W9-ANY-724

THEODORE GRAY
Reactions
Photographs by Nick Mann

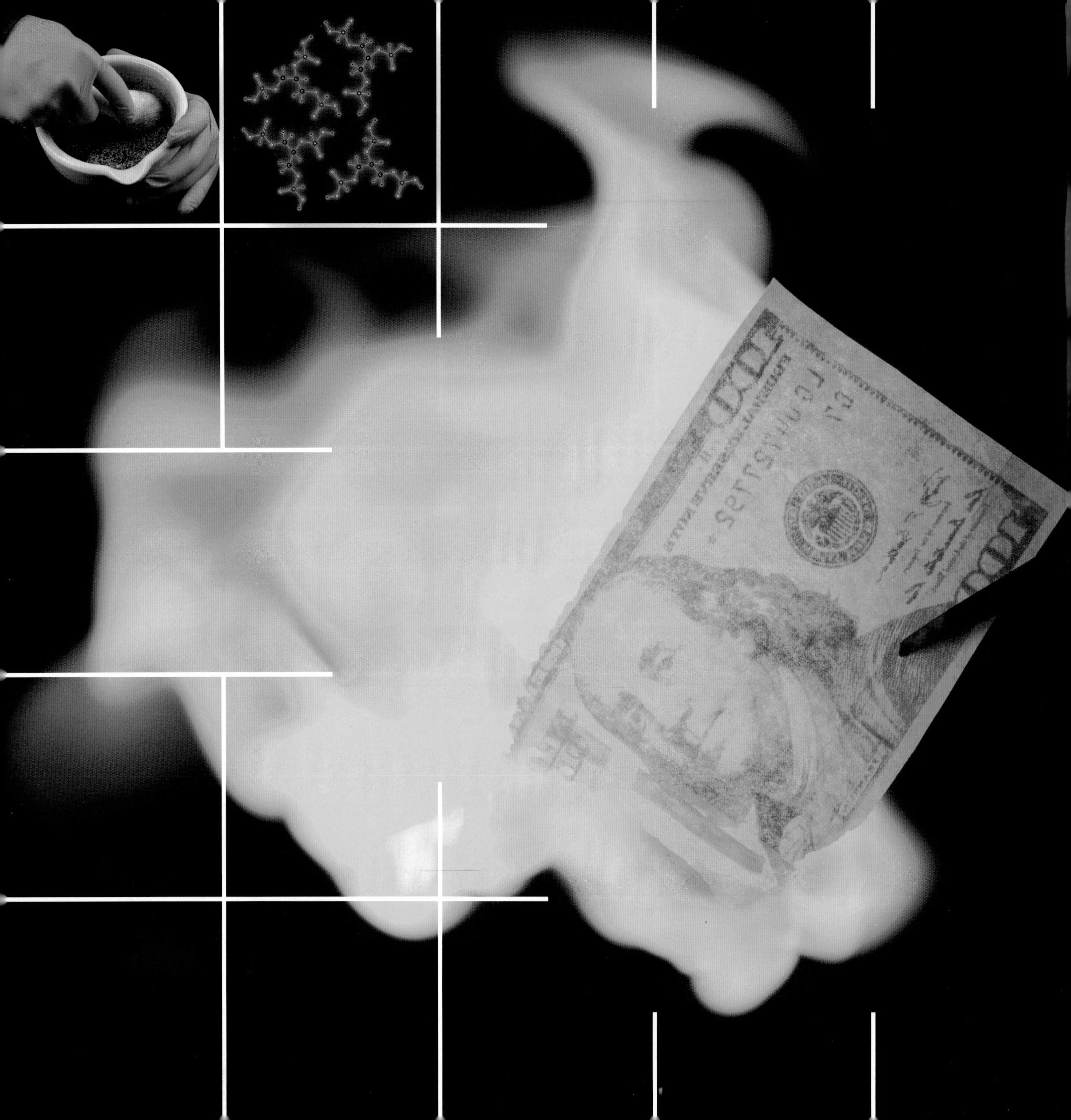

Reactions

An Illustrated Exploration of Elements, Molecules, and Change in the Universe

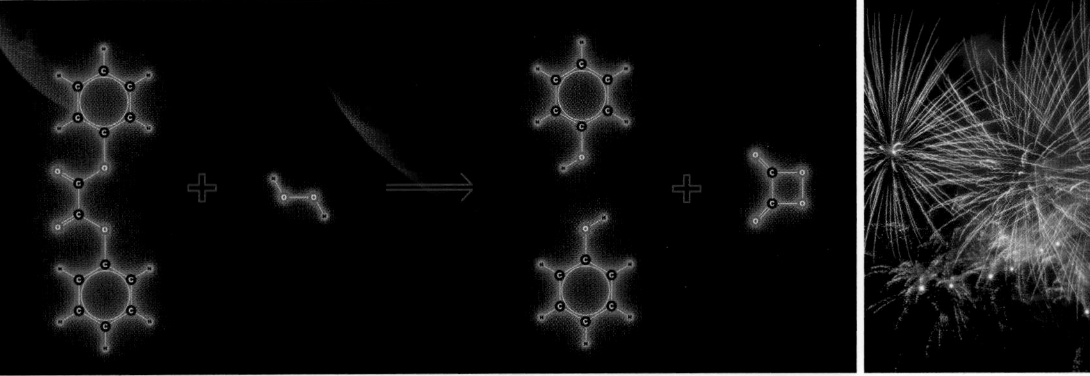

THEODORE GRAY
Photographs by Nick Mann

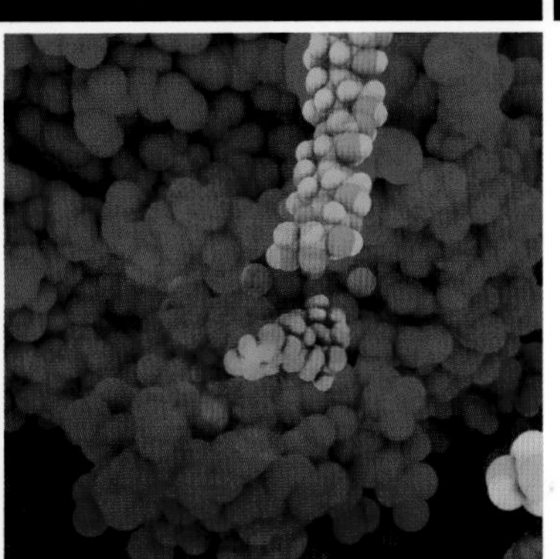

BLACK DOG
& LEVENTHAL
PUBLISHERS
NEW YORK

Black Dog & Leventhal Publishers
Hachette Book Group
1290 Avenue of the Americas
New York, NY 10104

www.hachettebookgroup.com
www.blackdogandleventhal.com
www.theodoregray.com

First Hardcover Edition: October 2017

Black Dog & Leventhal Publishers is an imprint of Hachette Books, a division of Hachette Book Group.
The Black Dog & Leventhal Publishers name and logo are trademarks of Hachette Book Group, Inc.

The publisher is not responsible for websites (or their content) that are not owned by the publisher.

Additional copyright/credits information is on page 210.

Print book interior design by MRCokeley Design, LLC.

Library of Congress Cataloging-in-Publication Data

Names: Gray, Theodore W., author. | Mann, Nick, photographer.
Title: Reactions : an illustrated exploration of elements, molecules, and
change in the universe / Theodore Gray ; photographs by Nick Mann.
Description: New York, NY : Black Dog & Leventhal Publishers, Hachette Book Group, 2017.
Identifiers: LCCN 2017020093| ISBN 9780316391221 (hardcover) | ISBN 9780316508742 (ebook)
Subjects: LCSH: Molecules. | Molecular structure. | Chemical elements. |
Chemistry.
Classification: LCC QC173 .G658975 2017 | DDC 530--dc23
LC record available at https://lccn.loc.gov/2017020093

ISBNs: **978-0-316-39122-1** (hardcover), **978-0-316-50874-2** (ebook);

Printed in China

1010

10 9 8 7 6 5 4

Contents

Introduction

WHEN I FIRST STARTED writing my book *The Elements* in 2008, I had the idea that it should be the first in a three-part series—*Elements, Molecules*, and *Reactions*. Together these three books would represent a tour of the world of chemistry. You start with elements, because everything starts with elements. Then you put them together into molecules. Then you send those molecules at each other in a sort of nanoscale fight club.

It's been nearly ten years (*ten years*!), but the trilogy is now finished! In the time it's taken to write these books, and live the life necessary to make them possible, I feel that I have been transformed as much as any of the molecules I talk about. My children have grown and my hair has shrunk. It's been worth it.

I hope you enjoy reading any or all of these books, as much as I've enjoyed writing them.

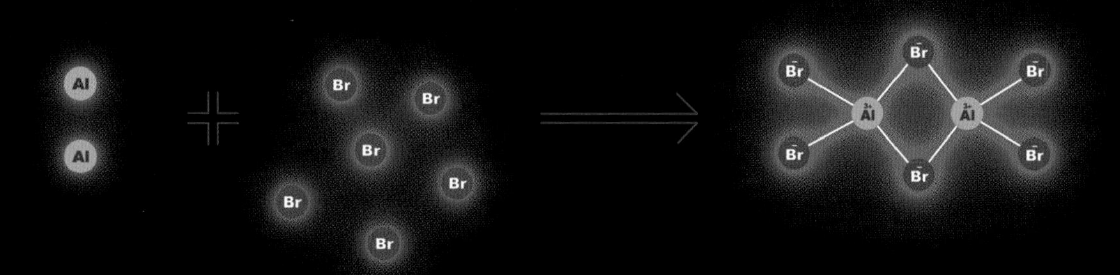

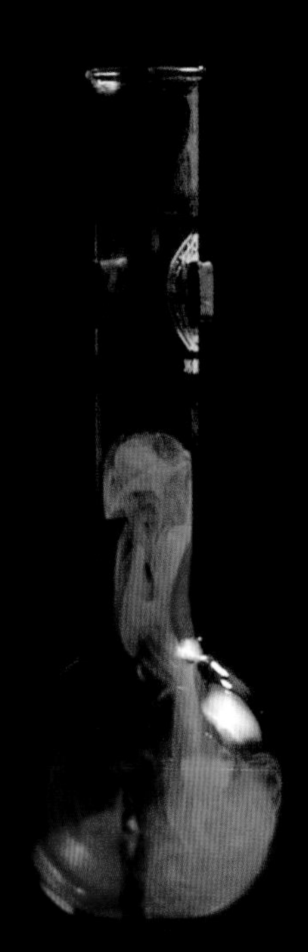

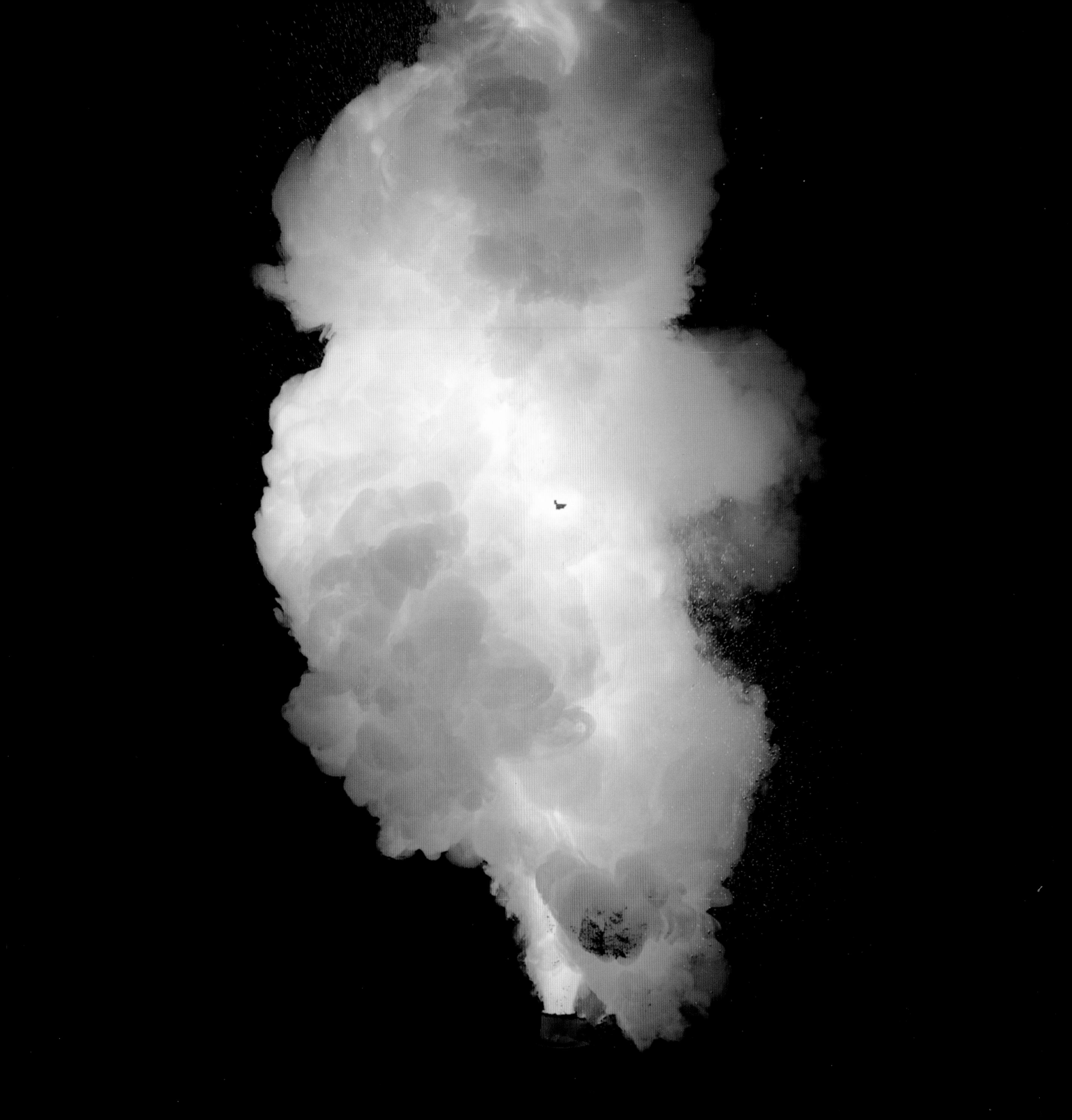

Chemistry Is Magic

IT IS IMPORTANT to stand in awe of nature, and to be no less dazzled by the energies and forces of the world just because we have learned to understand and control them.

Push two strong magnets against each other and feel the invisible force that repels one from the other. Go on—do it now if you can. Come back when you are filled with a sense of miracle and wonder that such things exist and that you have the supreme privilege of holding not one, but *two* of them.

Do not be misled by the fact that magnets are dirt common, or that we know exactly how they work and how to make more of them. A magnet is an object from another world, like a moon rock or meteor that fell from another star. It is a visitor to our human world that carries with it the knowledge and power of its home world.

That home is not another planet, but another scale. It is the vanishingly small world whose native inhabitants are the quantum forces that control the nature of matter and energy.

Quantum magnetic forces exist in all matter, all the time, but normally they work in opposite directions, cancelling each other out. They are hiding in plain sight. But when we create a strong magnet, we align a vast number of individual quantum forces in the same direction, coaxing this astonishing force up into our world where we can feel it push and pull on our hands.

The quantum world of the very small is also the home world of chemistry. When we see a fire burn or a leaf change color, this is the action of atoms in numbers beyond imagination, all working together to create an effect visible on the human scale.

This is the world we will explore in the chapters to come.

Consider the humble glow stick. You break an inner capsule to mix two solutions together, and suddenly the whole thing starts to glow! How on earth is that even possible? I insist that you be amazed by this object, even though it is available at gas stations for less than the price of a bottle of water.

Where does the light come from?

Light is made of countless individual photons—packets of energy that travel through space at the speed of light. There are many different ways of making photons. The way it's done in a glow stick is one of the most complicated: each photon is lovingly handcrafted by its own single-use chemical machine.

We designed and built these machines from scratch, so we know exactly how they work, and I can explain them to you. (Don't worry if many of the words in this explanation are not familiar: we'll come back to them one at a time later in the book.)

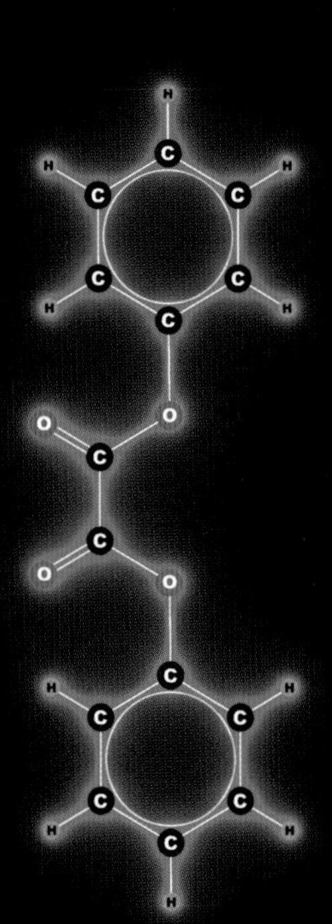

◁ The mechanism starts with a molecule of diphenyl oxalate, which has an unusual combination of four oxygen atoms in the center between its two mirror-image wings. Countless copies of this molecule are present in the liquid in the outer shell of the glow stick.

△ When the inner capsule is broken, it liberates molecules of hydrogen peroxide to mingle with the much bigger diphenyl oxalate molecules.

△ When the two molecules meet, a chemical reaction happens, destroying the hydrogen peroxide and splitting apart the diphenyl oxalate.

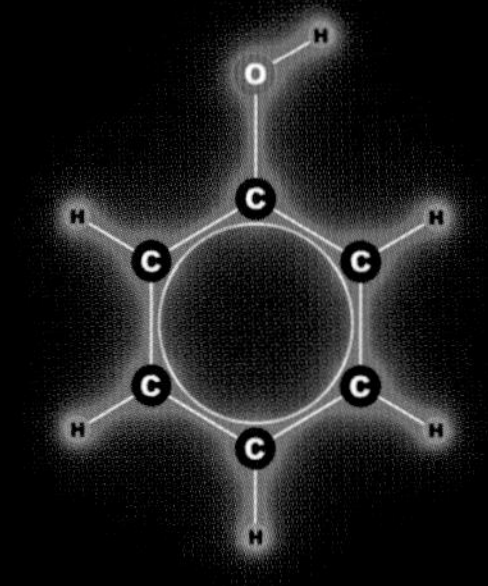

△ The two wings of the diphenyl oxalate become two identical phenol molecules. We don't care about these.

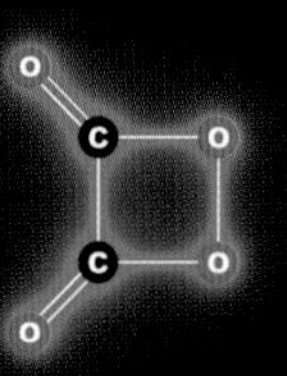

△ It's the small molecule from the middle, peroxyacid ester, that carries the action forward. The square, four-member ring in this molecule is unusual and indicates a high degree of stress. This molecule is a wound-up spring ready to unleash its pent-up energy.

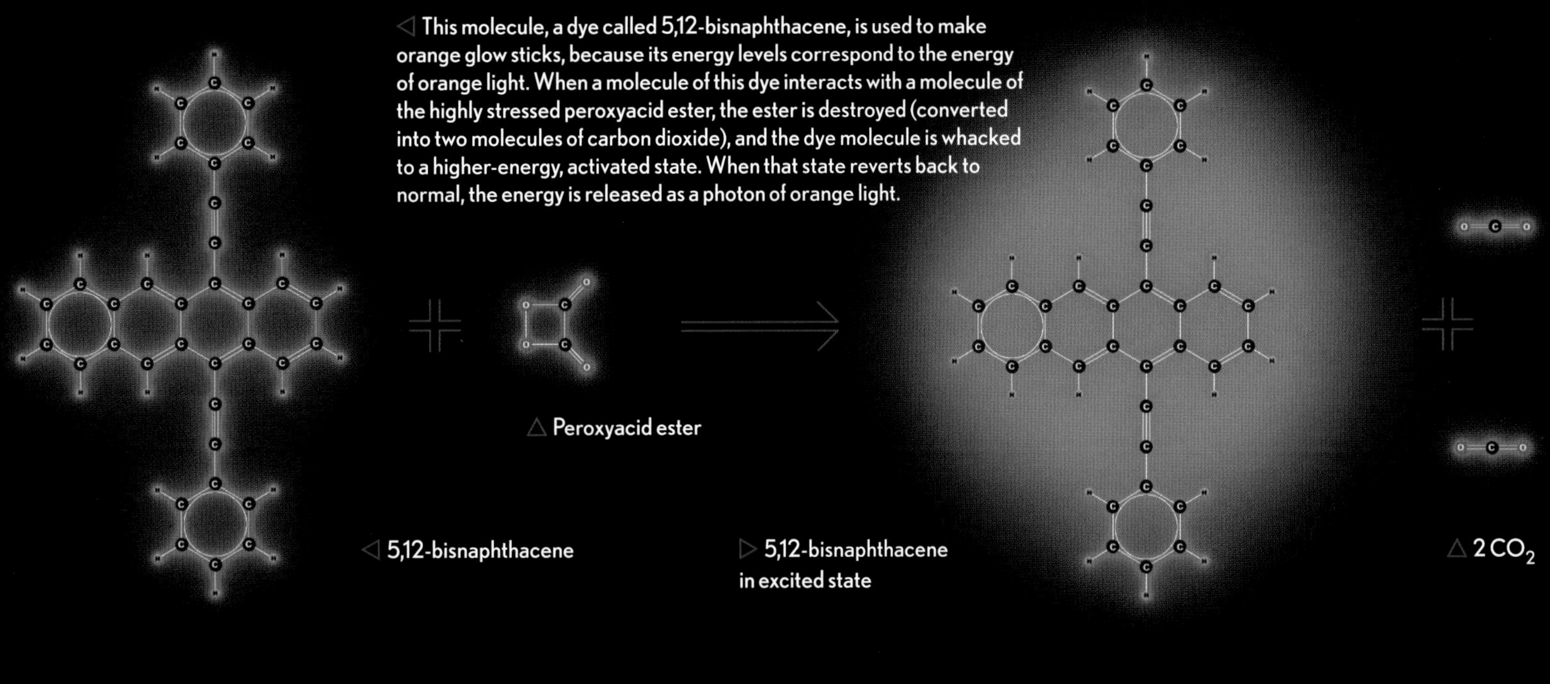

This molecule, a dye called 5,12-bisnaphthacene, is used to make orange glow sticks, because its energy levels correspond to the energy of orange light. When a molecule of this dye interacts with a molecule of the highly stressed peroxyacid ester, the ester is destroyed (converted into two molecules of carbon dioxide), and the dye molecule is whacked to a higher-energy, activated state. When that state reverts back to normal, the energy is released as a photon of orange light.

To get light from a chemical mechanism, you need a molecule that is able to absorb and emit photons of light with the right amount of energy to be visible (too little energy and the photons are invisible infrared light, too much energy and they are invisible ultraviolet light).

The dye molecule in this reaction can be reused indefinitely, but the other molecules can only be used once: Their destruction supplies the energy needed to create the photon. All the work of this elaborate machine goes into creating *one single photon* of light. If you want another photon you have to use up another set of molecules.

A typical glow stick activates and destroys about 10,000,000,000,000,000 (ten thousand million million) of these machines *every second* to produce the light we see coming from it.

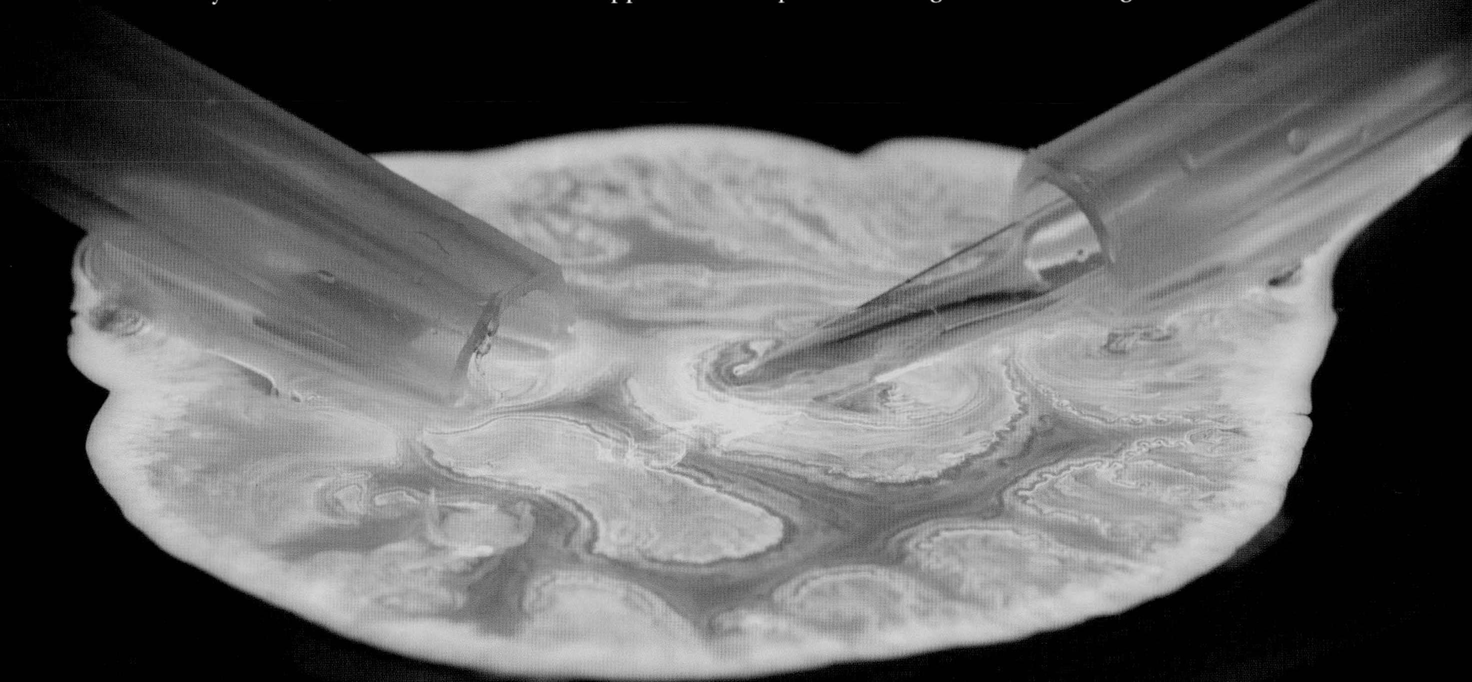

Is That Really Magic?

THE TITLE OF this chapter is "Chemistry Is Magic," yet I just told you exactly how one particular bit of chemical "magic" works. Doesn't that mean it's not magic?

I like to use the word *magic* the way it's used by stage magicians. A magic trick is "magic" because it *seems* to be supernatural, until you understand how the trick is done. Some people enjoy the mystery of not knowing, but I always find it far more satisfying to learn the trick. The cleverness and skill involved are much more interesting than the end result.

Professional magicians, and ancient conjurers, carefully guarded their secrets as a matter of commercial necessity: if everyone knew how they did their tricks, they'd be out of a job. Scientists are the opposite. When they discover a particularly clever trick, something that makes the seemingly impossible suddenly possible, they want to tell the whole world about it.

I am definitely in the science camp: I am eager to tell you about all the clever but invisible chemical magic going on all around you. This doesn't make it any less magical. It just means you will have the opportunity to tell someone else how the trick is done—next time you see someone with a glow stick, you can tell them about the tiny machines inside making the light one photon at a time.

△ A phone. You can't tell one single thing about how it works from the outside, and not much more even if you take it apart.

Any sufficiently advanced technology is indistinguishable from magic.

–Arthur C. Clarke

Machines from a hundred years ago are wonderfully understandable: All the moving parts actually move, and they really are separate parts that you can see and dismantle. These machines are marvelous, but you can plainly see that they are not magical.

But imagine if you showed someone from the Steam Age a modern phone. It would seem like magic, right?

The thing that makes phones seem magical, even today, is that you can't see how they work *at all*. There's no part of their mechanism that is visible. If you take one apart, all you see is some little plastic blocks with tiny metal pins coming out. If you take those chips apart, there doesn't seem to be anything interesting inside. All the action is happening in circuits that are literally smaller than the wavelength of visible light—they are not just microscopic, they are smaller than light itself.

"Magic" machines whose parts are too small to see are a modern invention, but chemistry has always been this way, with its mechanisms of action completely invisible. Today, chemistry only seems magical: we know it actually isn't. But in ancient times no one had the slightest idea how it worked, which left a lot of room for both mystical nonsense and serious study.

▷ Every single part of this beautiful old machine is immediately visible. You can practically see it working just by looking at it.

Ancient Magic Was Mostly Chemistry

△ *The Three Witches from Shakespeare's Macbeth* by Daniel Gardner

SOME PEOPLE MAKE a big deal about ancient "secret knowledge," the supposed keys to unlocking great powers, lost in the crass modern world of fast food and stupid television. The fact is that most of this secret knowledge was either pretty simple by modern standards, or plain wrong. There's more interesting new knowledge discovered in an average week today than in all the secret books of the ancients.

What I find interesting is that much of what passed for magic and secret knowledge in the ancient world was actually chemistry, or at least attempted chemistry. And *all* of the ancient magic that actually works is chemistry. Exactly none of the spells and incantations work, but a few of the *potions* do.

The potions that worked became modern chemistry. The potions that didn't work live on in the quack medical products, homeopathic remedies, and bogus "age defying" creams that remain as popular as ever.

> Eye of newt, and toe of frog,
> Wool of bat, and tongue of dog,
> Adder's fork, and blind-worm's sting,
> Lizard's leg, and howlet's wing, –
> For a charm of powerful trouble,
> Like a hell-broth boil and bubble.
>
> **–William Shakespeare**

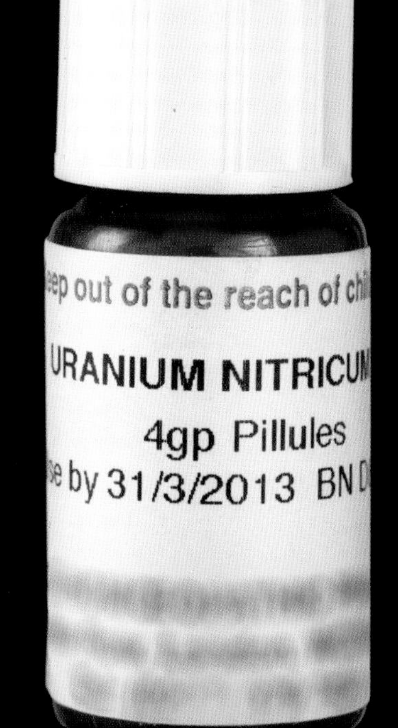

▷ The best modern example of commercially available witch's brews are homeopathic "medicines." They list a bunch of exotic ingredients, but don't actually contain any of them (yes, this is legal due to a stupid law authorizing lying for these specific products). To make them "properly" the manufacturer has to dilute the solution a certain number of times, and with each dilution shake and tap the solution in a certain direction a certain number of times. It's all complete nonsense, and would be funny if not so widely sold for so much money to so many people who are being completely ripped off.

It's easy to mix together a bunch of ingredients that sound exotic, boil them for a while, say some silly words from a book of spells, and hope that the result is a powerful potion. But potions made with eye of newt do not work, no matter how bad they smell, and no matter how passionately you ask the forces of darkness to make them work for you.

Fortunately not everyone in the ancient world based their potion-making on wishful thinking. Some of them knew that nature is very picky and really doesn't care about your pleadings. These people knew that in order to make a functional potion, they were going to have to work hard, study the world, follow up on lucky breaks, and try a lot of variations before homing in on a combination that gave them something interesting. In other words, they were scientists long before that word had been invented.

The ingredients for gunpowder: charcoal (black), saltpeter (white), and sulfur (yellow).

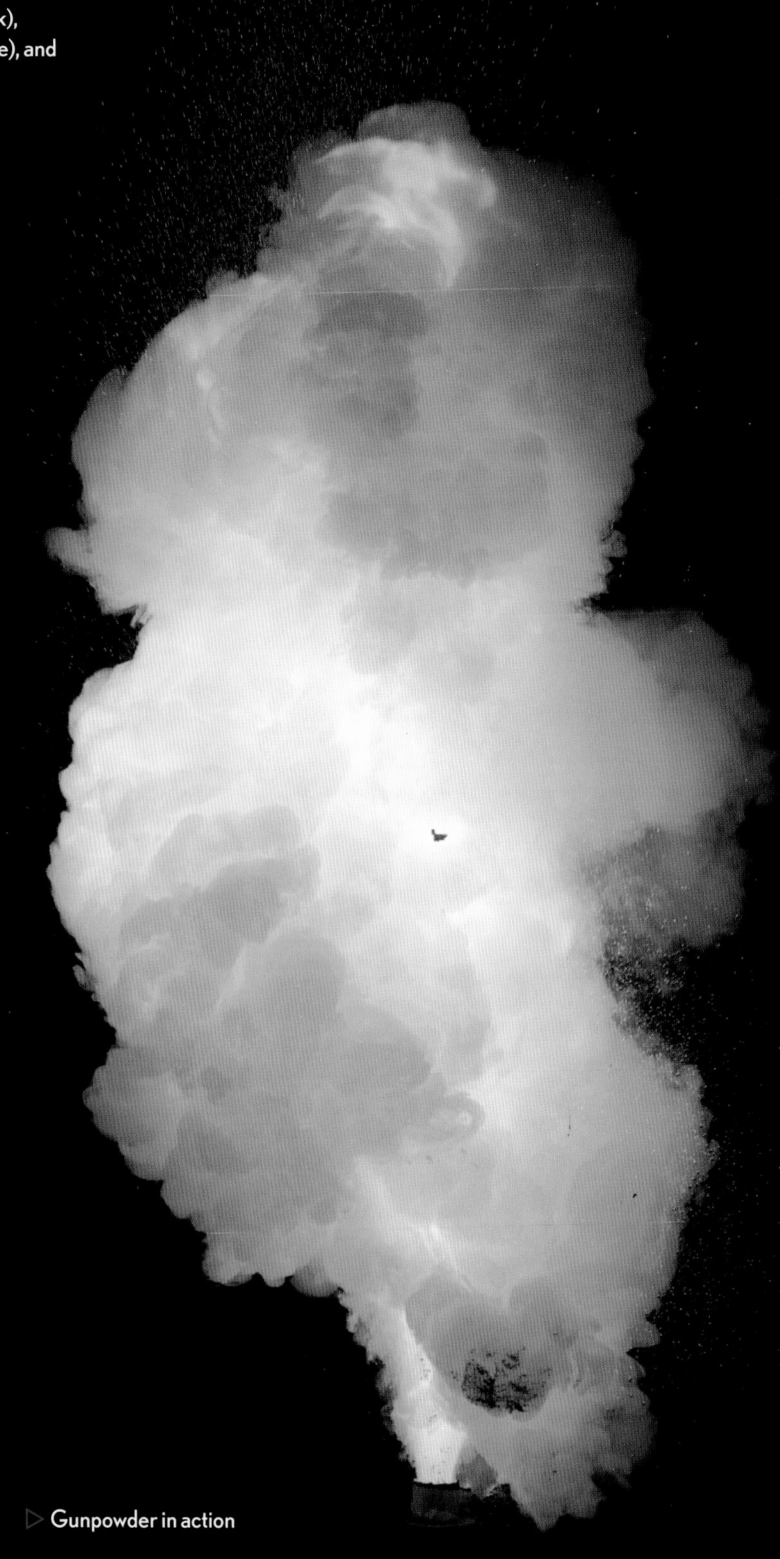

Gunpowder in action

If you pick three powders at random and mix them together, it's very unlikely that the result will have any special properties. But if you pick *these three specific powders*—this particular white one, this particular black one, and this particular yellow one—and if you mix them in just the right proportions, you get a powder that vanquishes your enemies, consumes them in the fire of damnation, or blows them to bits on your command.

In other words, these are the ingredients for gunpowder.

Gunpowder is like a magic potion—a really strong one—in that it does amazing things and unleashes great power in the world. But it's not like a magic potion in that you don't have to say any silly words while mixing it, and if you get the mixture right, it works every time.

This is the key: we don't usually call gunpowder "magic" only because it actually works. And because it works, you can *use* it. It leads to new inventions, like fireworks and guns. It is an invention that carries true power, unlike the non-functional dead ends you find in witches' brews (both ancient and modern). We're going to learn how gunpowder can be arranged in beautiful ways (see page 134), about a better substitute (see page 148), and finally, on page 193, we will learn the details of the chemical reaction that makes it work.

Finished gunpowder

Ancient alchemists are sometimes made fun of because their obsession with trying to do things that we now ow are impossible, like turn lead into gold or find a tion for eternal life. But this criticism is unfair, and nfuses alchemists with the mystics and charlatans of eir age.

The alchemists worked diligently to identify and derstand powerful substances—chemicals—that they uld experiment with and combine with each other in attempt to advance toward their goals. They got a long y toward unraveling the nature of chemical change, cluding the notion of immutable elements (which they re interested in studying precisely because they wanted find an exception that allowed them to make gold from se metals).

They were wrong about a lot of things, but they were so right about a lot of other things. More importantly, ey based their work on reality. They believed in testing eir ideas through experiment—a thoroughly modern, ientific way of doing things. They believed in evidence, udy, and confirmation. When the modern field of emistry emerged as a solid science in the 1700s, it was ilt on a platform of alchemy.

△ *The Alchemist* by Newell Convers Wyeth, 1937

◁ Phosphorus burning in the air

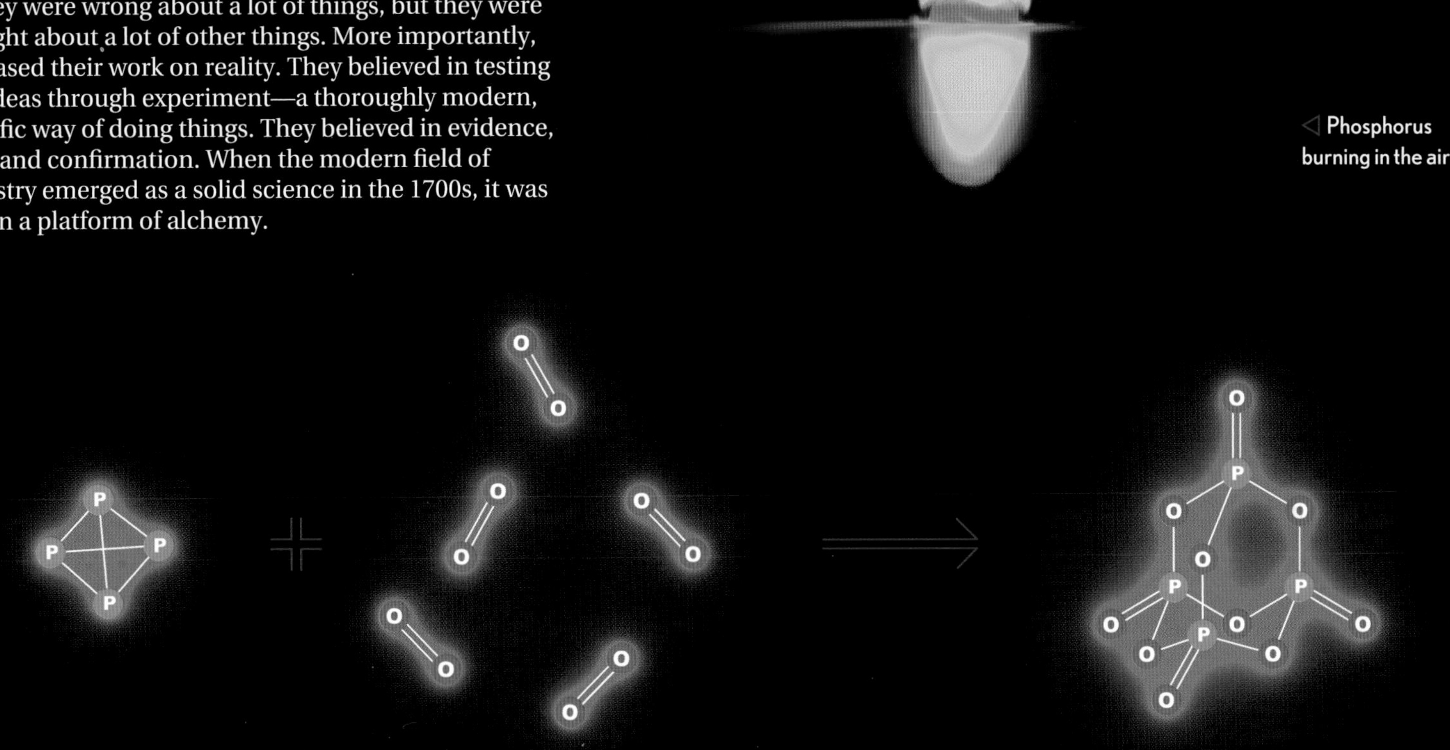

△ P_4 △ $5\,O_2$ △ P_4O_{10}

△ The reaction between white phosphorus and oxygen (from the air) creates lovely 3-dimensional molecules, each containing four phosphorus and ten oxygen atoms.

Glowing chemicals endlessly fascinated the alchemists. When Hennig Brand discovered phosphorus in 1669, he was absolutely convinced that it must be the key to making gold, because he derived it from pee (which is yellow like gold) and because it glows in the dark (which he thought proves that it must have inherited some kind of life force from its origins). He was correct that white phosphorus is powerful stuff, just not in quite the way he hoped it would be.

▷ The "phosphorus sun" demonstration is a favorite of those comfortable handling highly toxic, spontaneously flammable chemicals in front of crowds. Here we see my intrepid colleague Prof. Hal Sosabowski performing the demonstration, in which a small chip of white phosphorus is suspended in a ball of pure oxygen.

▷ Hennig Brand discovered phosphorus while working in Hamburg, Germany. This is what Hamburg looked like 270 years later, after Allied forces in World War II dropped thousands of bombs, many of them white phosphorus incendiaries, on the city. Power is not about good or evil. Phosphorus is powerful, and whether you use that power to grow plants or level a city is up to you.

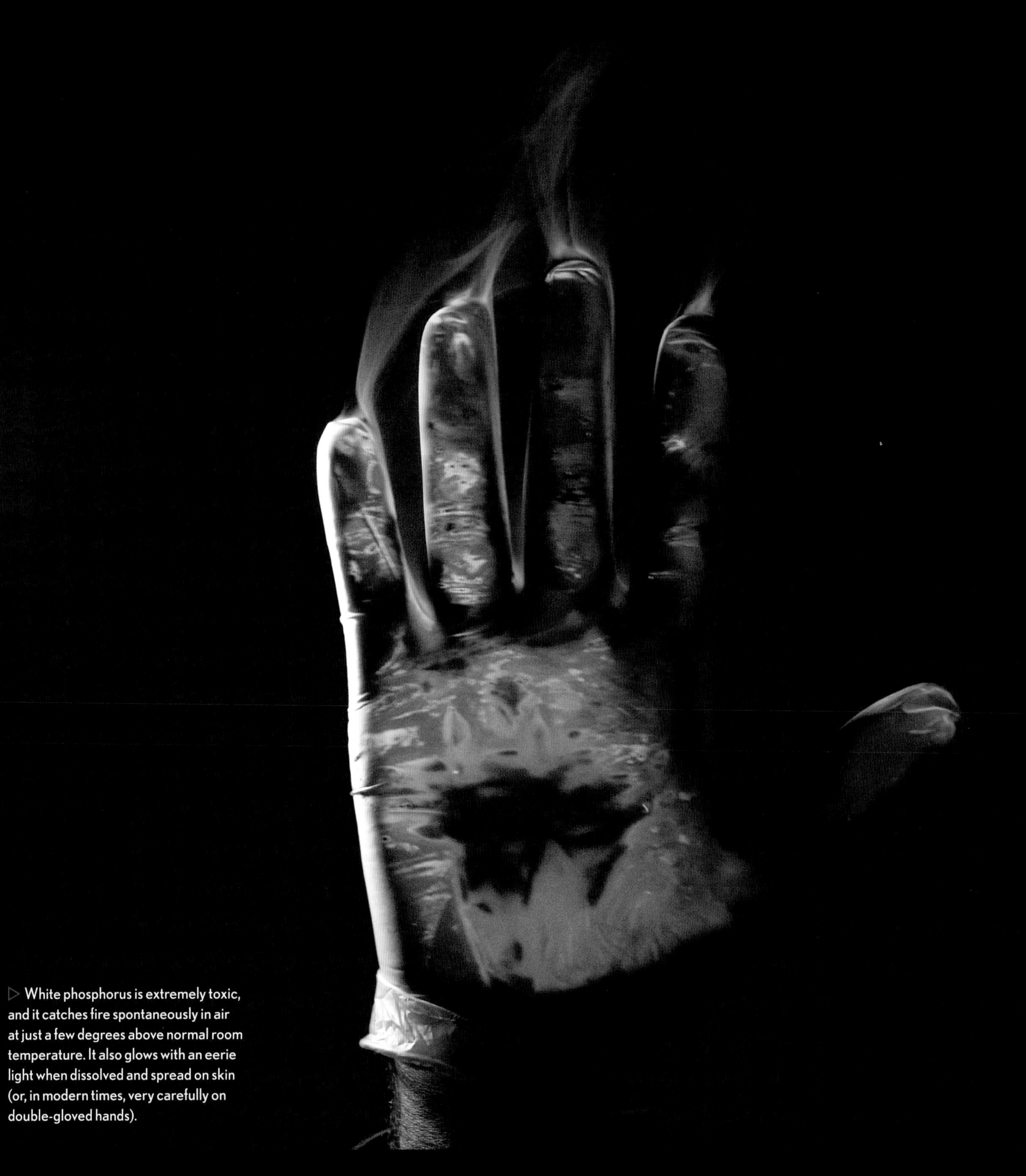

▷ White phosphorus is extremely toxic, and it catches fire spontaneously in air at just a few degrees above normal room temperature. It also glows with an eerie light when dissolved and spread on skin (or, in modern times, very carefully on double-gloved hands).

Lycopodium powder was a favorite of alchemists wishing to put on a good show, or convince a king of their power to control nature. Throw a handful into a candle and you get a powerful burst of flame that vanishes as quickly as it arrived, leaving no trace of smoke.

Lycopodium seems almost like gunpowder (when thrown in the air), but it's actually the spores of club moss: It's a plant, not a mixture of simple chemicals. But its incredibly high surface area makes it burn so rapidly that it seems very "chemical" in nature.

▷ When you see this stuff in action you can really understand why people of earlier times thought it was true magic!

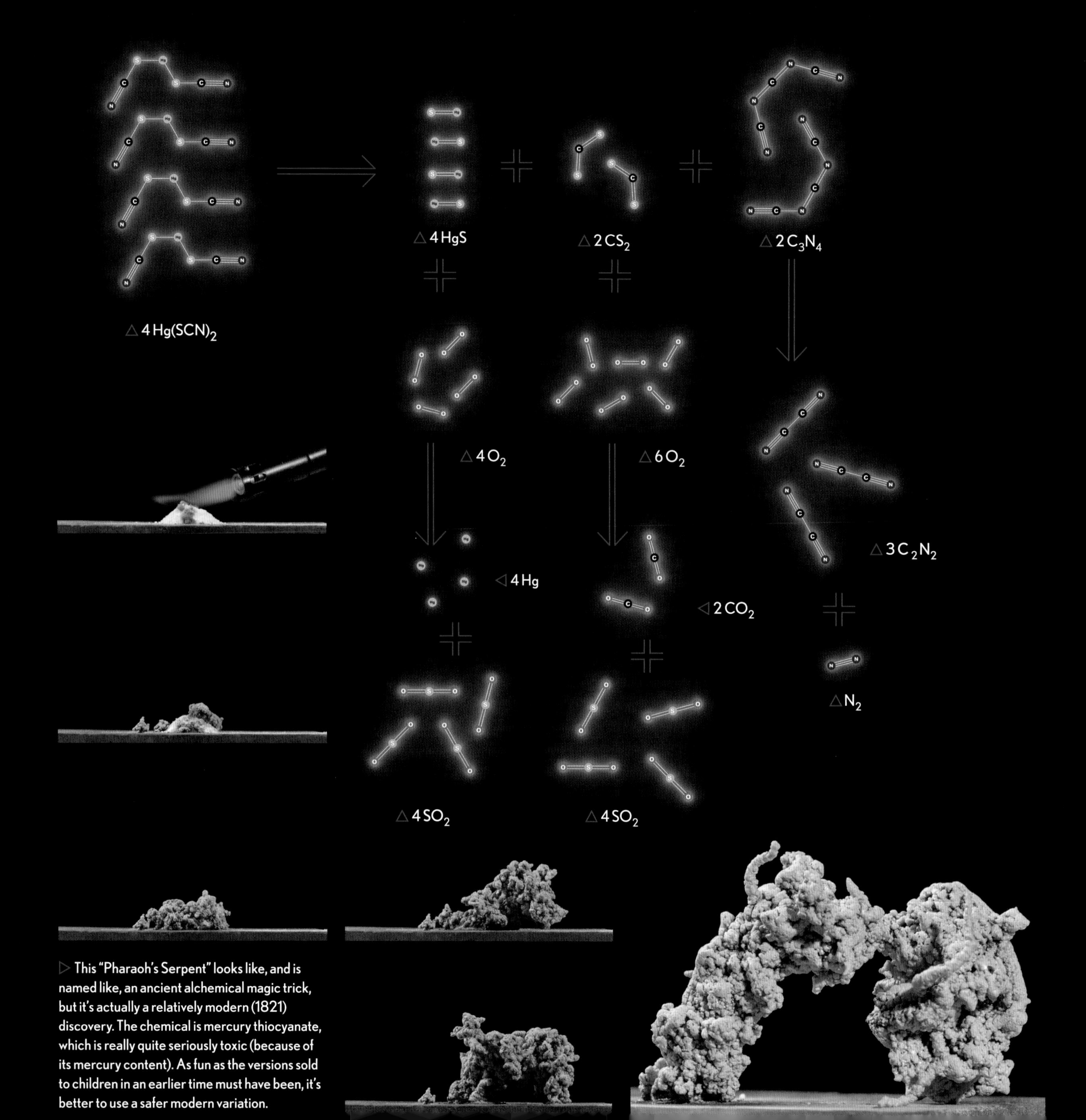

▷ This "Pharaoh's Serpent" looks like, and is named like, an ancient alchemical magic trick, but it's actually a relatively modern (1821) discovery. The chemical is mercury thiocyanate, which is really quite seriously toxic (because of its mercury content). As fun as the versions sold to children in an earlier time must have been, it's better to use a safer modern variation.

▷ These very cheap kids' Black Snake fireworks nicely replicate the Pharaoh's Serpent without the heavy metal poison. The ingredients in commercial versions are trade secrets, but include something that decomposes into carbon (often linseed oil), something that burns (perhaps naphthalene), and just enough oxidizer (typically potassium nitrate) to make it burn at a good rate, not too fast and not too slow. (Normally you light one pellet at a time, but here we have ground up several hundred of them, because it looks really cool when you do that.)

The Reactions I Remember from When I Was but a Wee Lad

I'VE LIKED CHEMICALS since I was little, for pretty much the same reason people a thousand years ago were fascinated by them. Unlike a lump of clay or the eye of a newt, the right chemicals actually *do something* when you mix them up. It took the first alchemists hundreds of years of messing around to find just a few really cool chemicals, but as a wee lad I had the advantage of all their experience and more—captured in a pre-internet encyclopedia and later at a university—to guide me.

The day I found a list of the ingredients for gunpowder (with percentages!) was a major thrill for me. I can remember it like yesterday, taking the "G" volume off the high shelf, flipping through the pages, getting closer and closer…and there it was, just as plain as day: 75% potassium nitrate, 15% charcoal, and 10% sulfur.

These proportions are pretty flexible: Historical versions of gunpowder vary them by plus or minus 10% or more. Different ratios give you different burning rates, useful for different applications. (For example, the gunpowder in a rocket engine needs to burn over at least a few seconds, while the gunpowder in a gun has to expend itself in a tiny fraction of a second. Much more on the speed of gunpowder on page 193.)

The flexibility of the gunpowder formula is no doubt why it was possible for the Chinese to have discovered it long before there was any understanding of why it works. You don't have to get the mixture right at all to start seeing something interesting. In fact, all you really need is the potassium nitrate: add it to just about anything that burns, even ordinary paper, and the result is something that burns *more*. If you're paying attention to the world, this is a clear sign that potassium nitrate is something worthy of further study. It's not a huge leap to try adding it to other things that burn, including charcoal, sulfur, or—bingo—both of them at the same time. From there it's just a question of systematic tests to find the best proportions.

◁ (Full disclosure: This is actually a photo of paper soaked in the closely related chemical strontium nitrate, not potassium nitrate—frankly for no other reason than that this picture came out way more attractive than the one using potassium nitrate, and they both demonstrate basically the same phenomenon.)

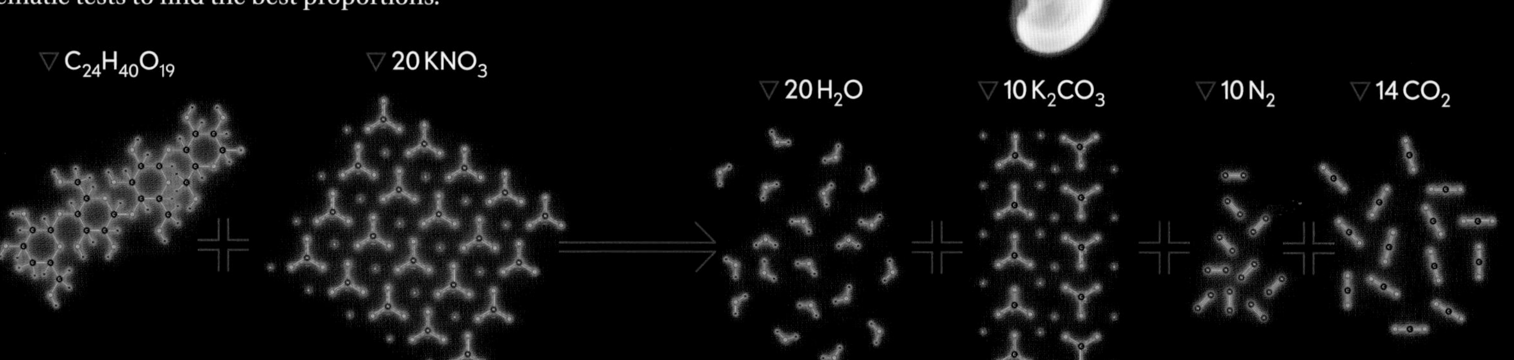

△ The hardest part of making gunpowder is the mixing and grinding. Making gunpowder that actually explodes in the traditional sense requires grinding for several hours and is usually done with an electrically driven ball mill (see page 194). If you do it wrong, the powder explodes while you're mixing it. Fortunately all I had was a mortar and pestle, and I didn't have the patience to grind the powder long enough. (Fortunately, because inadequately ground-up ingredients make gunpowder much less dangerous!)

▷ With badly handmade gunpowder like mine was, you get what might be described as "vigorous burning," not a real explosion. It's great for making sparkle cones, terrible for making a cannon. I'm sure that when people first stumbled on this mixture of ingredients, this is the kind of reaction they got. It's not an explosion, it's not obviously useful for anything, but it sure is *interesting.* It would have grabbed their attention and encouraged further experimentation. Slowly, as they worked out what made the powder work better, they developed it into something that could be used for rockets and cannons.

We'll see an example of how gunpowder is used in commercial fireworks in Chapter 4, and learn about how fast it burns in Chapter 6.

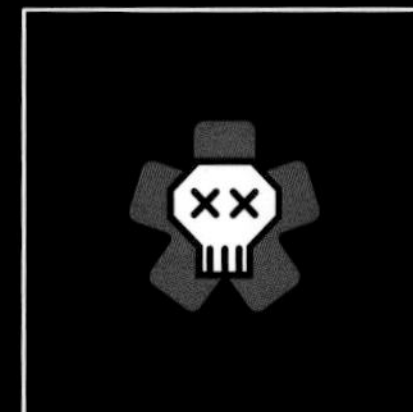

▷ This missing photograph represents the flash powder rockets I made as a kid. There is no picture because I am not willing to make another one. It's just too dangerous. What I did, and would never do again, was to tightly pack special effects flash powder (a mixture of aluminum powder and potassium perchlorate) into thick-walled cardboard craft straws. I'm not sure where I got the idea to do this, but it actually worked: packed tightly in a thin tube, the flash powder did not "flash," but rather burned quickly, generating enough thrust to lift the rocket 20 or 30 feet into the air. Sometimes, for added stupidity, I added a firecracker on the end. Yes, it was a really dumb thing to do, and I even knew that at the time, but what can I say, kids are dumb. None of the rockets ever exploded in my face while I was packing the powder, which could very easily have happened. One reason I'm careful around chemicals to this day is the feeling I get in the pit of my stomach when I think back to when I was sitting at the dining room table, looking out the back windows with an open bottle of flash powder, a straw, and a nail, thinking to myself, "Why am I jamming this nail repeatedly down into a straw with enough friction to set off the flash powder, when I know that's very dangerous?" Seriously, kids are very dumb sometimes. If you're currently a kid like I was, remember that. You'll get smart, but in the meantime, please don't blow yourself up.

I love this stuff! It's a clear liquid, but if you put a few special drops in it, an hour later it looks exactly the same except now it's a clear solid. Magic. Tremendous fun to embed things in. This is polyester casting resin, commonly available in hobby shops for craft projects. The special drops are usually called the catalyst, but this is an incorrect use of the term. They are actually a free-radical initiator which kick-starts a chain reaction that eventually consumes the entire mass. Vast numbers of small molecules end up linked together into a small number of very large ones. A dense web of cross-links connecting all parts of these giant molecules makes the resulting material hard and strong.

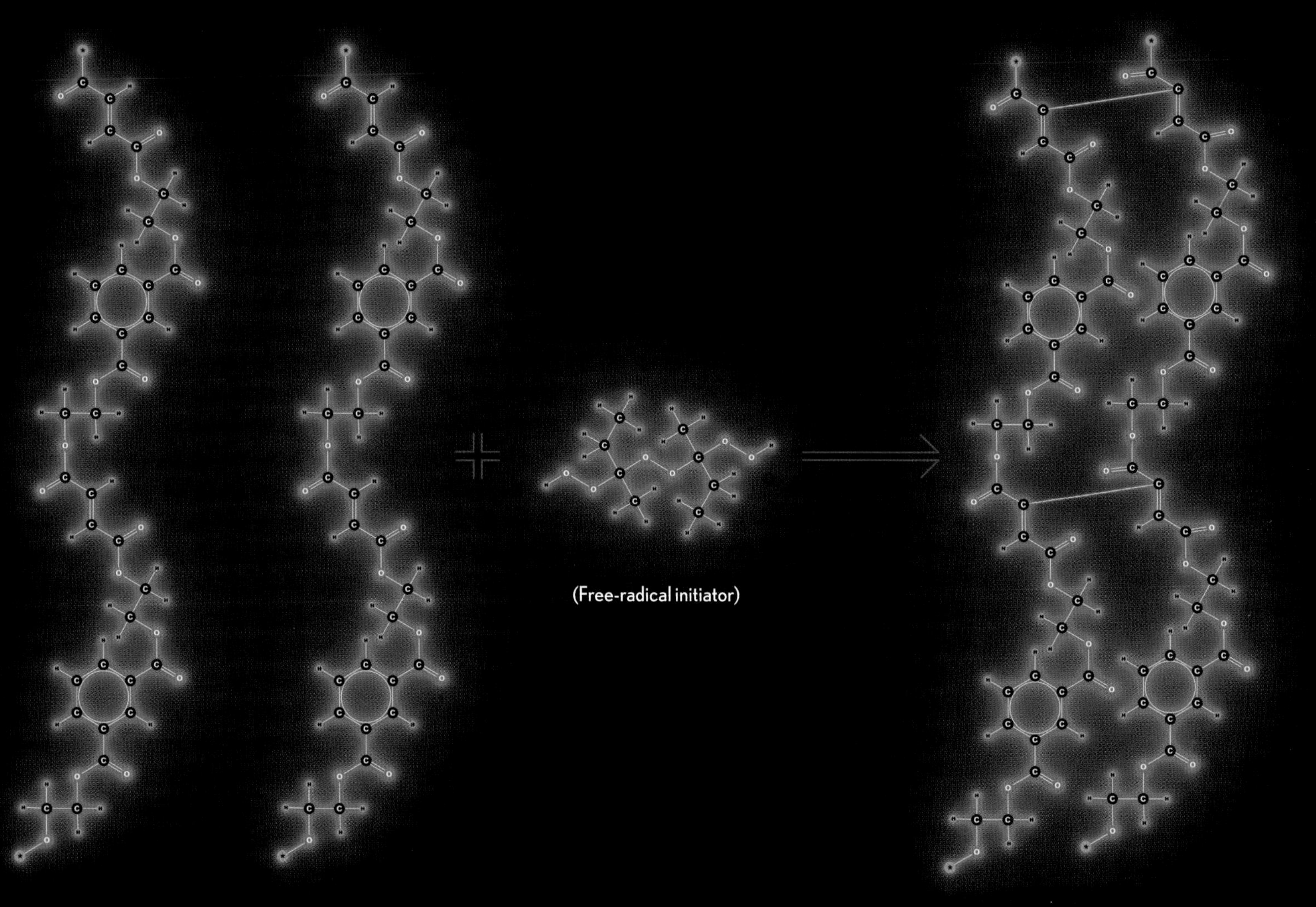

One of the ground rules we expect objects in our daily experience to obey is that just looking at them isn't supposed to cause them to change form. But this stuff, a type of light-cured epoxy glue (see page 155 for more on the chemistry of epoxies), turns rock solid in a few seconds when you do nothing more than shine some blue or UV light on it! Keep it in the dark and it stays liquid. Try to look at it under too strong a light source, and suddenly it's a solid. (Fortunately there is a wide middle ground where you can work with it under normal indoor lighting without setting it off. Convenient kits include a tube of glue and a blue/UV LED flashlight to solidify it on command.)

▷ Light-cured glues have existed for decades, but it's only in recent years that it has become widely available in retail packaging. It was worth the wait (and the fairly high cost).

▷ Making salt from scratch is something I'd wanted to do for many, many years before I finally got a chance. It still scares the pants off me on the rare occasions I film it again for one reason or another.

I like it because it's such a primal demonstration of elemental chemistry in action. You blow chlorine gas (a pure element that kills quickly and painfully) at sodium metal (a pure element that explodes on contact with water) and the result is a fire whose smoke is common table salt (sodium chloride, NaCl).

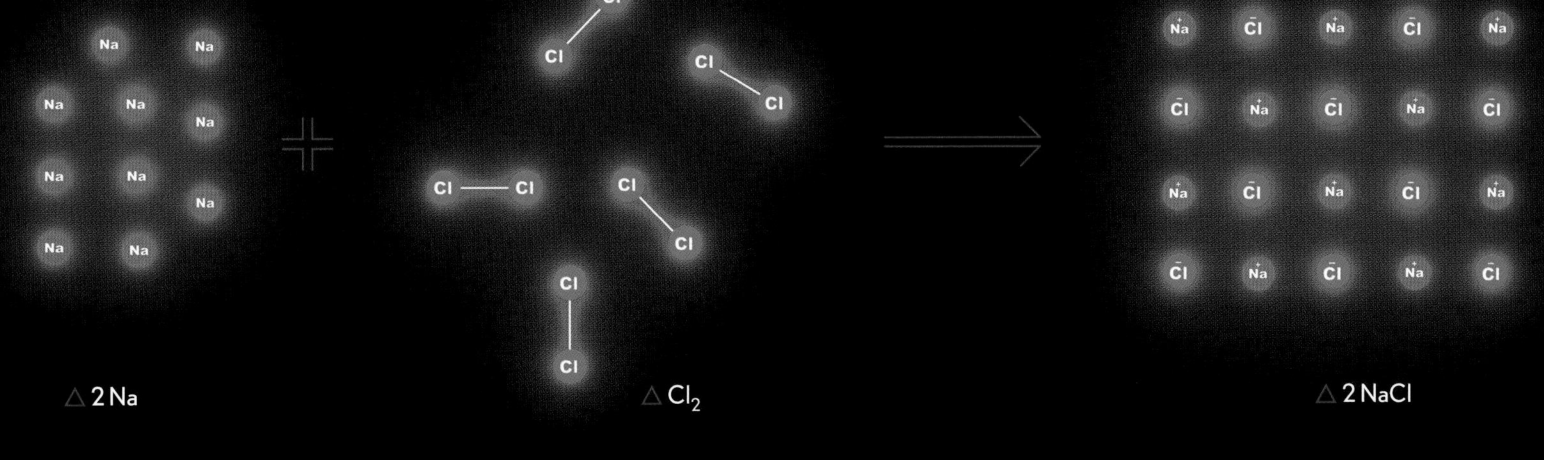

△ Sometimes in this book I draw chemical reactions with more copies of the atoms or molecules than are needed. But the text labels usually show the smallest numbers that give the right ratio. In this example, two sodium atoms react with each Cl_2 molecule, but I've drawn ten sodium atoms reacting with five chlorine molecules (containing ten chlorine atoms). The numbers are different, but the important part, the 2:1 ratio, is the same. I did it in this case so that I could then show a reasonably-sized block of NaCl as the reaction product. In any real reaction there would of course be trillions of trillions of these atoms and molecules reacting, but the ratio stays the same.

△ The first time I did this demonstration it was to salt a plate of pasta.

This ancient retort (▷), an evolution of the even older alembic, was made in the 1700s. It's an example of the kind of apparatus alchemists designed to further their study. Hennig Brand used one in his discovery of phosphorus (see page 7).

These direct descendants of the retort are what most people think of when they hear the words "chemical laboratory." If you showed a setup like this to a medieval alchemist, they would be astonished at the skill of the glass blower, but not fundamentally mystified by the purpose of the various parts. They would recognize boiling flasks, condensation columns, and receiving pots. Mostly I imagine them admiring the clever advances in design, and the incredible precision in manufacture. And I imagine that they would take pride in knowing that it was their original invention of the alembic and the retort that led to this magnificent creation. (There are even modern glass versions of the retort design, thought they are rarely used these days.)

This kind of apparatus is how chemistry students first encounter the world of classical chemistry. If chemistry had a "golden age," it surely was the late 1800s to mid 1900s, when a tremendous number of discoveries and advances in techniques were made in synthetic organic chemistry. Today it's possible to create, from scratch, molecules with nearly any chemically possible configuration of atoms you like. That's due to generations of work by chemists who figured out thousands of individual reactions that can be used to transform one molecule into a slightly, or completely, different one.

▷ Glass retorts

△ Ground glass "T" connectors

△ Fritted glass filter funnel

△ 3-necked reaction vessels

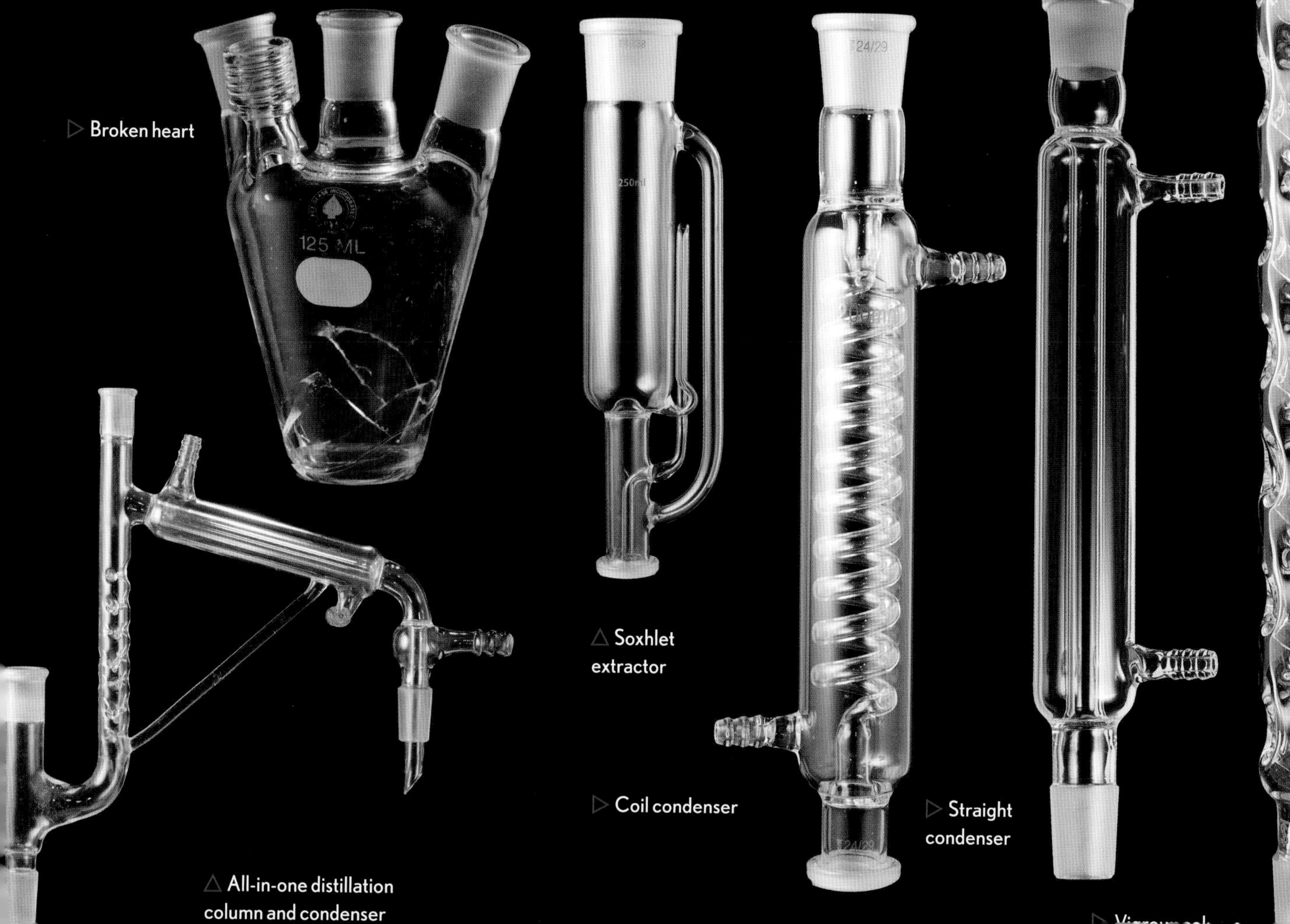

▷ Broken heart

△ Soxhlet extractor

▷ Coil condenser

▷ Straight condenser

△ All-in-one distillation column and condenser

▷ Vigreux col

This is Percy Julian, and on the next full-page spread
ı can see his molecule, physostigmine (along with the
ɔs needed to make it from scratch, which he figured
: in 1935). So significant, and so difficult, was this "total
thesis" of physostigmine that it made Julian famous and
to a long and lucrative career as an industrial chemist.
u know you've had a financially successful career when
ɪre's a university building named after you. Generally
ɐaking, they name these things after whoever donated
ɛ money to build them.)

The leap from alchemists stumbling into the discovery
ɔhosphorus, to Julian's systematic engineering of a
ɪthesis of this complex molecule, is as great as the leap
m Orville and Wilbur Wright's bicycle shop to the first
ɔon landing.

◁ The Julian Science Center at DePauw University celebrates Julian's synthesis of physostigmine as one of the great achievemen of that university. Synthetic organic chemistry was, and is, serious business: If you can figure out a way to make a commercially or medicina important molecule more cheaply, you can change the world. Millions of lives have been saved and improved by Julian's discoveries of ways to make physostigmine and other important molecules.

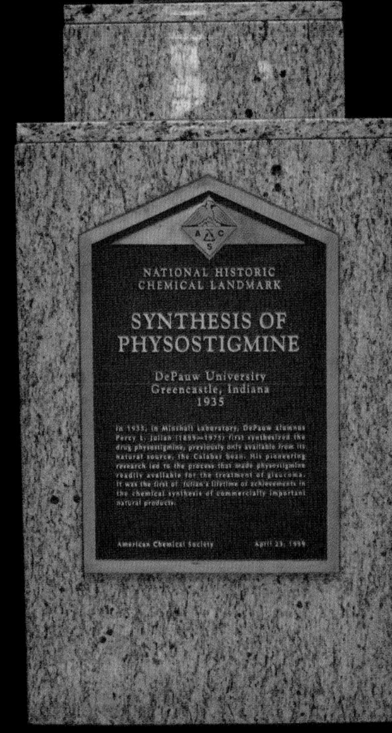

▽ In 2003 my friend Max Whitby and I instal a periodic table display in the Julian Science Center at DePauw University. It remains a tou attraction of sorts in the small Indiana town of Greencastle. So in a very indirect way, Max a too are beneficiaries of Julian's discoveries.

◁ The other tourist attraction of sorts in Greencastle is the impossibly out-of-place German V-1 "b bomb" on display in the town square. I almost don't want to know how on earth a Nazi terror weapc (designed to randomly bomb London during WWII) ended up being proudly displayed in front of th courthouse of a small town in rural Indiana. So many possibilities for how that could have happene

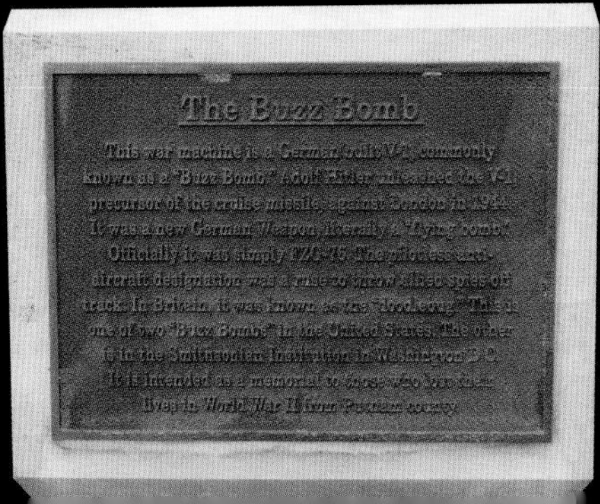

▷ Each element in the DePauw University periodic table display is given a little diorama highlighting its forms and uses.

O
8
oxygen
Al
13
aluminum
Cu
29
copper
Nb
41
niobium

The Total Synthesis of Physostigmine

"Total synthesis" means that Dr. Julian found a series of reactions that builds the target molecule, physostigmine, from scratch (which in this case means the very simple molecule phenol, available in large amounts from crude oil). The first four steps, up to phenacetin, were provided by earlier researchers, then Julian took it from there.

Phenol

4-Aminophenol

N-methyl-N-(2-bromopropanoyl) phenetidine

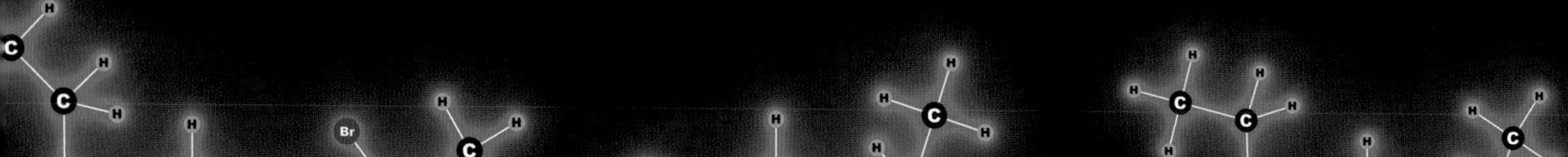

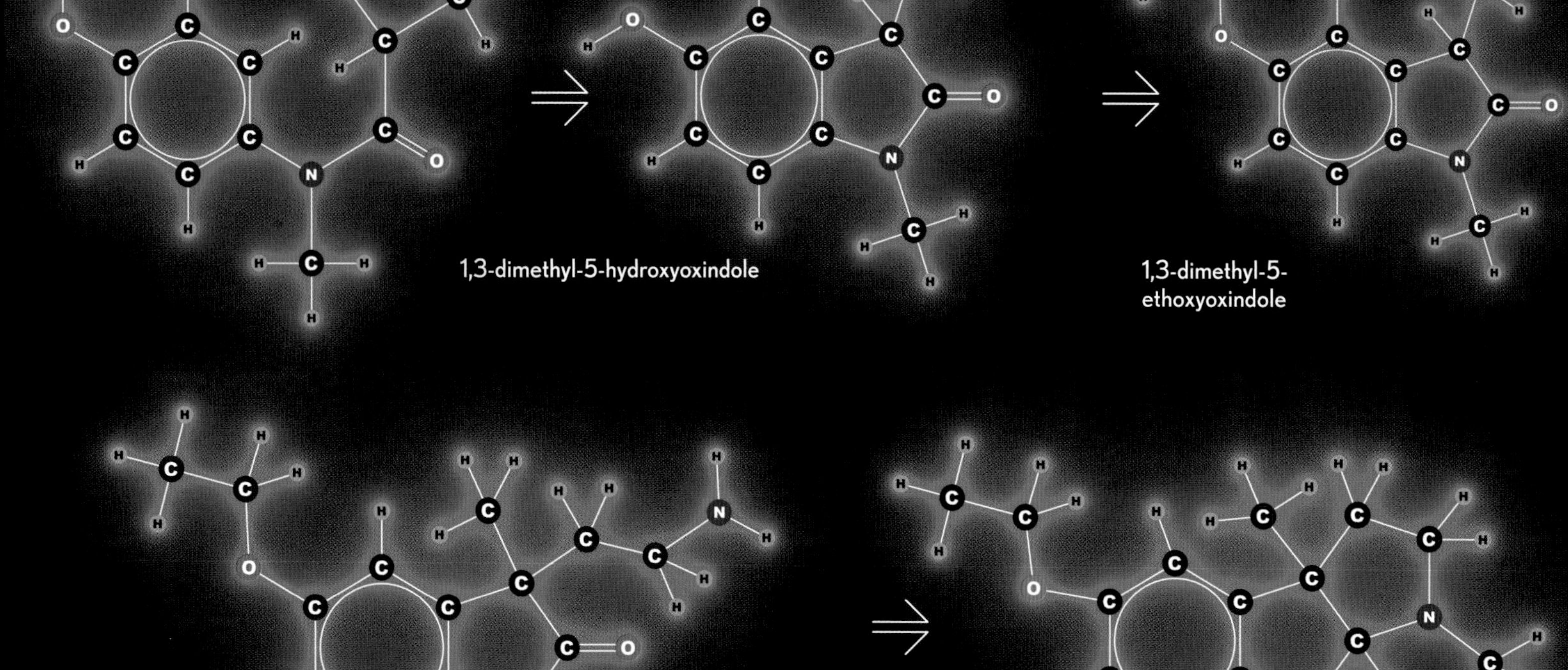

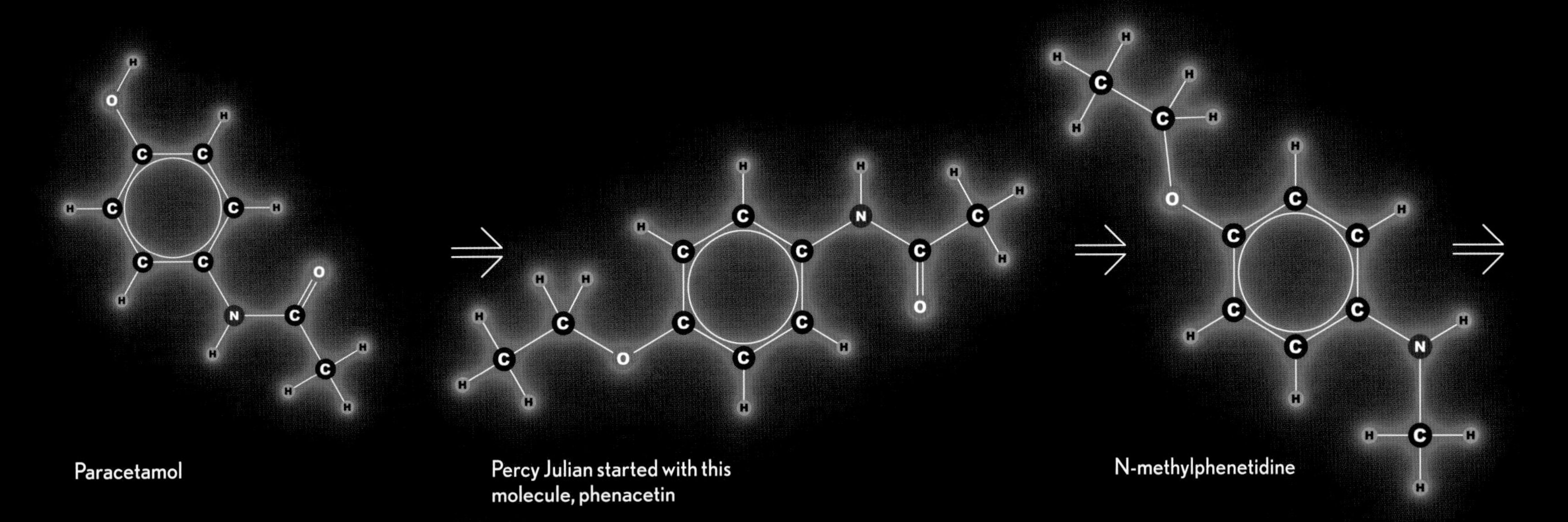

Paracetamol

Percy Julian started with this molecule, phenacetin

N-methylphenetidine

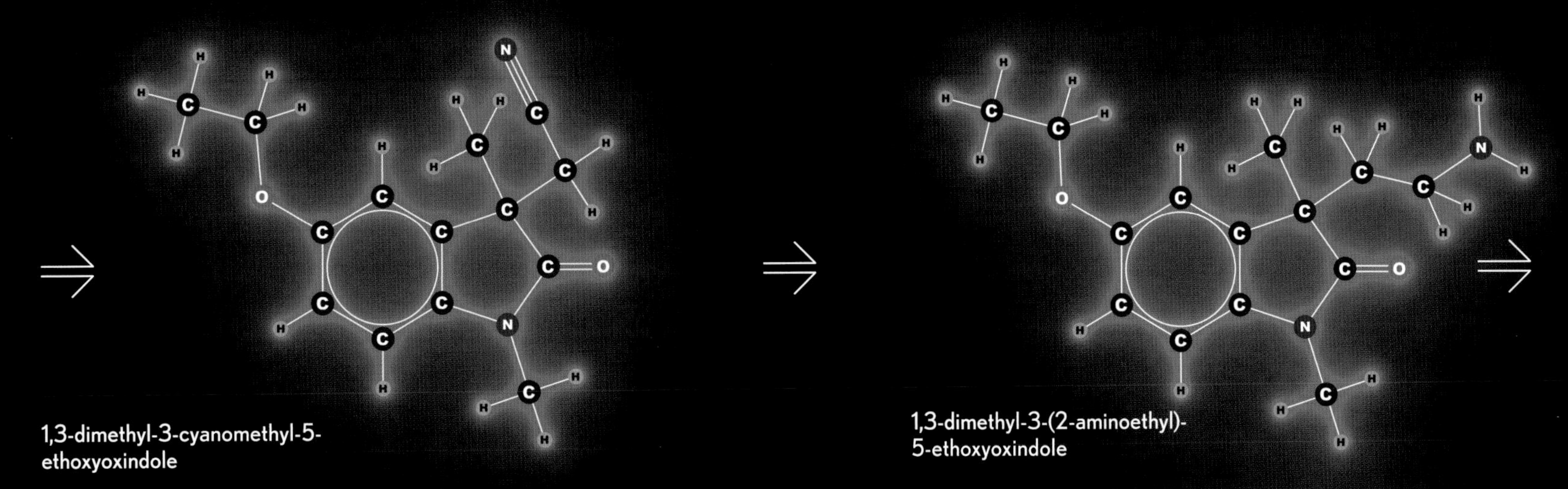

1,3-dimethyl-3-cyanomethyl-5-ethoxyoxindole

1,3-dimethyl-3-(2-aminoethyl)-5-ethoxyoxindole

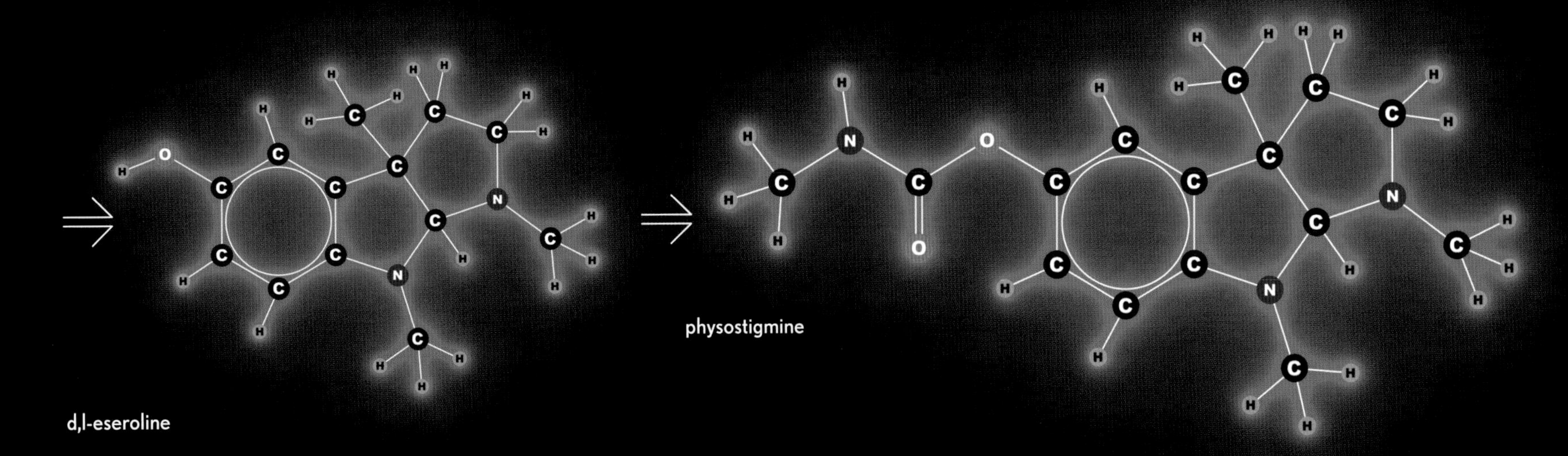

d,l-eseroline

physostigmine

When I was taking an advanced organic chemistry lab class (the one that weeds out future chemists from the tourists), the old stockroom attendant told stories of the good old days before "safety" and "not poisoning the students" became such an obsession.

I thought he was kidding when he said they used to do a lab exercise where they would feed a large dose of a certain chemical to all the students, then have them collect their pee overnight. The next day in class the students would distill and extract a reaction product that the students' own bodies had synthesized from the precursor they had been fed.

Only many years later did I track down the truth. He wasn't kidding. This biosynthesis (a chemical synthesis carried out within a living creature) of hippuric acid was actually done by (or should I say done on?) students for many years. I am *so* tempted to try it myself, but the procedure is actually quite involved, and the stuff probably tastes terrible, so I will be content to reproduce this paragraph from the book, published in 1935, that gives the procedure without the slightest irony.

increase in the hippuric acid [illegible] benzaldehyde (C_6H_5CHO), toluene ($C_6H_5CH_3$), cinnamic acid ($C_6H_5CH = CHCOOH$), and other similar compounds.

The ingestion of small amounts of sodium benzoate appears to have no harmful effects, but the amount which can be tolerated is of course not indefinite. Experiments with doses as large as 50 g. per day have shown that if the benzoic acid is increased beyond a certain point it is not all conjugated but is excreted as such in the urine (sometimes causing diarrhea). It has been concluded that the body has available for conjugation a maximum of 13 g. of glycine per day (corresponding to 30 g. of sodium benzoate).

Human urine contains a considerable quantity of urea (see Experiment 23), together with various acids (hippuric, uric, phosphoric, sulfuric, and other acids) present largely in combination with basic substances (ammonia, creatinin, purine bases, sodium and potassium salts). The hippuric acid is probably present in the form of a soluble salt, and the simplest method of isolating it is to acidify the urine and to add an inorganic salt to further decrease the solubility of the organic acid. Extraction processes are more efficient but also more tedious.

Procedure.[1] — Ingest a solution of 5 g. of pure sodium benzoate in 200–300 cc. of water (or increase the hippuric acid output through special, measured diet), and collect the urine voided over the following twelve-hour period. If it is to be kept for

[1] Plimmer, "Practical Organic and Bio-Chemistry," p. 172 (1926); Adams and Johnson, "Laboratory Experiments in Organic Chemistry," p. 321 (1933). For an alternate procedure, see Cohen, "Practical Organic Chemistry," p. 344 (1930).

Experiments in Organic Chemistry by Louis F. Fieser, D.C. Heath and Company, 1935

Why am I telling you so much about organic syntheses and complex laboratory glassware? Because that class is when I realized I didn't want to be an organic chemist. Not because of the pee, but because I realized that I enjoyed the names of the chemicals—the beautiful, mathematical logic of their construction—more than the chemicals themselves. Don't get me wrong: I love chemicals! But if you're a serious chemist working in a lab, you don't get to play with the fun chemicals and blow stuff up. I also wasn't particularly good in that class: too much careful measuring and weighing things out.

I went another way in life, but this only makes me admire more the skill and dedication of the people who have developed the art and science of wet chemistry to such a high degree.

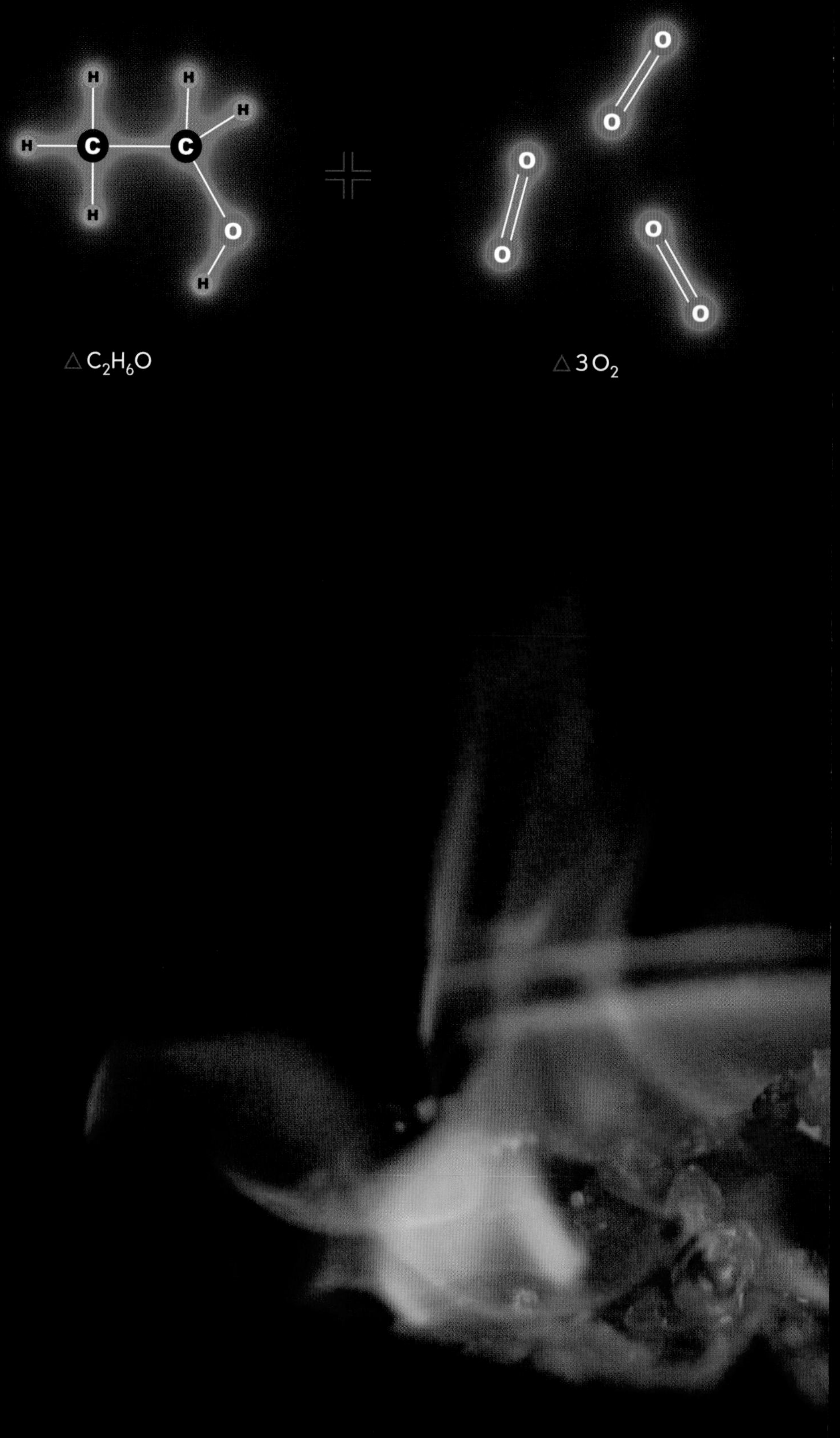

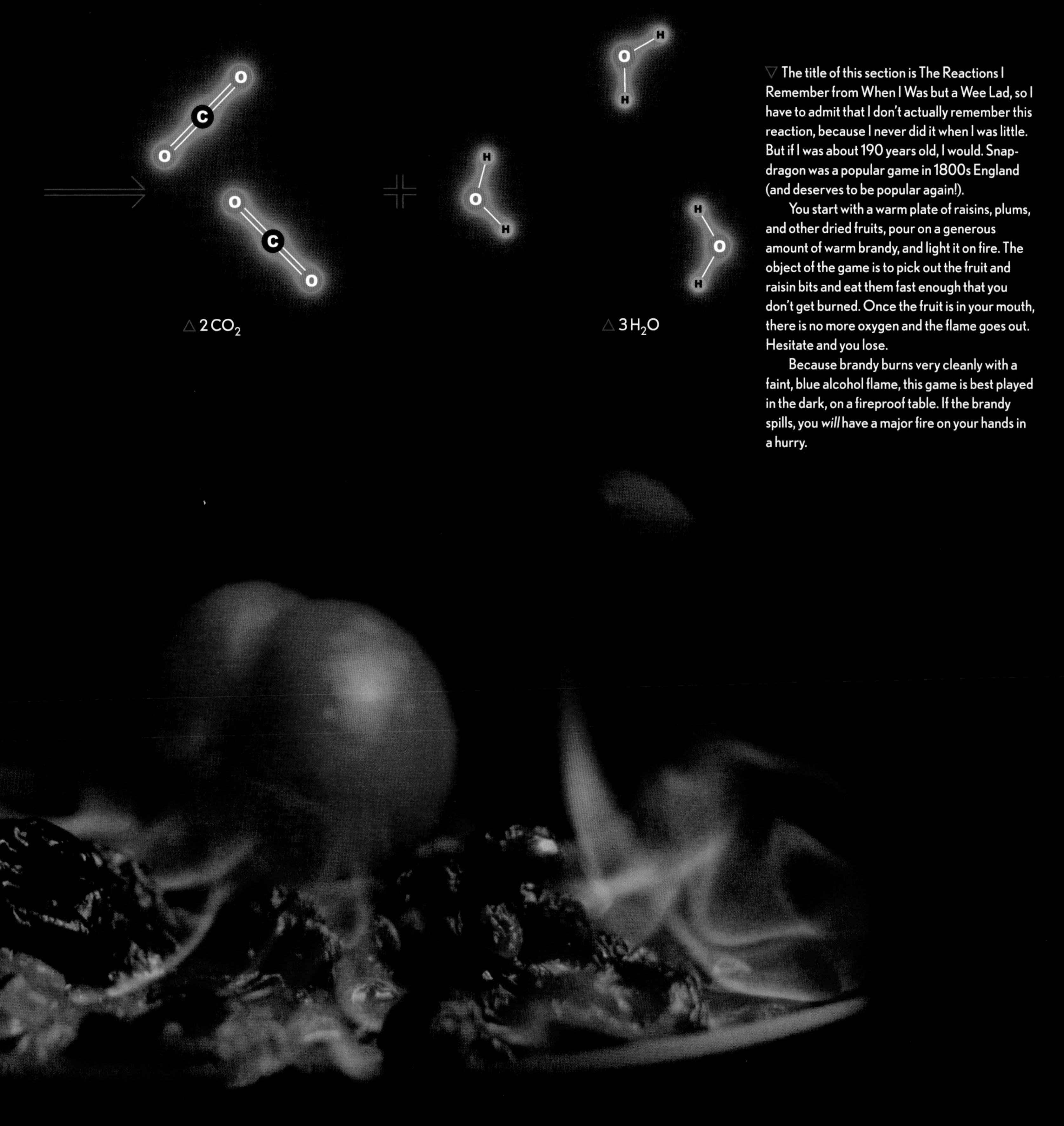

▽ The title of this section is The Reactions I Remember from When I Was but a Wee Lad, so I have to admit that I don't actually remember this reaction, because I never did it when I was little. But if I was about 190 years old, I would. Snapdragon was a popular game in 1800s England (and deserves to be popular again!).

You start with a warm plate of raisins, plums, and other dried fruits, pour on a generous amount of warm brandy, and light it on fire. The object of the game is to pick out the fruit and raisin bits and eat them fast enough that you don't get burned. Once the fruit is in your mouth, there is no more oxygen and the flame goes out. Hesitate and you lose.

Because brandy burns very cleanly with a faint, blue alcohol flame, this game is best played in the dark, on a fireproof table. If the brandy spills, you *will* have a major fire on your hands in a hurry.

Atoms, Elements, Molecules, Reactions

ALL OF THE WORLD is made of elements—copper, cobalt, calcium, and the others. All these elements are made of atoms, each according to its own kind. Iron is made of iron atoms, carbon is made of carbon atoms, and so on, and so it has always been.

These atoms are ancient and unchanging, lasting, most of them, for the better part of eternity. The cores of heavy atoms are held in padded rooms, walled off by dozens of electrons, for spans of time measured against the age of the universe. These serene cores feel nothing of the wider world, save only for vague magnetic tugs that whisper of the goings on around them.

But surrounding the atom's calm interior is a maelstrom of activity. Outer electrons are in constant flux, coming and going, trading places, gathering and spreading, on a timescale a thousand million million times faster than our everyday world. This is the world of reactions.

Elements, and the molecules they combine into, are the nouns of the physical world. They are the stuff of which you, I, and everything else is made.

Chemical reactions are the verbs of this world, the action words. For the most part, when something interesting happens in the world—when a tree grows, a fire burns, or a life begins—it is the result of a chemical reaction.

The computer that contained these words as they were being written operates largely without chemical reactions. It's an electrical creature. But the mind that is reading them now—and perhaps doubting them—is nothing more and nothing less than an intricate dance of chemical reactions. The thoughts that form in your head are elaborate patterns, waves, and pulses of electricity completely controlled and shaped by chemical reactions.

To make you, it takes chemical reactions. To make reactions, it takes molecules, and to make molecules, it takes atoms.

So what exactly are atoms, and where do they come from?

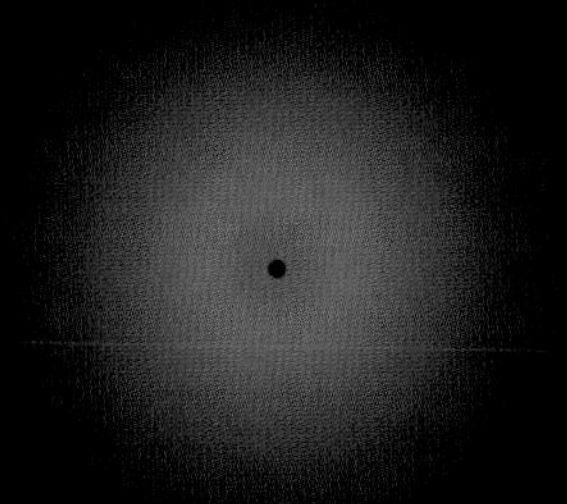

An atom cannot be seen, at least not in the conventional sense of the word: Atoms are much smaller than light. Clever instruments do now let us resolve individual atoms, but the best way to understand them is still to look at drawings, like this one, of how they are put together.

At the center of every atom is a nucleus, which contains protons and neutrons (known as "subatomic particles" simply because they are smaller than atoms). Surrounding the tiny nucleus is a cloud of electrons (another subatomic particle). You will sometimes see these electrons drawn as if they were tiny planets orbiting around the nucleus, but this is incorrect. Electrons in an atom are not really in any one place, but rather exist as a sort of wave of probability filling the space around the nucleus. I only ever draw them as the clouds that they are.

As we will soon learn, what matters most about an electron isn't so much where it is or what it looks like, but rather the *energy* of its position.

△ Ninety percent of the atoms in the universe are hydrogen atoms, each with one proton surrounded by one electron. Nearly all of them were formed immediately after the big bang, and haven't changed since, other than to trade their electrons with each other during their infrequent meetings in the deep cold of space. They have never been part of a star, fallen as rain, or helped shape a molecule of DNA. Participation in the excitement of a reaction is a rare privilege in the universe.

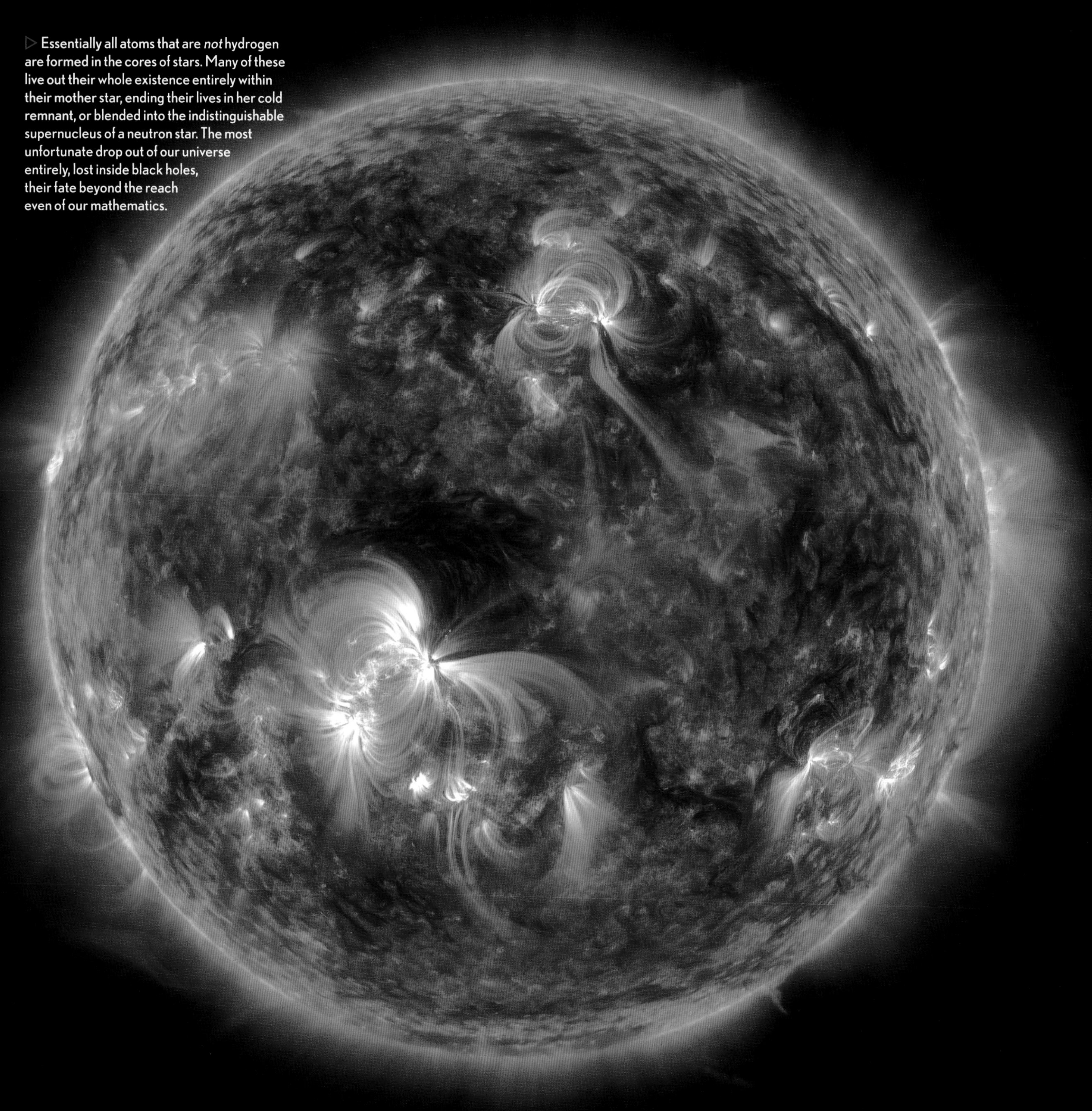

▷ Essentially all atoms that are *not* hydrogen are formed in the cores of stars. Many of these live out their whole existence entirely within their mother star, ending their lives in her cold remnant, or blended into the indistinguishable supernucleus of a neutron star. The most unfortunate drop out of our universe entirely, lost inside black holes, their fate beyond the reach even of our mathematics.

▷ A few lucky atoms are born inside stars destined to end life in the spectacle of a supernova, an explosion of such scale and power that words are largely useless in describing it. Traveling at the speed of light, 186,000 miles (300,000 km) per second, it would take you five and a half *years* to get from one side of this thing to the other. Yet it is nothing more than the puff of smoke put out by a single supernova. (We call it the Crab Nebula.)

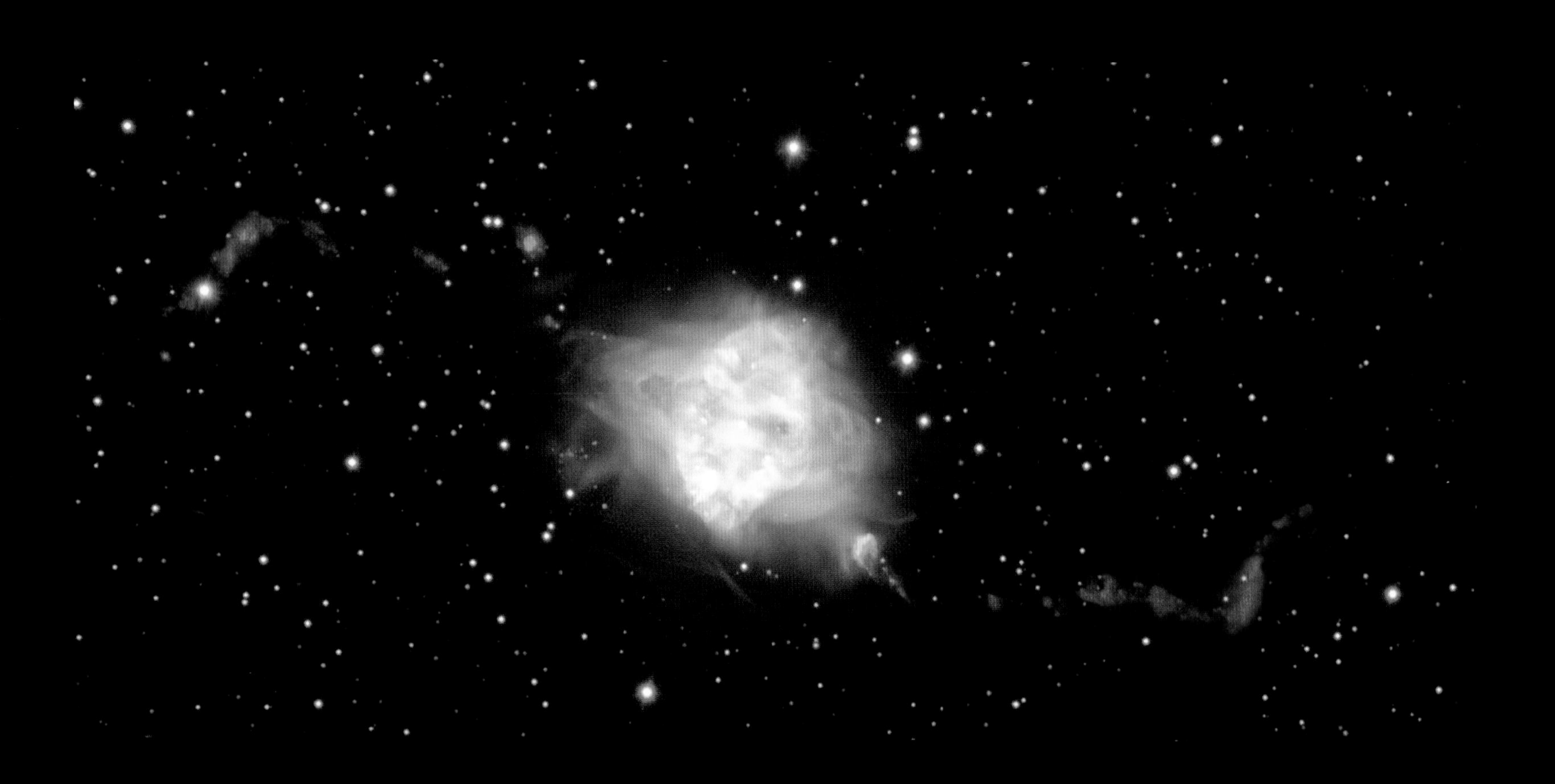

△ Matter thrown out of a supernova is given a second chance to do something interesting. These atoms of helium, carbon, oxygen, sodium, calcium, and so on, eventually form both new stars and the planets that surround them. As planets slowly coalesce out of their accretion disks, the star dust they are made of comes together to form their hot cores, rocky crusts, rolling oceans, blue skies, and curious inhabitants. Look at your hands. Stars died to make you. Never forget that.

When a planet forms in just the right way and at just the right distance from its host star, life may evolve on it. If that life is clever and hardworking, it will eventually figure out all of the previous, and construct the ultimate catalog of matter, the periodic table of the elements. The details might differ a bit from planet to planet, but the fundamental shape and logic of the table will be the same everywhere, because the laws of physics that determine it are the same.

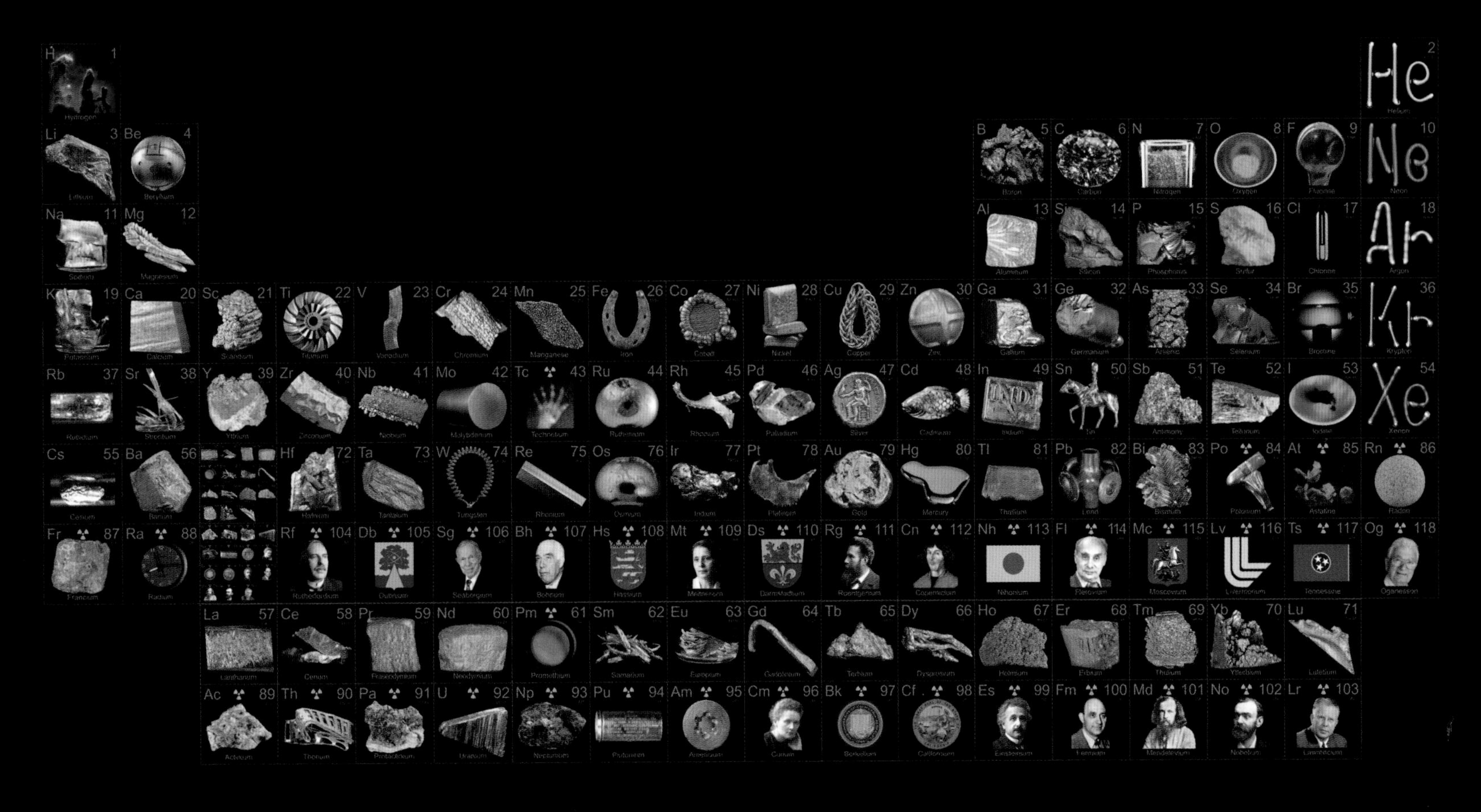

Each element is like a LEGO block of a particular shape, able to combine with itself and with other "blocks" only in certain ways. Each column is like a class of blocks with similar shapes. For example, any block from the first column can always combine with any block from the next-to-last column. The rules for other columns get more and more complex and subtle the closer you go toward the middle. Because this isn't a textbook, we're not going to go into those rules in detail, but just know that they exist, and they can be learned. There is no mystery about how atoms of different elements behave, but there is a great deal of rich behavior that rewards the diligent student.

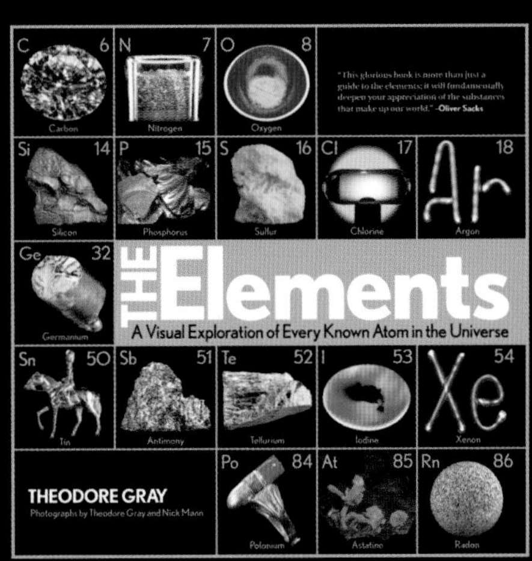

If you want to learn more about the elements, both in general and individually, I wrote a whole book about them. *The Elements: A Visual Exploration of Every Known Atom in the Universe* has two pages about each and every element. (Except the silly ones, elements 101 through 118, which I don't like because it's not possible to collect them.)

Instead of spending a lot of time talking about elements, let's move on to what happens when you combine atoms together to create *molecules*.

What Is a Molecule?

MOLECULES ARE MADE by connecting two or more atoms to each other with chemical bonds. The smallest, simplest molecule is H_2, made of two hydrogen atoms connected by a single bond. Many common, important molecules are made of just a few atoms, or a few dozen. Sugar molecules, for example, are made of 45 atoms each—12 carbon atoms, 22 hydrogen atoms, and 11 oxygen atoms, connected in the particular way shown in this diagram.

H—H

▷ The properties of molecules are usually completely different from the properties of the pure elements they are made of.

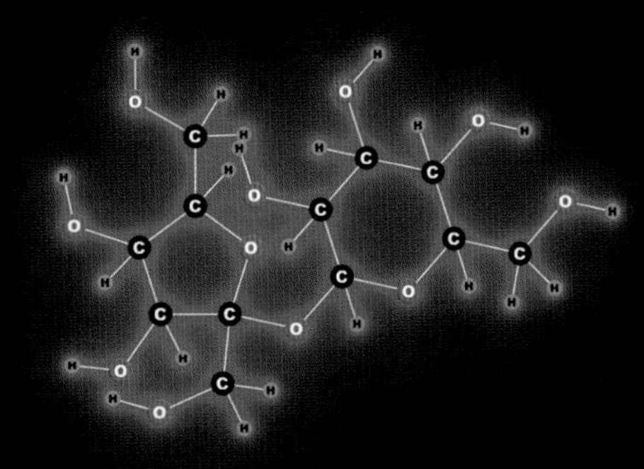

The four pictures on this page are all examples of elements in their pure forms.

Three of these four elements are seriously dangerous.

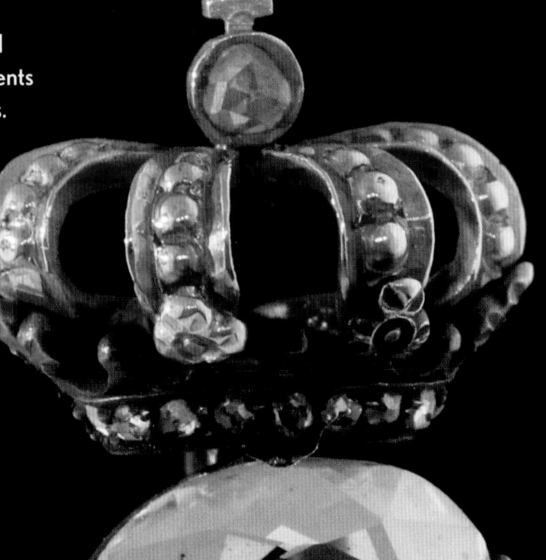

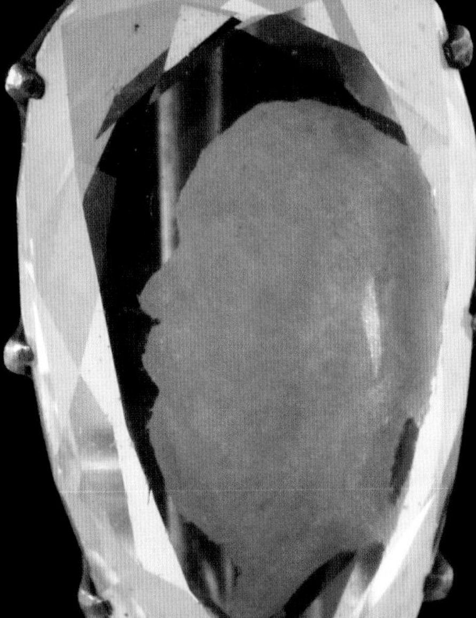

▷ A diamond carved with the image of a king is pure carbon.

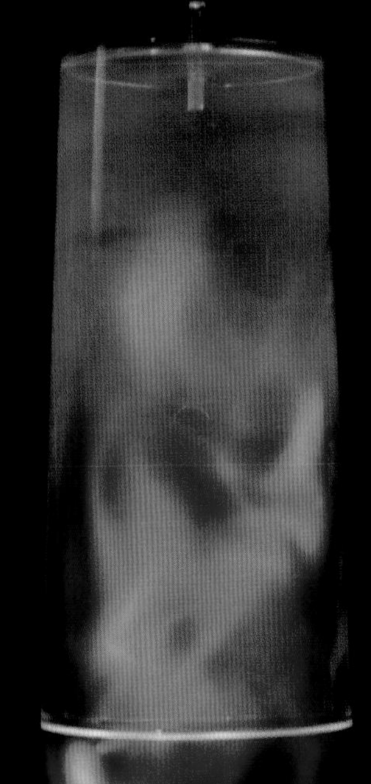

◁ Hydrogen fills a jar only when it's upside-down, because hydrogen floats upward in the air. It also explodes when lit.

▽ Sodium metal explodes beautifully when merely dropped in a bowl of water.

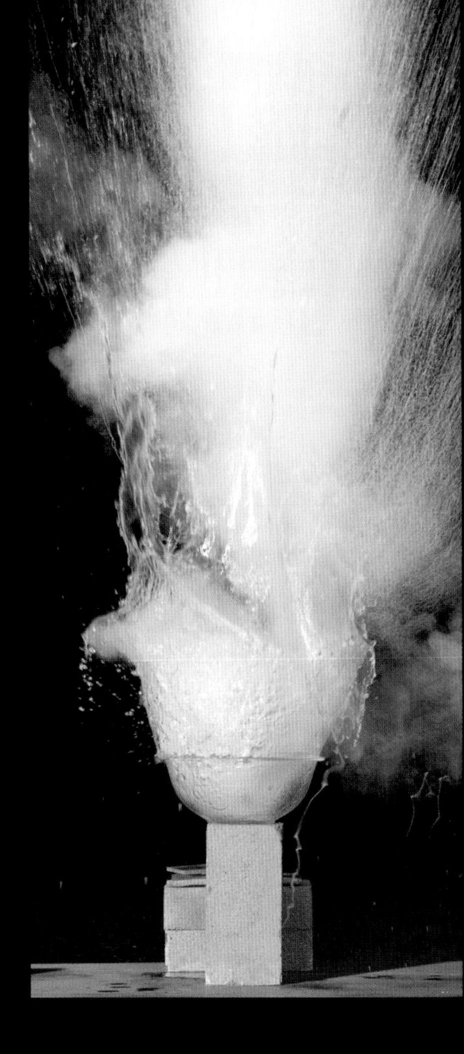

▽ Chlorine gas condenses into a pale yellow liquid when passed through a coil cooled with liquid nitrogen.

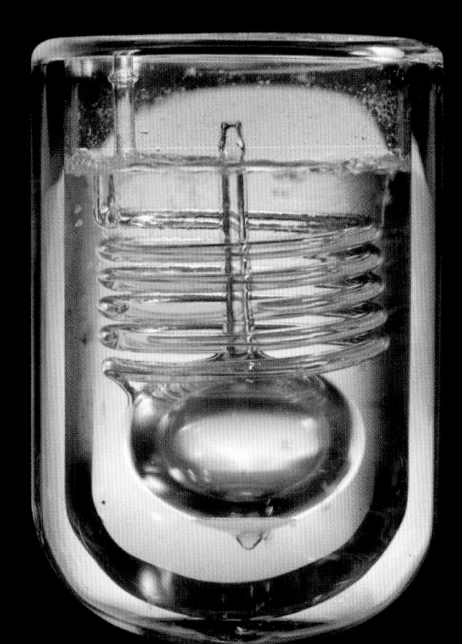

▽ Different combinations of these four elements create a huge variety of compounds whose properties are nothing like those of the elements. Here are just a few examples.

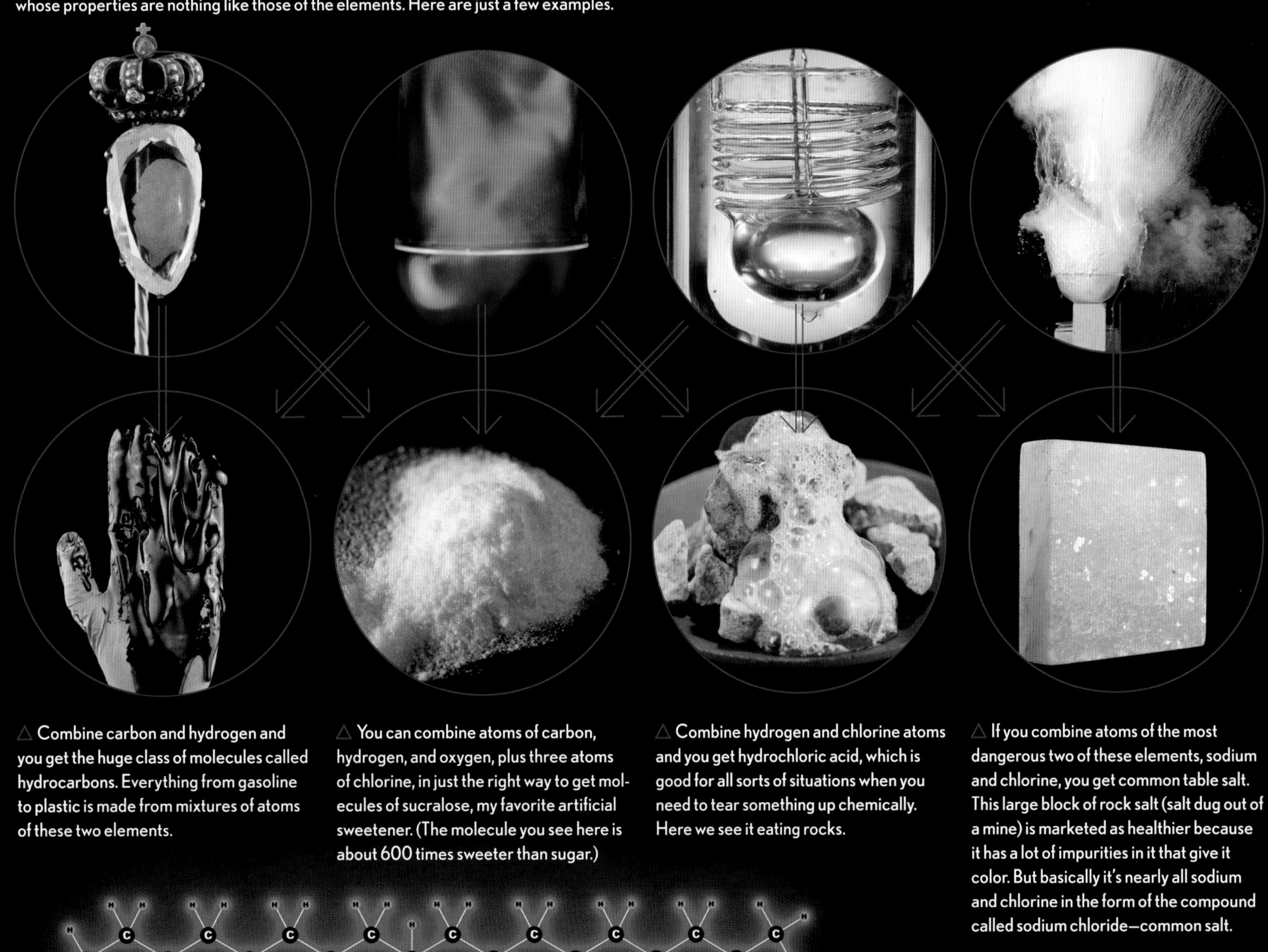

△ Combine carbon and hydrogen and you get the huge class of molecules called hydrocarbons. Everything from gasoline to plastic is made from mixtures of atoms of these two elements.

△ You can combine atoms of carbon, hydrogen, and oxygen, plus three atoms of chlorine, in just the right way to get molecules of sucralose, my favorite artificial sweetener. (The molecule you see here is about 600 times sweeter than sugar.)

△ Combine hydrogen and chlorine atoms and you get hydrochloric acid, which is good for all sorts of situations when you need to tear something up chemically. Here we see it eating rocks.

△ If you combine atoms of the most dangerous two of these elements, sodium and chlorine, you get common table salt. This large block of rock salt (salt dug out of a mine) is marketed as healthier because it has a lot of impurities in it that give it color. But basically it's nearly all sodium and chlorine in the form of the compound called sodium chloride—common salt.

▷ 9-heptyloctadecane is made of carbon and hydrogen. It's sticky.

▷ Sucralose is made of carbon, hydrogen, oxygen, and chlorine. It's ridiculously sweet.

▷ Salt is made of three-dimensional crystals of alternating sodium and chlorine atoms.

What Force Holds Molecules Together?

WHAT DO THE LINES in the diagrams you've just seen actually mean? Each one represents a chemical bond. These bonds are a result of forces between electric charges. So, to understand bonds, we need to understand electric charge.

It's not hard to find clues that electricity is in everything. Literally just walking across a carpet can trigger the release of electricity. When you shuffle across a carpet, or rub a balloon on your hair, electrically charged particles called electrons are transferred to your body or to the balloon, where they become trapped. That's called *static electricity*.

"Static" in this context means "not moving"—water in a bucket is static water, water flowing in a river is a current of water. Electrons trapped in your body are static electricity, electrons flowing through a wire are an electric current. When you touch a doorknob and get a spark, that is a short-lived electric current created when the electrons trapped in your body escape and flow into the doorknob.

It's also not hard to see that electric charges can create a force, which we call the electrostatic force. When you stick a balloon on the wall after rubbing it, the force holding it to the wall is this kind of electrostatic force. (When talking about polyester clothes, for example, the term "static cling" is short for "electrostatic force causing your dress to look like it's stuck to your butt".)

Electric charges come in two kinds: positive and negative. Charges with the same sign (both positive or both negative) push each other apart (electrostatic repulsion). Opposite charges (one positive, one negative) pull toward each other (electrostatic attraction). A balloon stuck to the wall is stuck there because negative charges on its surface are attracted to nearby positive charges in the wall.

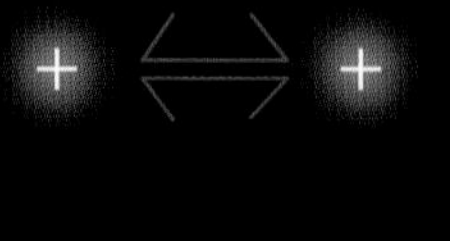

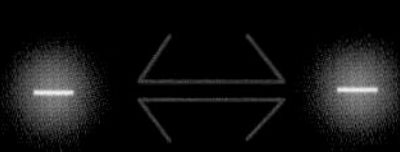

▷ The core of every atom contains one or more protons, which have a positive electric charge. The electrons smeared out around the outside of the atom have a negative electric charge. The electrons stay close to the nucleus because there is an attraction between their negative charge and the positive charge in the nucleus.

In a normal "neutral" atom, there are exactly the same number of electrons on the outside as protons in the inside. Each electron has exactly the opposite amount of charge as each proton, so as a whole the charges exactly cancel each other out, leaving the atom with zero charge overall.

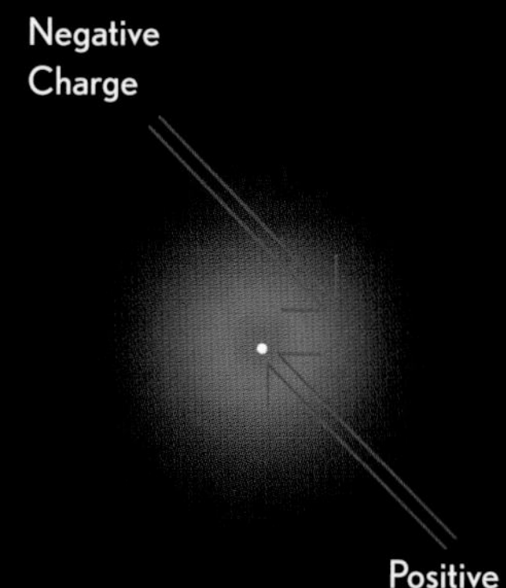

How do these forces combine to create a chemical bond? Very cleverly.

▽ Imagine two positive charges a short distance apart. If you let them go, they are going to fly apart, because like charges repel each other.

▽ But what if you put a negative charge between the two positive charges? The plus charge on the left is attracted to the minus charge in the middle, and so is the plus charge on the right. Those two attractions are stronger than the repulsion between the two plus charges, because the plus and minus charges are closer to each other than each plus charge is to the other plus charge. This collection of charges as a whole will pull itself together, not fly apart.

This is almost exactly how chemical bonds are formed.

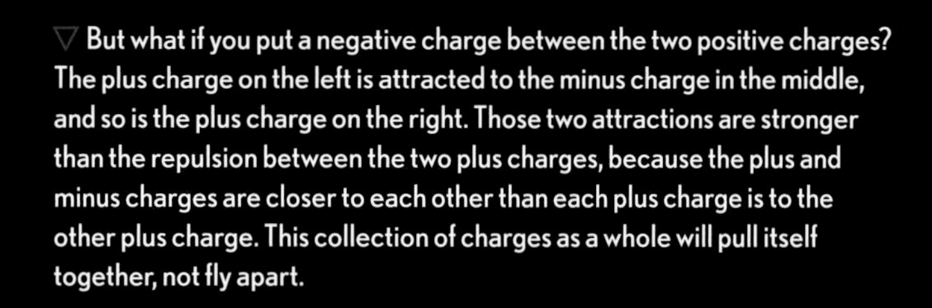

◁ On the left we see what happens when two hydrogen atoms come close together. Their electrons form a pool of negative electric charge (represented by the purple glow) between the two hydrogen nuclei. This collection of particles is called an H_2 molecule, which is a good, stable molecule.

▷ On the right we see what happens when two helium atoms come together. See the purple glow to the left and right of the pair of nuclei? That is negative electric charge that is pulling the atoms apart. Helium atoms bounce off each other without forming any bond.

Knowing whether you'll get a bond or not is a big part of what people learn when they study chemistry. It's more complicated than what I can cover here but, rest assured, it is knowable, and it makes sense once you have studied the quantum mechanics that underlies all of the rules for bonding.

(I know it's frustrating for me to say "it's complicated," but frankly, it is complicated. Look at it this way: It's an opportunity for you to learn so much more than I can possibly put in one book. It's wonderful and amazing how much we know about how protons and electrons interact with each other. There's no way around the fact that there's a lot to know, and it takes a while to learn it all. But what a boring place the world would be if I could explain it all in one book!)

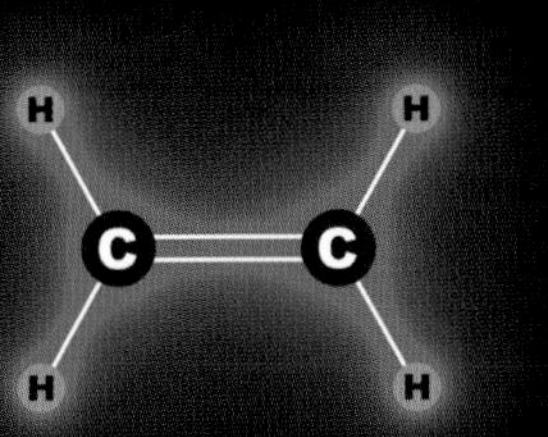

◁ This example of a molecule, ethylene, shows how I draw molecules in this book. It's made of two carbon atoms and four hydrogen atoms (the disks labeled C and H). Each white line represents one pair of electrons that are participating in a chemical bond, pulling two of the atoms together by electrostatic force. Reading the diagram, we see that each hydrogen atom is bonded to a carbon atom by two electrons (one pair, one line). The two carbon atoms are bonded to each other by four electrons (two pairs, two lines, known as a double bond). Very rarely you will see a triple bond (six electrons), and sometimes you will see an interesting circle drawn inside a ring of six carbon atoms: This circle represents a total of six electrons that are shared equally among all six of those atoms. For this book, you don't need to worry much about these details.

I use purple glows to represent the negative electric charge of the electrons that form molecules. In the two examples on the opposite page showing hydrogen and helium atoms approaching, the glow is mathematically correct—it represents the accurately calculated "electron density" around those atoms at those positions.

But in the other molecules in this book (and my other books), the glow is purely metaphorical. I calculate it using a simplified formula mainly designed to look pretty. (It would be pointless to use the real thing when the molecules are not actually flat anyway.) It's there because I like purple and it lets black carbon atoms stand out from the black background. And, in a gentle way, it reminds us that there is a fuzzy, diffuse cloud of electrons spread out around the whole molecule, particularly near the centers of the atoms and between the pairs that are bonded to each other.

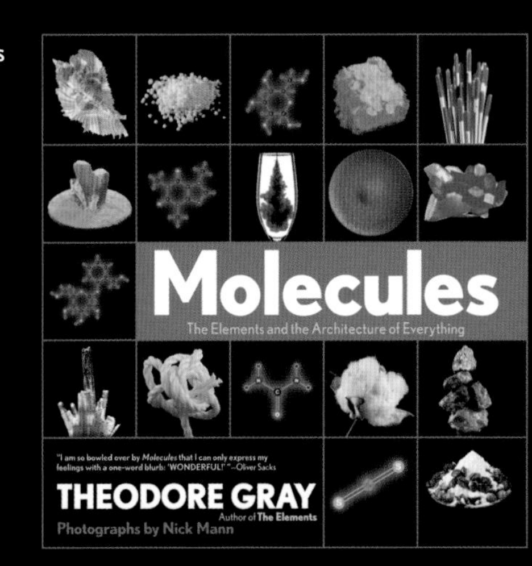

△ If you're interested in all the different ways atoms can be assembled into molecules, I've written a whole book about that. *Molecules: The Elements and the Architecture of Everything* contains hundreds of examples of molecules, showing both a picture of what they look like on a human scale, and a diagram of the way their atoms are connected.

Rather than spend a lot of time on molecules, we're going to go straight to talking about how molecules can be brought together to trigger *reactions*.

What Is a Chemical Reaction?

A CHEMICAL REACTION happens when chemical bonds are created or destroyed. Since these bonds are made of electrons, another way of saying it is that chemical reactions happen when electrons move to find new bonding opportunities between different sets of atoms.

Reactions, in short, are electrons in motion.

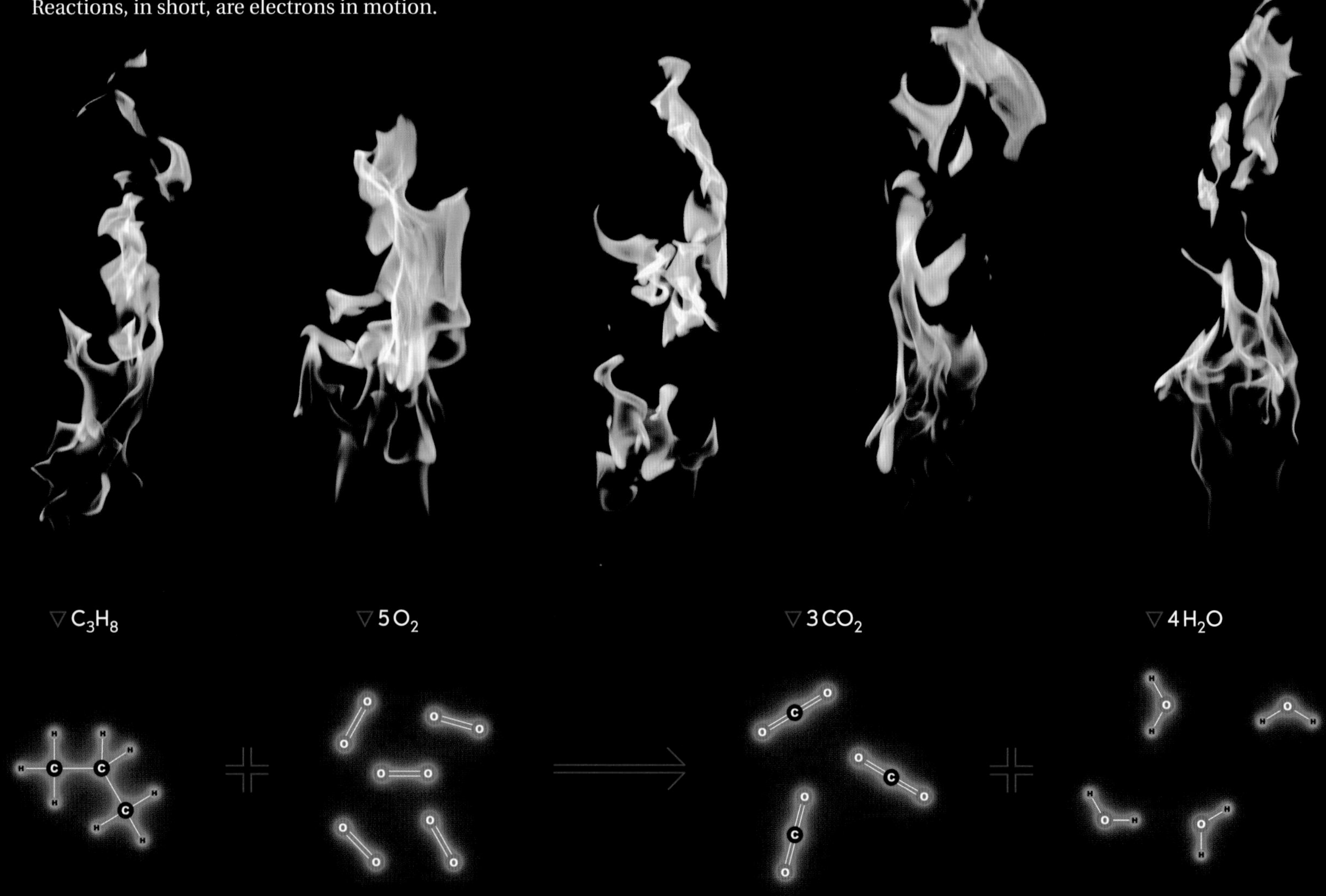

△ This "reaction expression" describes the reaction that happens when propane (C_3H_8, often used in camp stoves and rural homes), burns in the air, combining with oxygen (O_2) to form carbon dioxide (CO_2) and water (H_2O).

You read a reaction expression from left to right. On the left are the "reactants" you are starting with. On the right are the "products" you end up with. In the middle is the arrow, which represents the process that gets you from reactants to products: The arrow represents the reaction itself.

(Normally you see chemical reaction expressions written out in plain text form. For example, propane in text form is C_3H_8, because it's made of three carbon atoms and eight hydrogen atoms. So you would normally see this reaction written in plain form as $C_3H_8 + 5\,O_2 \rightarrow 3\,CO_2 + 4\,H_2O$. But in this book I am always going to write reactions using diagrams of the molecules involved, because that makes everything clear and real and it's pretty and my publisher is willing to pay for the extra paper.)

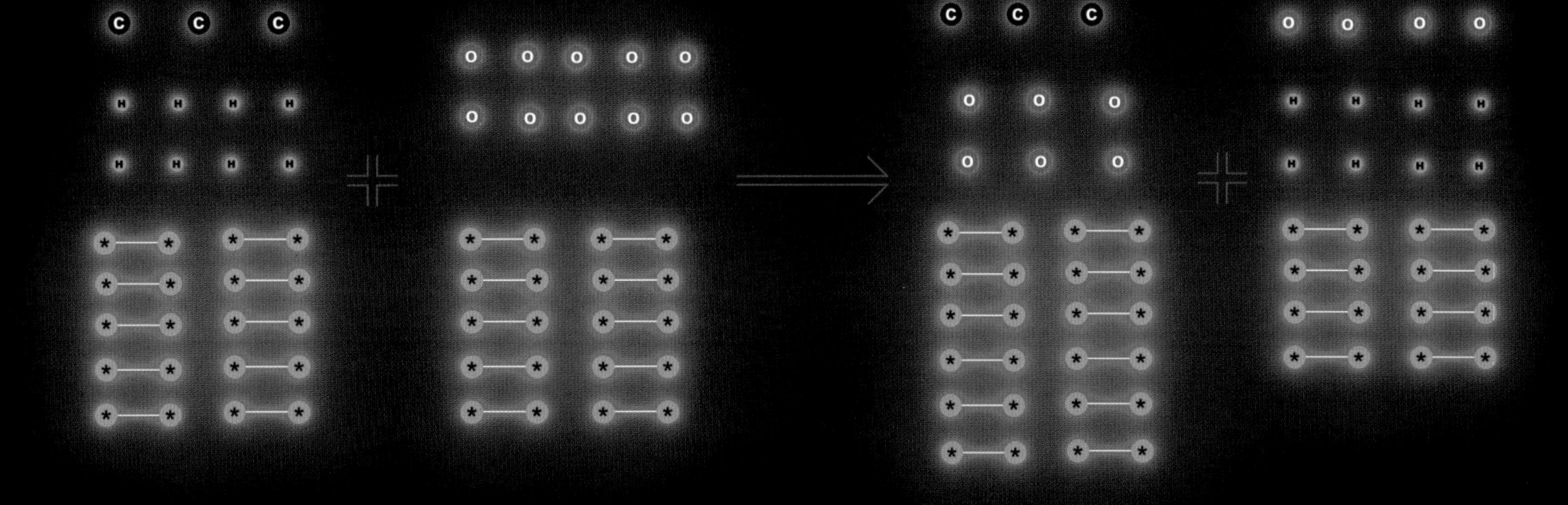

If you count the atoms of each kind on both sides of the reaction, you'll see that there are exactly the same number of each type: Three carbon atoms, ten oxygen atoms, and eight hydrogen atoms. That's *always* the case in every chemical reaction, period, no exceptions. Atoms are matter, and matter can never be created or destroyed by any chemical reaction. (To do that you need *nuclear* reactions, which are a whole other can of worms, and a whole other book.)

If you count the number of *lines* connecting the atoms you'll see that it's also the same: A total of twenty lines on each side. Since each line represents one pair of bonding electrons, that means there are forty electrons involved in holding the molecules together both before and after the reaction happens. That's less of an absolute rule, but it's often the case.

These two rules—keep the atoms the same and keep the line count constant—tell you a lot about what reactions can and can't happen. But they alone are not enough.

▽ Some silly results of propane combustion.

▽ C_3H_8 ▽ $5\,O_2$ ▽ C_3O_4 ▽ H_4 ▽ HO_2 ▽ H_3O_2 ◁ H_2O_2

To a chemist, this reaction expression is almost painful to look at, even though it obeys the rules we just learned. These molecules are just totally **wrong**. Wrong on so many levels.

There are many possible arrangements of these particular atoms with twenty lines, so what makes one arrangement happen spontaneously when propane is burned, and the other one make chemists laugh at the absurdity? To understand this we need to understand the two forces that drive all change in the world: energy and entropy.

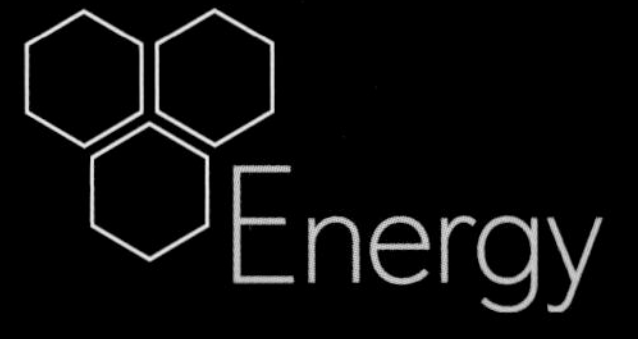

Energy

ENERGY IS LIKE an agitation, an itch that the universe needs to scratch. Something that contains energy is restless. It is either in motion (which is called kinetic energy) or it's coiled up, tense, and waiting, like in a compressed spring, or a rock balanced high on a mountain ledge (that's called potential energy).

Energy is "conserved," which means that, so far as we know, you can never create or destroy it, only move it or transform it from one kind to another. (Or hide it in the form of matter, which is a topic for a book on nuclear reactions.) This principle is called the *conservation of energy*.

People say that if you want to understand what's going on in business, politics, or crime, you should "follow the money." If you want to understand the natural world, you should instead follow the energy. Not much happens without the movement and transformation of energy.

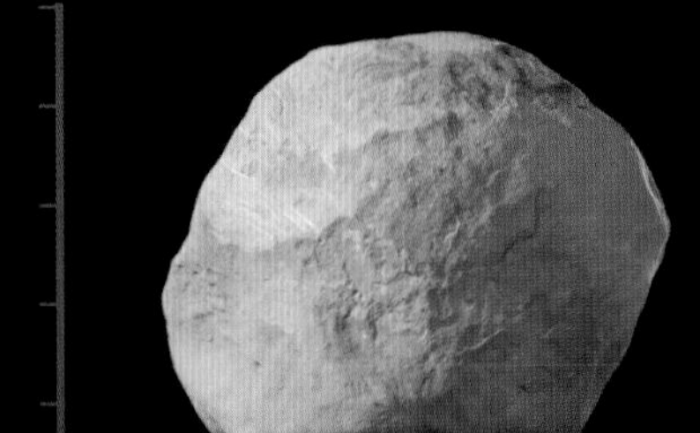

▽ A falling rock has less potential energy (because it's lower down), but more kinetic energy (because it's moving fast). Potential energy has been converted into kinetic energy.

▽ When the rock hits the ground, it stops, so now it's both low down (low potential energy) and not moving (no kinetic energy). Where did the energy go? It was converted into vibrations in the air (sound) and the ground, which ultimately ended up as heat: The falling rock warmed up the ground a little bit. Ultimately the energy will end up dispersed throughout the world as an imperceptibly tiny increase in temperature.

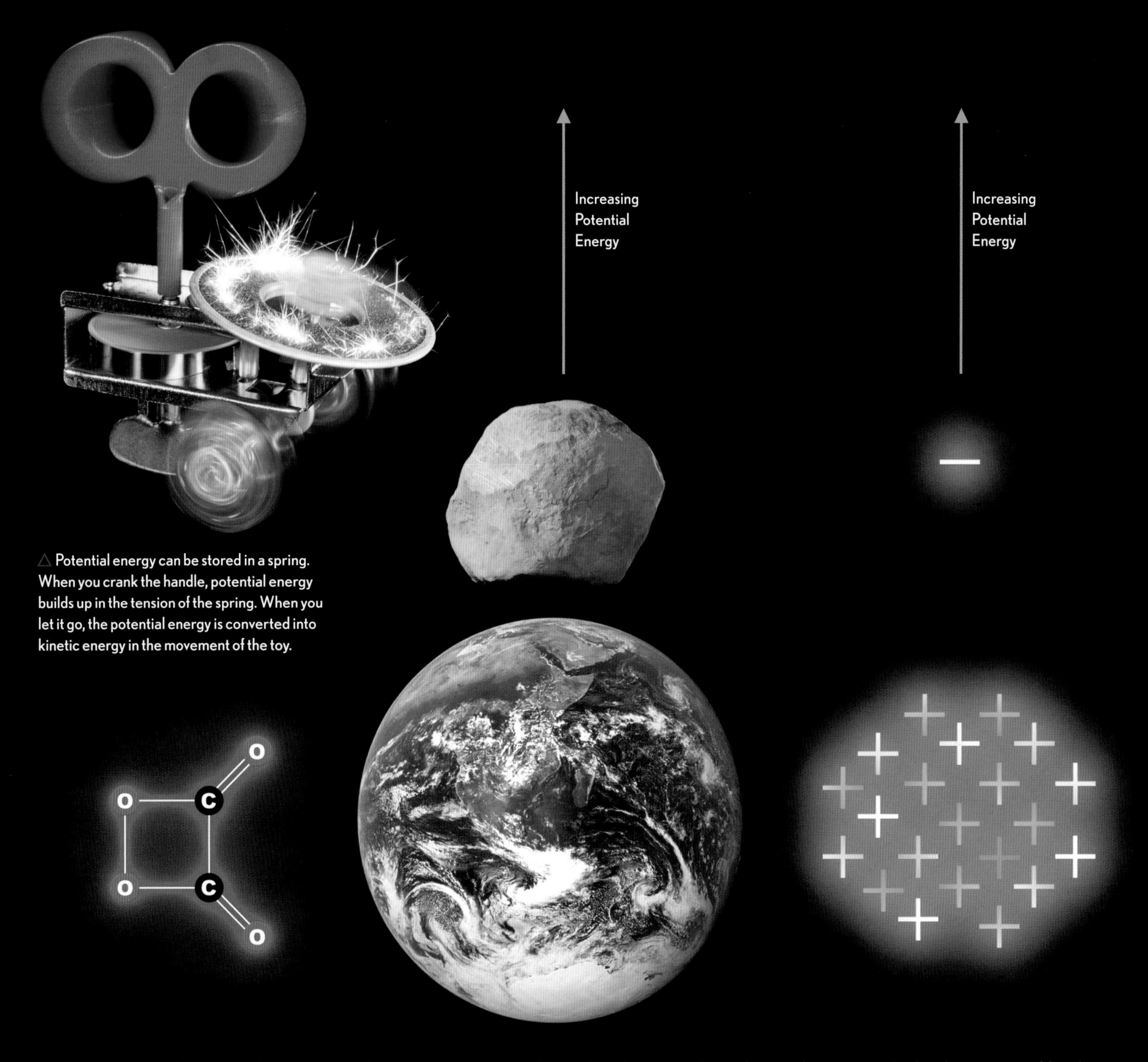

△ Potential energy can be stored in a spring. When you crank the handle, potential energy builds up in the tension of the spring. When you let it go, the potential energy is converted into kinetic energy in the movement of the toy.

△ We saw this molecule, peroxyacid ester, in the last chapter. Like a spring, it too contains potential energy in the wound-up, stressed condition of its bonds. All chemical bonds are a bit like springs: They can vibrate, and when you stretch, compress, or bend them, potential energy builds up, then gets released into the movement of the atoms they are connected to.

△ Chemical bonds can store chemical potential energy not only when they are stressed, but also by virtue of the potential energy of the electrons in them. Electrons are attracted to the positive charges in the centers of atoms, just as rocks are attracted to the center of the earth. Lifting a rock higher up, farther away from the center of the earth, gives it more potential energy. Similarly, pulling an electron farther away from a nucleus gives it more potential energy. When a rock falls closer to the earth, potential energy is released. When an electron falls in closer to a nucleus, potential energy is also released.

If you have a rock in a deep well (hole in the ground), it has low potential energy. Lifting it up to the surface takes a lot of work, because you have to give it a lot more potential energy to get it up to the level of the ground. The same rock in a shallower hole has relatively *more* potential energy compared to the one in the deeper hole, so it takes *less* work to raise it up to ground level.

Exactly the same principle applies to electrons. An electron very close to a nucleus is said to be in a deep "potential energy well." It takes a lot of work to pull it away from the nucleus. An electron farther away from the nucleus is easier to pull away, because it's starting from a higher potential energy level.

This is why some chemical bonds are stronger than others.

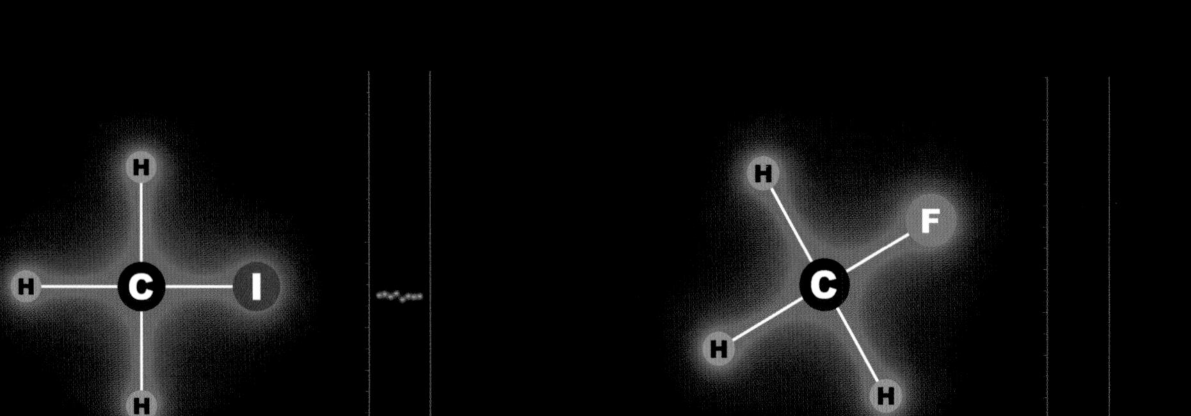

△ Different chemical bonds are made with electrons at different average distances from the centers of the atoms. For example, the bond between the carbon atom and the iodine atom in methyl iodide has the two bonding electrons relatively far away from the two atoms it is connecting. These electrons are in a shallow potential energy well, and are not hard to pull up and out of it. The C-I bond is weak.

△ The electrons connecting carbon and fluorine atoms in methyl fluoride, on the other hand, are very close to their two atoms. They are low in potential energy, in a deep well. The practical effect is that a C–F bond with its low potential energy electrons is much stronger than a C–I bond. It takes almost twice as much energy to pull the fluorine atom away from a methyl fluoride molecule as it does to pull the iodine atom out of a methyl iodide molecule.

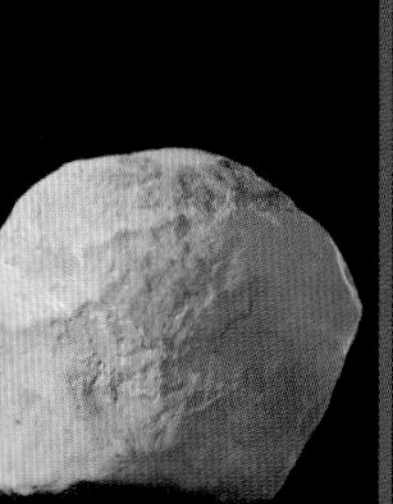

When a chemical reaction happens, some bonds are broken and new bonds are created. These bonds may have *different potential energy depths*. This is the source of the energy released by powerful chemical reactions, including fires, explosions, and the reactions that heat and power your body. The energy is released when electrons "fall" from high-potential-energy bonds into lower-potential-energy bonds, releasing the difference as kinetic energy, just like a rock falling from a high place to a low place (or from a shallow well down into a deeper well).

Let's look at how this works for a simple, and useful, way of generating heat with a chemical reaction: the combustion of methane (natural gas).

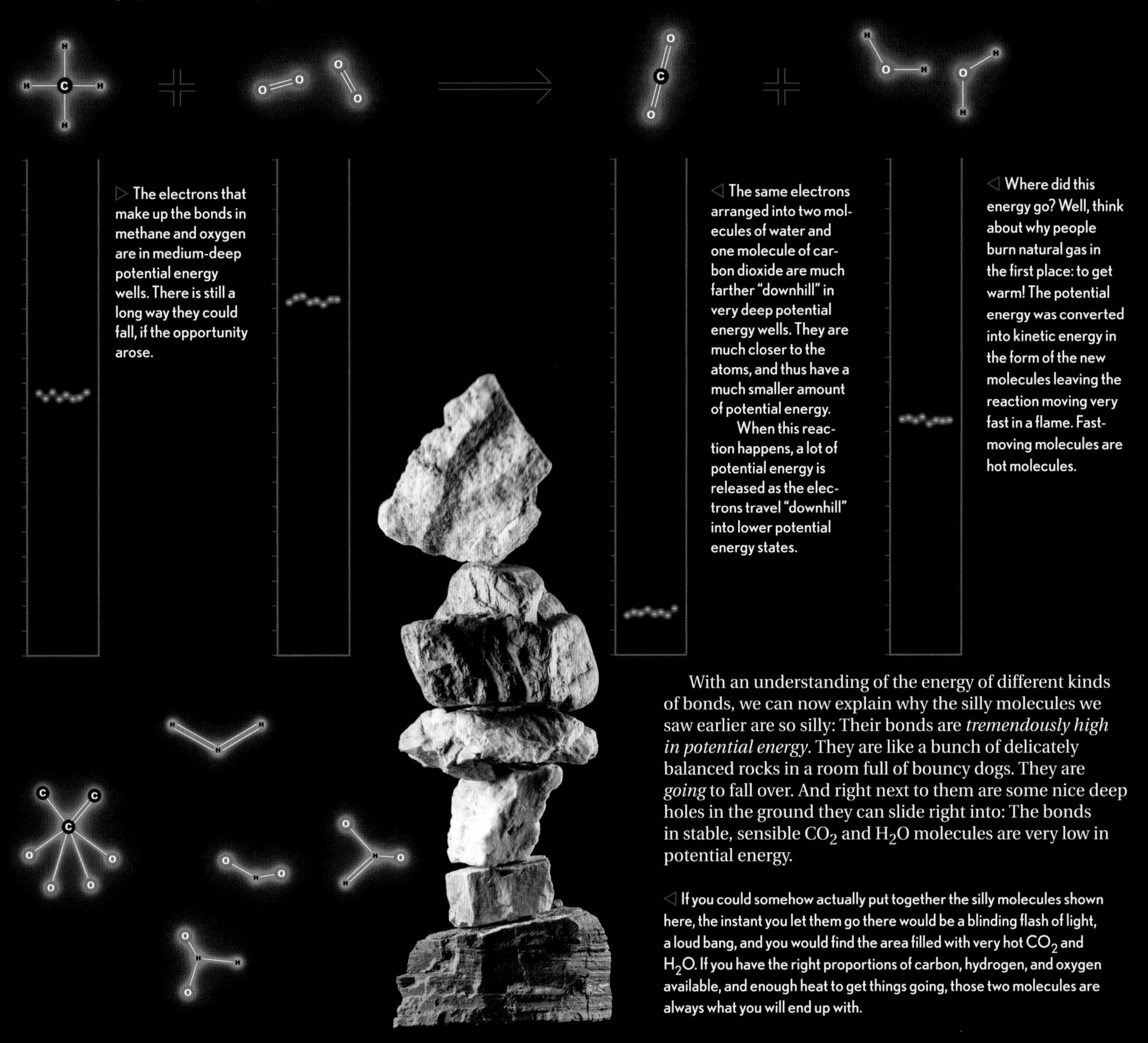

▷ The electrons that make up the bonds in methane and oxygen are in medium-deep potential energy wells. There is still a long way they could fall, if the opportunity arose.

◁ The same electrons arranged into two molecules of water and one molecule of carbon dioxide are much farther "downhill" in very deep potential energy wells. They are much closer to the atoms, and thus have a much smaller amount of potential energy.

When this reaction happens, a lot of potential energy is released as the electrons travel "downhill" into lower potential energy states.

◁ Where did this energy go? Well, think about why people burn natural gas in the first place: to get warm! The potential energy was converted into kinetic energy in the form of the new molecules leaving the reaction moving very fast in a flame. Fast-moving molecules are hot molecules.

With an understanding of the energy of different kinds of bonds, we can now explain why the silly molecules we saw earlier are so silly: Their bonds are *tremendously high in potential energy*. They are like a bunch of delicately balanced rocks in a room full of bouncy dogs. They are *going* to fall over. And right next to them are some nice deep holes in the ground they can slide right into: The bonds in stable, sensible CO_2 and H_2O molecules are very low in potential energy.

◁ If you could somehow actually put together the silly molecules shown here, the instant you let them go there would be a blinding flash of light, a loud bang, and you would find the area filled with very hot CO_2 and H_2O. If you have the right proportions of carbon, hydrogen, and oxygen available, and enough heat to get things going, those two molecules are always what you will end up with.

Follow the Energy

LET'S LOOK AT an example of how you can "follow the energy" when you put gas in a car and drive it down the road.

Gasoline is a lot like the methane and propane we just talked about. It too is a hydrocarbon (compound containing only carbon and hydrogen atoms), but the molecules in gasoline are larger, which allows them to stay liquid at room temperature. Methane is CH_4, propane is C_3H_8, while gasoline is a mixture of pentane (C_5H_{12}), hexane (C_6H_{14}), heptane (C_7H_{16}), octane (C_8H_{18}), and various other related compounds. With each new carbon atom, the molecule gets bigger and heavier, with a higher boiling point.

As in methane, the electrons in these larger hydrocarbons are relatively high in potential energy, and you can release energy by allowing them to combine with oxygen from the air, forming CO_2 and water. But how did those electrons get to be so high up in potential energy in the first place? For that, we need to go way back.

▷ Some of the sunlight that reaches the earth has the good fortune of running into one of these molecules—chlorophyll. Clusters of chlorophyll molecules are arranged in a protein scaffold that holds them in place. Primitive bacteria have them in rings, while higher plants use more complicated arrangements. When light hits these molecules, the energy in it is transferred through a complex (but remarkably well understood) electrical and chemical mechanism to ultimately end up as high-energy chemical bonds in a molecule called ATP.

Chlorophyll is, in a sense, the reverse of the glow stick we saw at the very beginning of this book. Glow sticks turn chemical energy into light. Chlorophyll turns light into chemical energy.

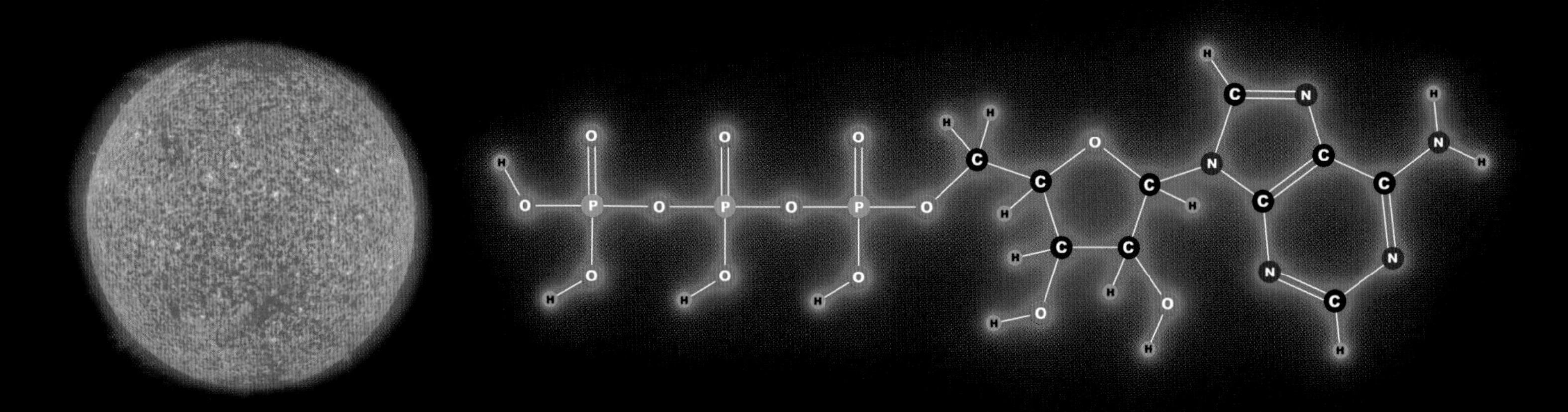

△ ATP, adenosine triphosphate, is the near-universal carrier of chemical energy in all living things.

All chemical potential energy on earth ultimately comes from one place: the sun. Deep inside the sun, nuclear reactions are turning hydrogen into helium. An atom of helium is a bit lighter than the hydrogen atoms it came from, and the difference in mass is released as kinetic and electromagnetic (light) energy according to Einstein's famous formula, $e=mc^2$ (e for energy, m for mass, and c^2 for the speed of light squared).

A tiny fraction of the energy released this way reaches us in the form of sunlight. (The rest, whatever doesn't hit one of the other planets, streams out into space and becomes what someone on another world sees when they look at our sun as the distant star it is to them.)

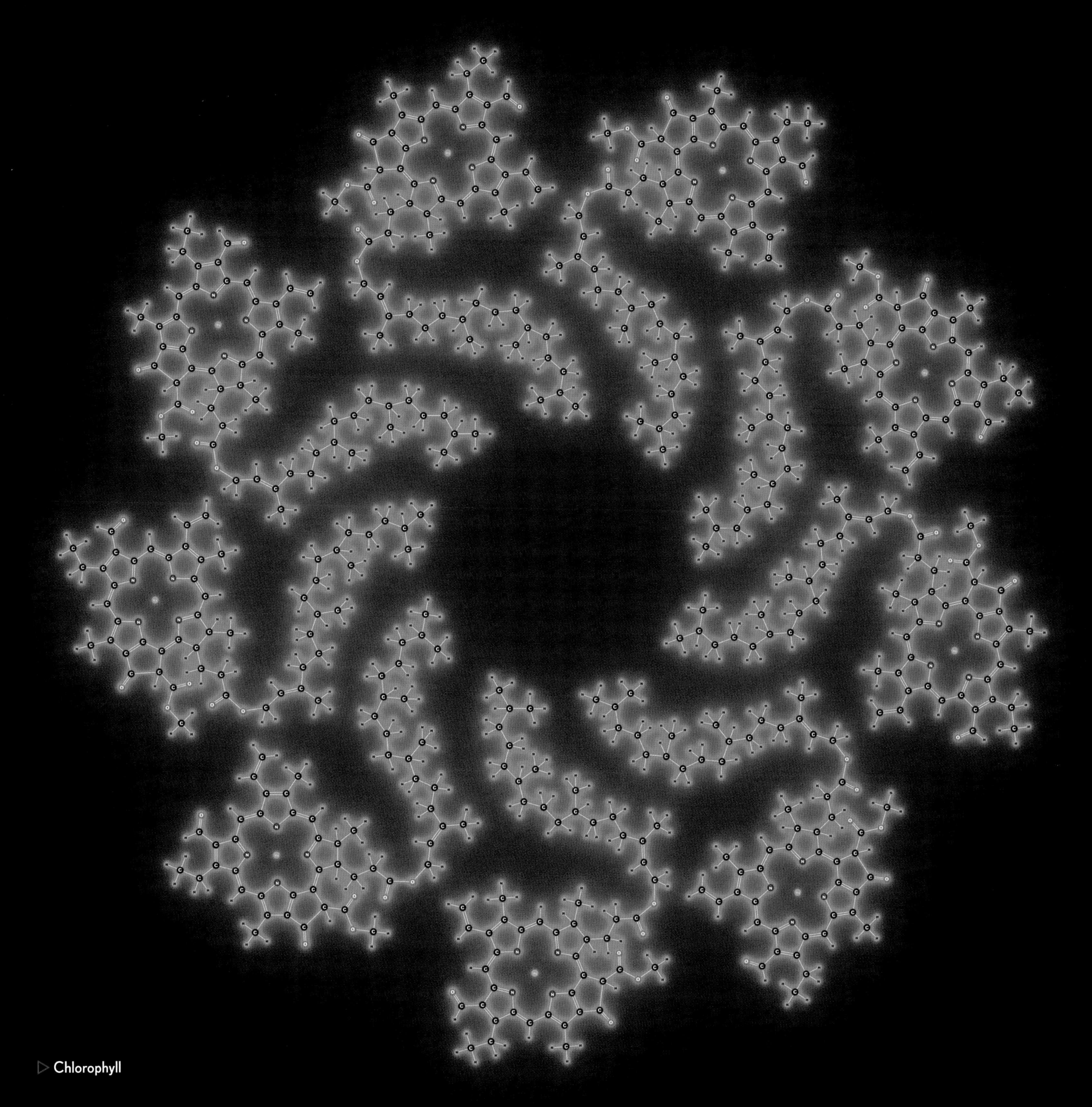

▷ **Chlorophyll**

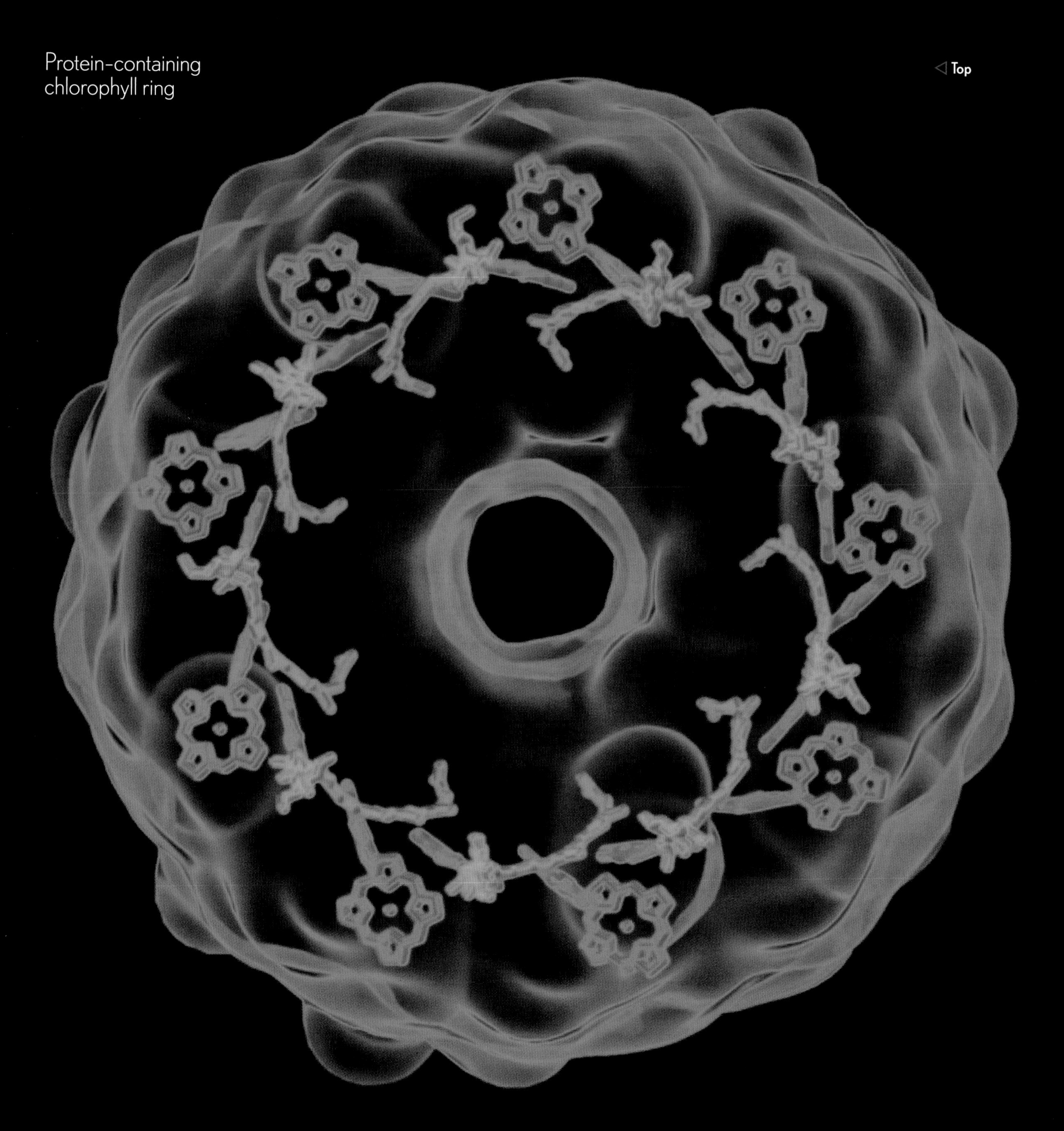
Protein-containing
chlorophyll ring
◁ Top

▷ Side

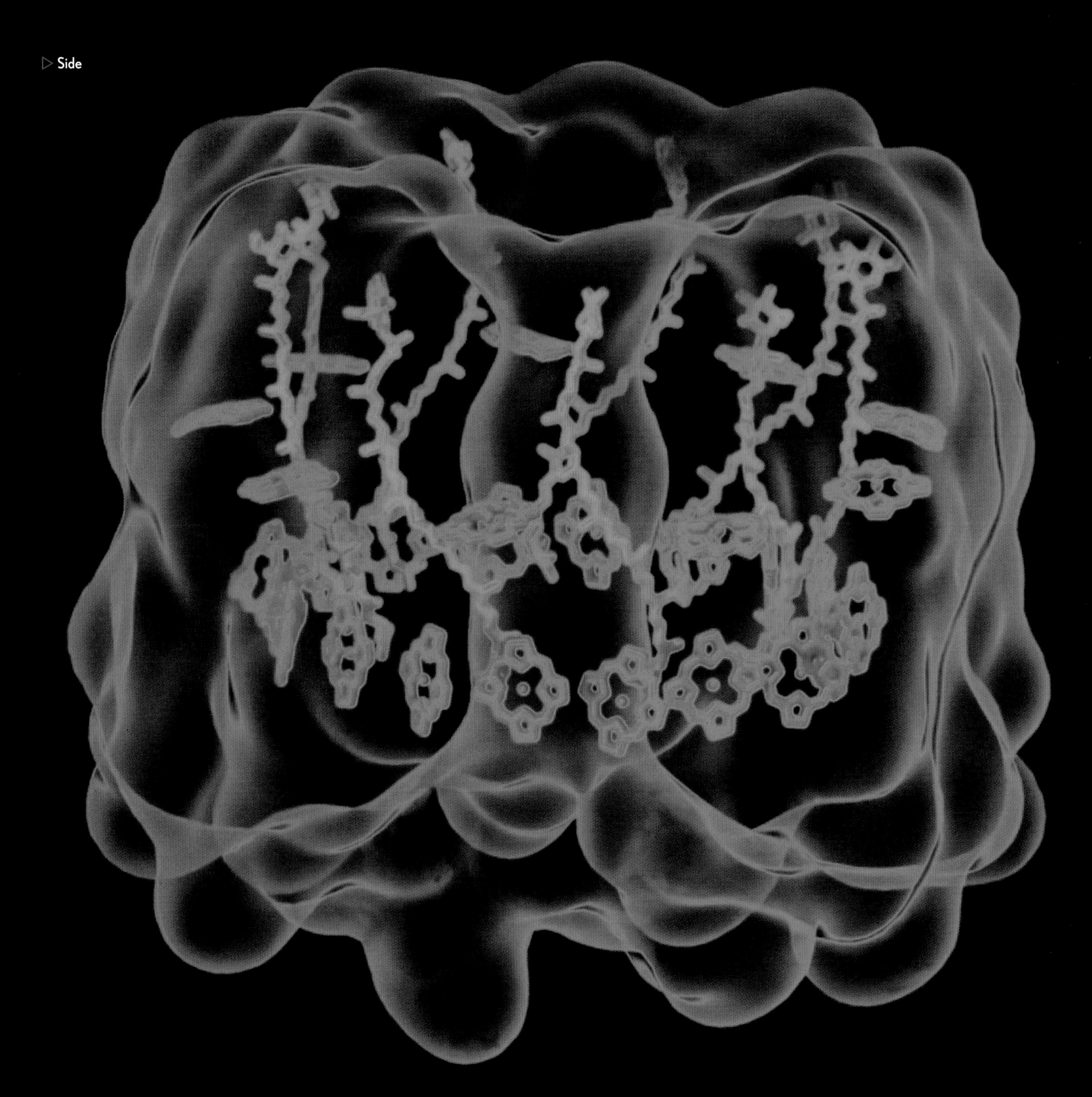

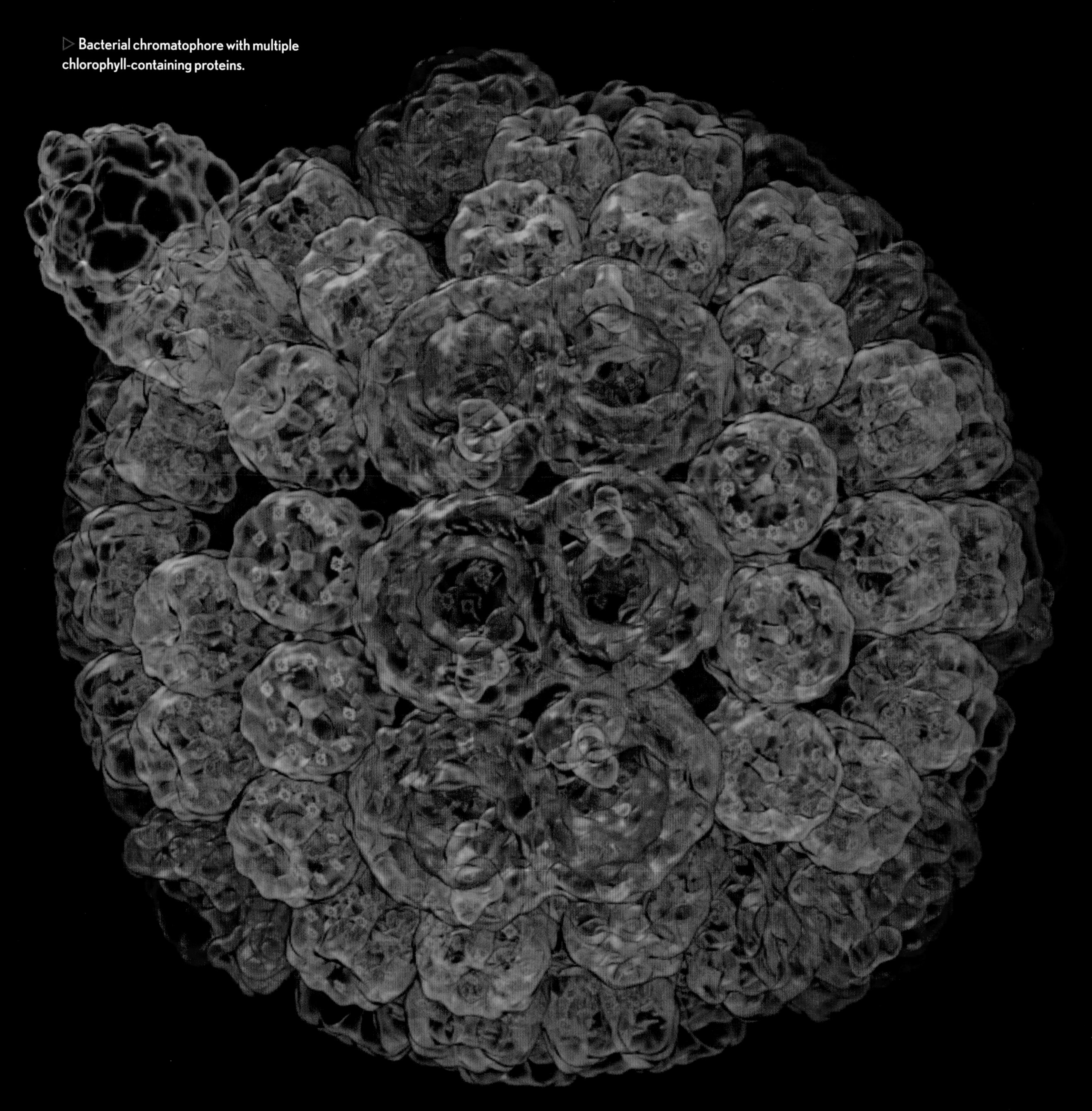

▷ **Bacterial chromatophore with multiple chlorophyll-containing proteins.**

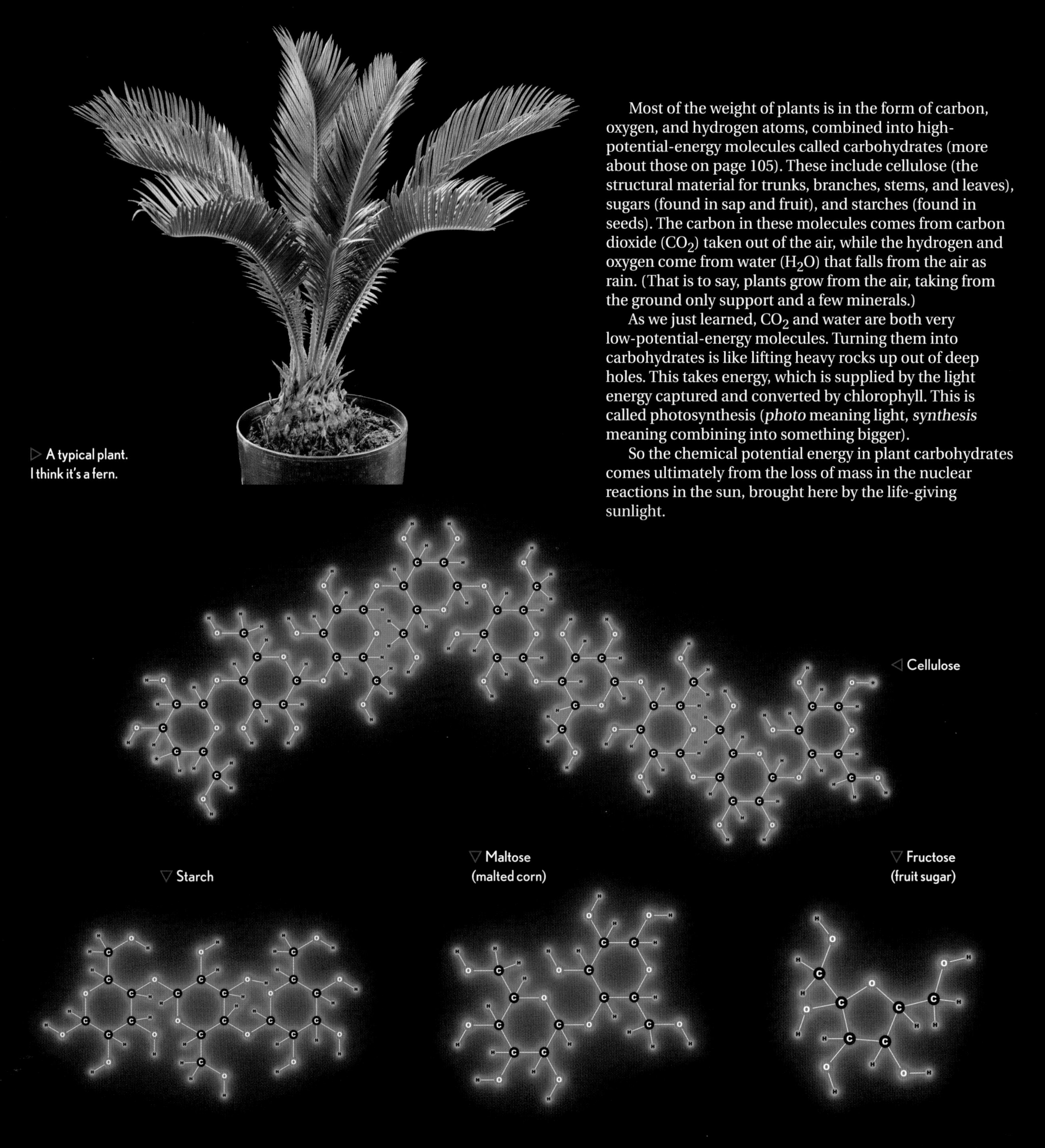

▷ A typical plant. I think it's a fern.

Most of the weight of plants is in the form of carbon, oxygen, and hydrogen atoms, combined into high-potential-energy molecules called carbohydrates (more about those on page 105). These include cellulose (the structural material for trunks, branches, stems, and leaves), sugars (found in sap and fruit), and starches (found in seeds). The carbon in these molecules comes from carbon dioxide (CO_2) taken out of the air, while the hydrogen and oxygen come from water (H_2O) that falls from the air as rain. (That is to say, plants grow from the air, taking from the ground only support and a few minerals.)

As we just learned, CO_2 and water are both very low-potential-energy molecules. Turning them into carbohydrates is like lifting heavy rocks up out of deep holes. This takes energy, which is supplied by the light energy captured and converted by chlorophyll. This is called photosynthesis (*photo* meaning light, *synthesis* meaning combining into something bigger).

So the chemical potential energy in plant carbohydrates comes ultimately from the loss of mass in the nuclear reactions in the sun, brought here by the life-giving sunlight.

◁ Cellulose

▽ Starch

▽ Maltose (malted corn)

▽ Fructose (fruit sugar)

Sometimes when plants die, they end up buried underground for a very long time. In decades they become something like peat moss. In a few thousand years they become more like bogwood, black, but still clearly of plant origin. With more time, millions of years, with heat, and with pressure, plant matter is transformed into coal. Lignite, or brown coal, is the freshest (if you can call something many millions of years old "fresh"). Bituminous and anthracite are progressively more compacted and transformed more fully from plant carbohydrates (carbon, hydrogen, and oxygen) to hydrocarbons (carbon and hydrogen only).

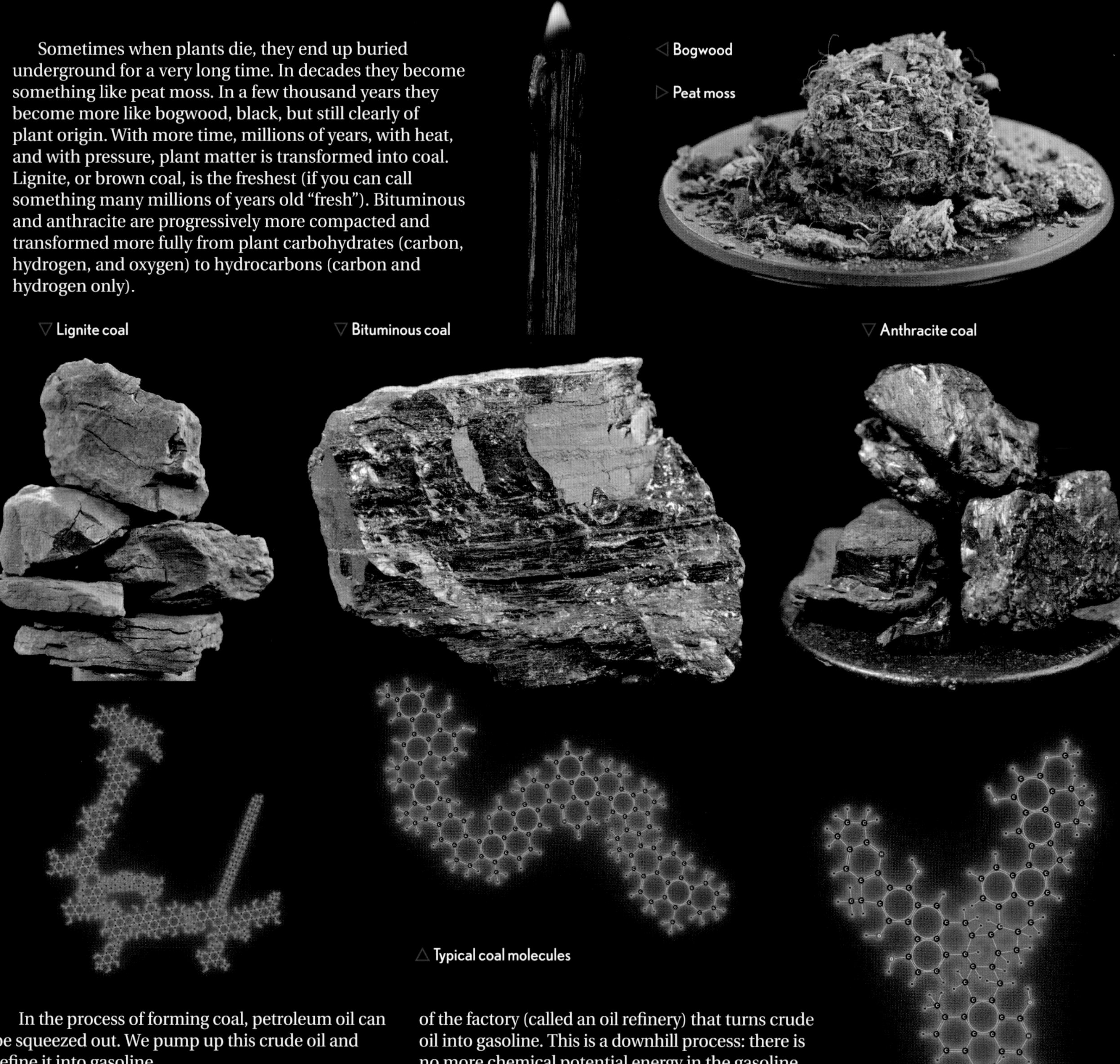

◁ Bogwood

▷ Peat moss

▽ Lignite coal

▽ Bituminous coal

▽ Anthracite coal

△ Typical coal molecules

In the process of forming coal, petroleum oil can be squeezed out. We pump up this crude oil and refine it into gasoline.

If coal is left underground the process continues, with more heat and pressure creating graphite (pure carbon) and ultimately diamond (also pure carbon, just arranged in a more expensive way).

In Chapter 3 (see page 90) we will see a picture of the factory (called an oil refinery) that turns crude oil into gasoline. This is a downhill process: there is no more chemical potential energy in the gasoline than in the crude oil it came from. In fact, some of the energy is used up to run the factory. But a lot of it remains in the gasoline.

So the energy in gasoline is entirely the energy of sunlight, just very, very old sunlight.

▷ Fancy diamond

△ Graphite

▽ Crude oil

▷ Crude diamond

▷ Typical crude oil molecules

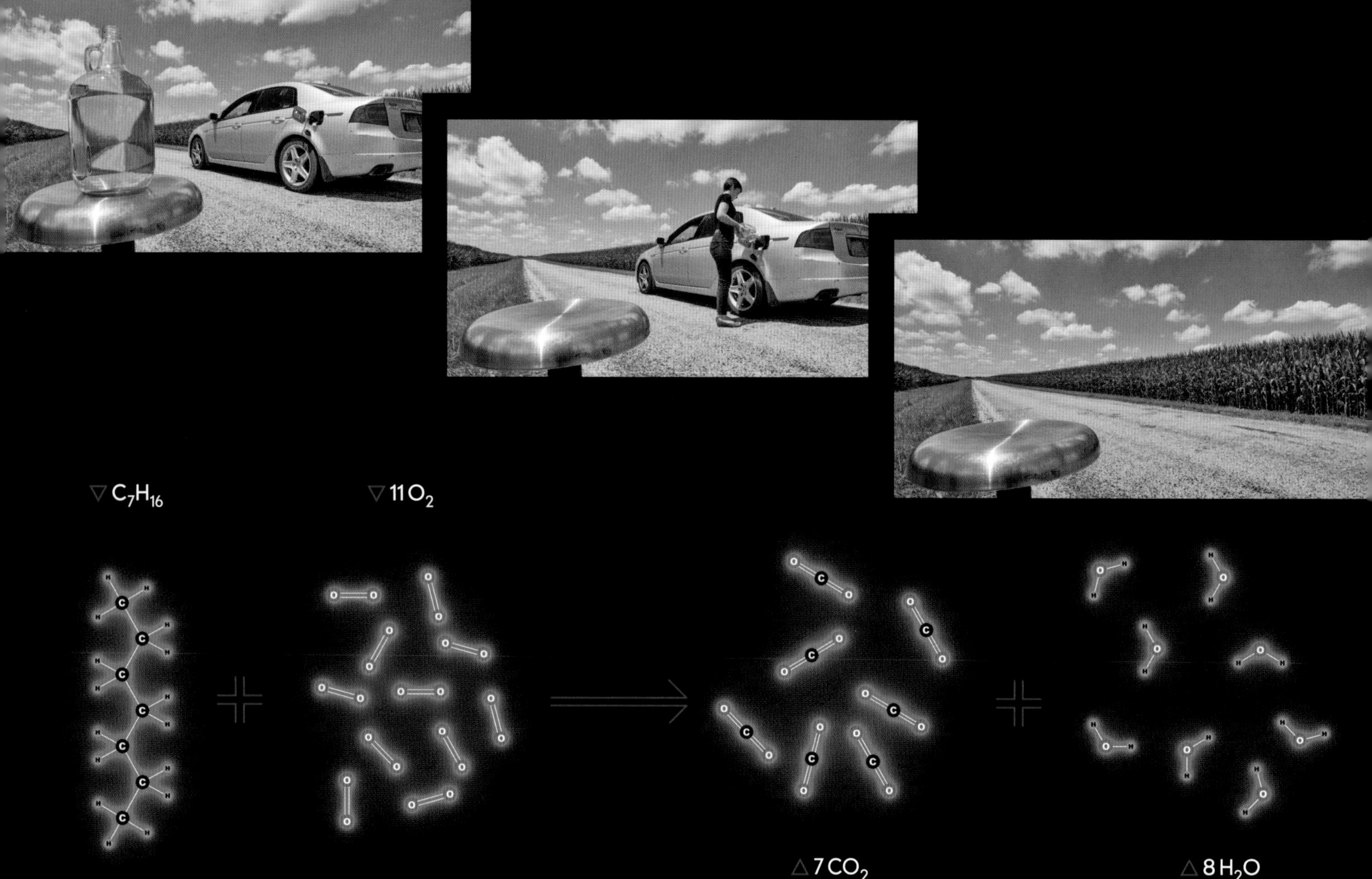

Now consider what happens when you put a gallon of that gasoline in a car and drive it down the road until the gallon is used up.

As the car is running, gasoline is combining with oxygen, releasing the stored-up sunshine and creating low-potential-energy molecules (CO_2 and water). The chemical potential energy is being turned into kinetic energy in the motion of the car. When the car is going fast, it's got a lot of kinetic energy, but when it comes to a stop eventually, the potential energy in the gasoline is gone, and so is the kinetic energy of the motion of the car. Where did all that energy go?

As when we dropped a rock on the ground, all the energy has ended up as thermal energy (heat) that raised the temperature of the road and the air around the car.

Some left the tailpipe in the form of hot exhaust gases. Some was conducted away from the engine and dumped into the surrounding air by the radiator. When the car came to a stop, the brakes got hot, taking the kinetic energy from the movement of the car and turning it into heat, which again leaked out into the air around the car.

When gasoline burns, a small part of the ancient sunshine that once warmed a younger earth now warms the earth again.

Conservation of energy guarantees that if you added up *all* of the heat released from the car, it would exactly equal the difference in chemical potential energy in the gasoline (plus the oxygen needed to burn it) and the CO_2 and water produced as the end result of the combustion.

You will find that the end result of just about *everything* is a bit of heat spread out widely in the surroundings. Every machine that does useful work is always also releasing heat. Every time something rolls, slides, falls, flies, or moves in any other way, when it comes to a stop, its kinetic energy content is converted into heat, which then leaks away. Every living creature, even a cold-blooded one, is constantly leaking heat into the environment around them with every movement.

Why is it that heat, particularly, is the end result of essentially all activity in the world? What's special about heat? The answer to this question is very closely related to the reason that time has a forward direction.

n one direction.

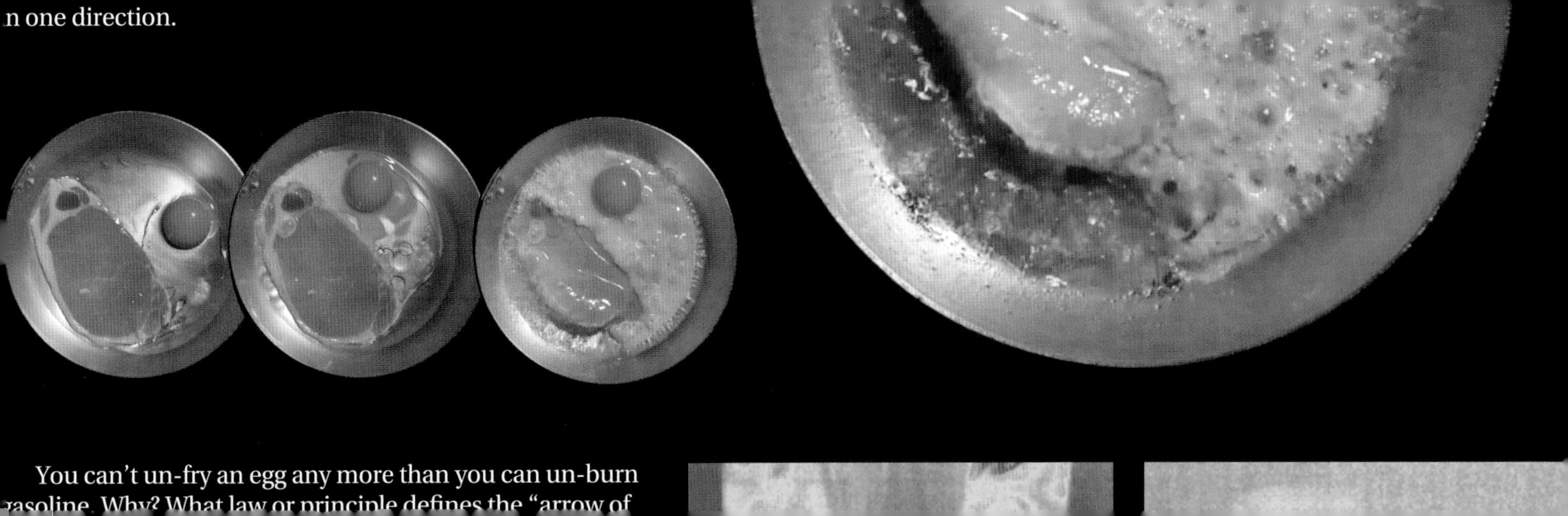

You can't un-fry an egg any more than you can un-burn gasoline. Why? What law or principle defines the "arrow of

Entropy

TWO IDEAS, FALLING DOWN to lower energy states and conservation of energy, have both failed to define the arrow of time. What's needed is an entirely new concept: entropy.

Entropy is perhaps the single most misunderstood scientific principle of all. Nearly everyone gets it wrong, including textbooks, teachers, working scientists, and nearly every student who's ever tried to understand it.

You, fortunately, are going to get an explanation that is at least not wrong, though it may be a bit difficult to follow. In short, entropy is a measurement of how *spread out* energy is. When energy becomes more spread out, entropy increases.

▷ Consider a "system" that consists of a 3x3 checkerboard with three checkers on it.

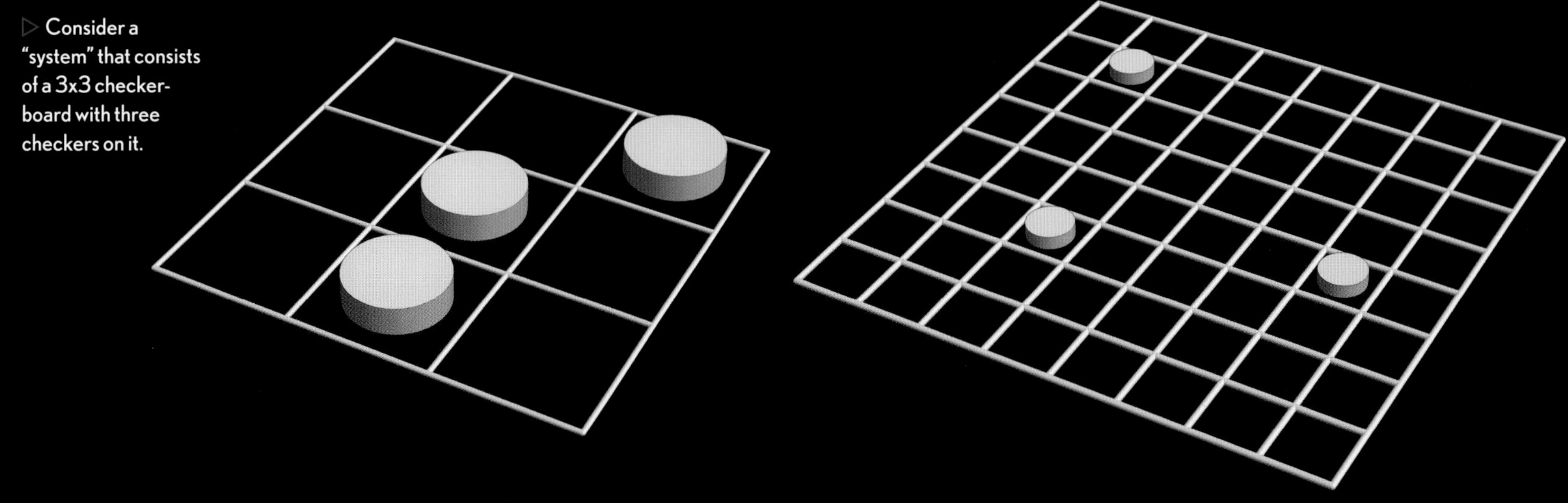

◁ There are 84 possible different ways these checkers can be arranged, all of which are shown here. The checkerboard is the physical space, and the list of 84 ways the checkers can be arranged is the state space.

△ If the physical space, the board, is enlarged to 8x8 squares, there are now 41,664 possible ways to arrange the same three checkers. (We're not going to try to draw all of them!) When we made the physical space a bit bigger, the state space got *much* bigger.

▽ Now suppose that instead of making the board bigger, we keep it the same physical size, but allow the checkers to be either lying flat or standing on edge. There are now 672 ways of arranging those same checkers on the same 3x3 board where there used to be only 84 ways. Without making the physical space any bigger, the state space has gotten much bigger—it's gone from 84 to 672 possible states, because we are allowing more flexibility, more possible states, for each checker.

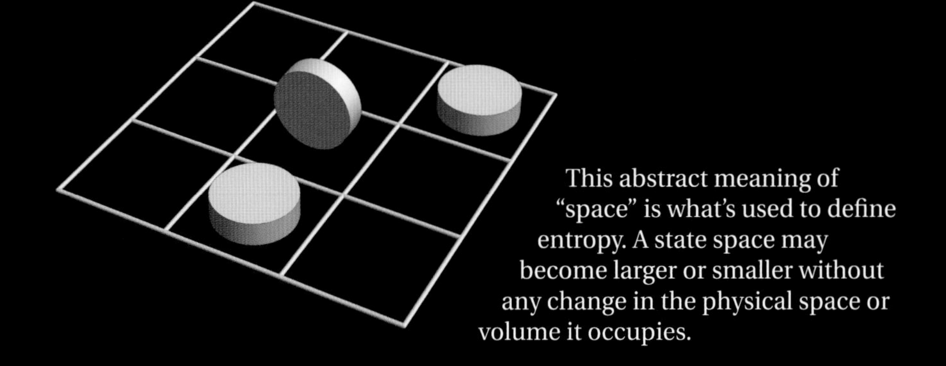

This abstract meaning of "space" is what's used to define entropy. A state space may become larger or smaller without any change in the physical space or volume it occupies.

How do entropy and state spaces work in chemistry? Instead of talking about ways of arranging checkers on a board, we talk about ways that energy can be distributed in a collection of atoms and molecules. There might be energy in the stretch of a certain bond, or energy in the movement of an atom in a particular direction, and so on. Each different mode or dimension where energy can be stored contributes to the possible ways of arranging the energy.

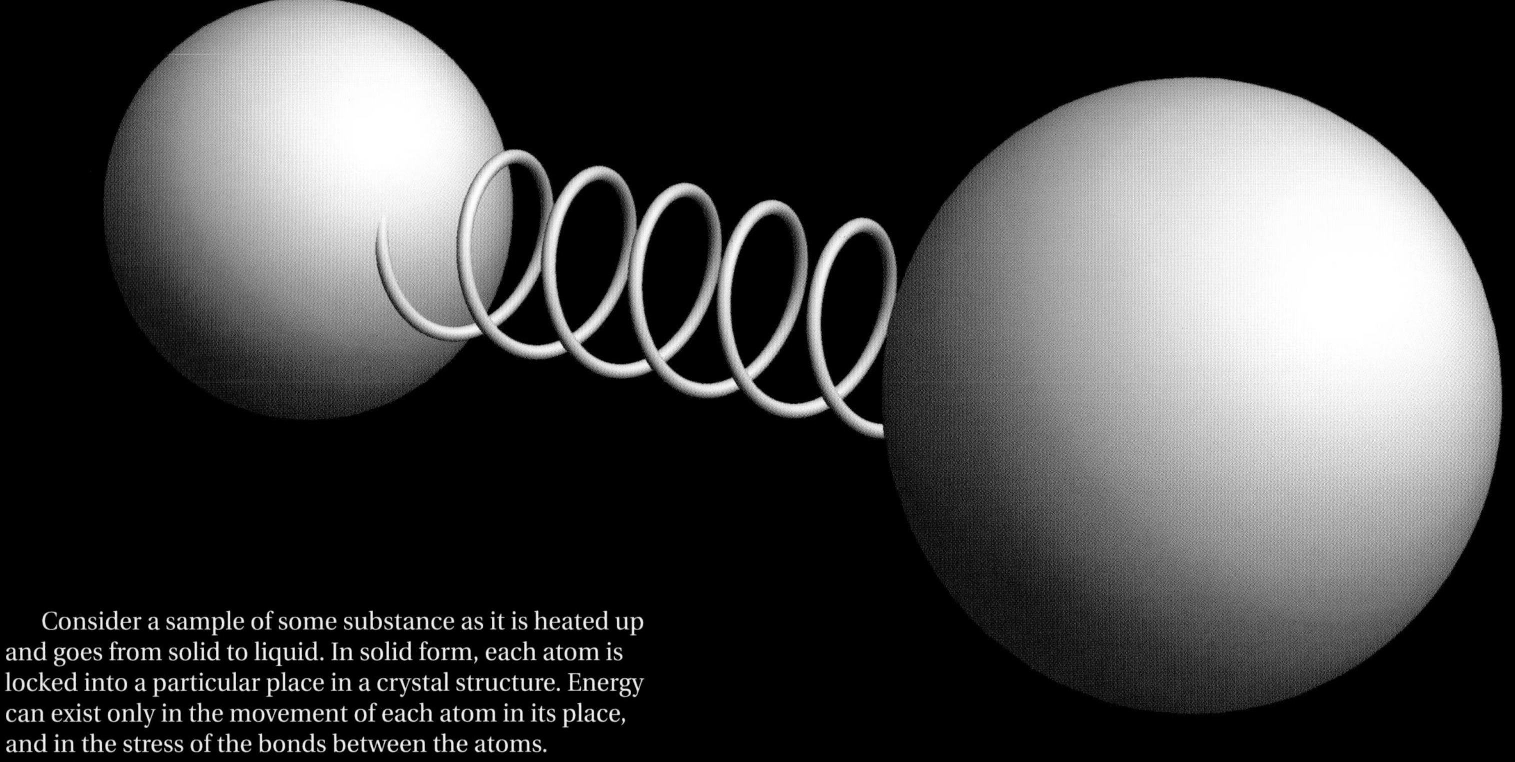

Consider a sample of some substance as it is heated up and goes from solid to liquid. In solid form, each atom is locked into a particular place in a crystal structure. Energy can exist only in the movement of each atom in its place, and in the stress of the bonds between the atoms.

In liquid form, the atoms are free to move around, so energy can also exist in the movement of atoms, and there are many more possible ways energy can be stored in the forces between the sliding atoms. Allowing the atoms to move is a bit like allowing the checkers to stand on edge: It's a new degree of freedom that we've given the atoms, allowing more variations within the same physical space.

The entropy of a system is a count of the total possible number of ways of distributing energy in that system (technically, the natural logarithm of that number).

So, all else being equal, a liquid state is a higher-entropy state—not because it's more random, not because it's any bigger, but strictly because there are more different ways the energy can be distributed within the material.

Having defined entropy, we can now reveal the most important thing about it, the thing that defines the direction of time and makes everything pointless: Entropy is *always increasing*.

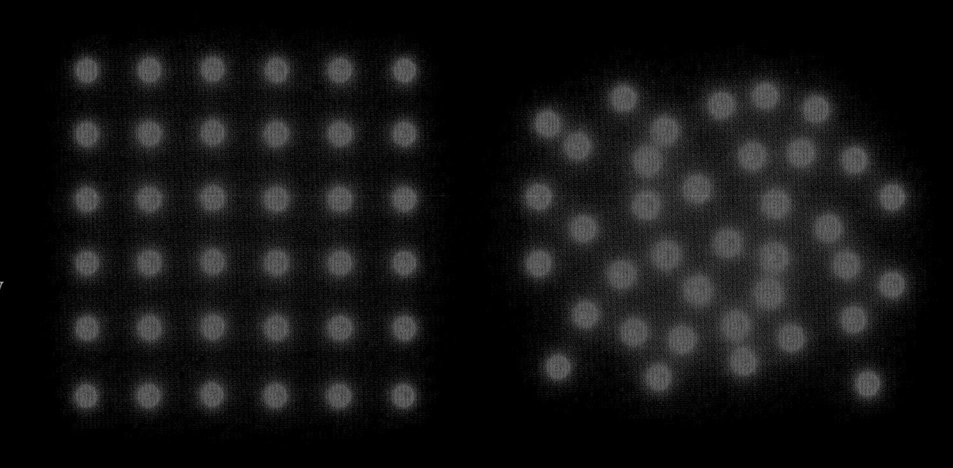

△ Solid

△ Liquid

Total Energy in an Isolated System Is Always Constant.

Total Entropy in a Closed System Always Increases with Any Change.

These two laws together determine what can and will happen spontaneously in the world.

The first one is usually called the law of conservation of energy, or the first law of thermodynamics. The second one is called the second law of thermodynamics. It is impossible to overstate how important these two laws are. They bring understanding to a vast range of subjects and situations, and when you have an intuitive grasp of how they work, much of the world will make sense that previously seemed mysterious.

The second law, that entropy is always increasing as time moves in the forward direction, is the physical law that defines the arrow of time. All our intuition about time, the sense that allows us to know instantly if a video is being played backward, boils down to us having an intuitive sense of entropy, and a deep, if subconscious, conviction that it only ever increases over time.

The law of conservation of energy cannot (currently) be explained. We know with great certainty that it is absolutely true in every situation that has ever been measured, but we don't know why. However, we *do* know why entropy is always increasing. This law is true by mathematical certainty.

The reason is as deeply satisfying as it is tricky to grasp. It is in a sense more absolutely true than any other law of physics, because it does not rely on any measurements or study of the world, but only on pure mathematics. You can hope to find exceptions or special sub-clauses to any physical law (even conservation of energy). But you cannot hope to escape the certainty of mathematics.

(Feel free to skip this section, by the way: it's really hard. It took me about 30 years and hours of conversation with a master teacher to finally get it straight. It's possible that I've written a magic chapter that will let you understand it overnight, but I doubt it. Nevertheless, I've done my best, so dive in if you like.)

These cubes represent the relative sizes of two state spaces, a small, low-entropy state and a large, high-entropy state. They *could* correspond to physical volumes, but really we are talking about the abstract listing-of-possible-configurations kind of space we described earlier.

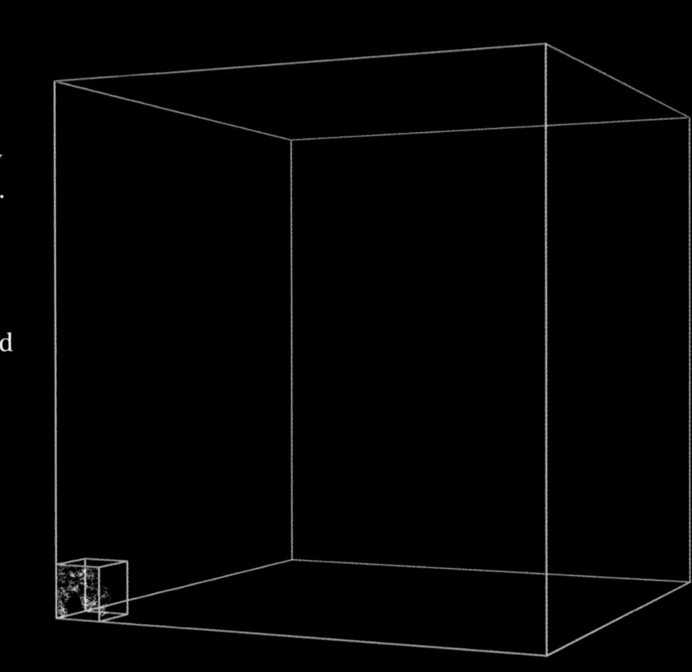

△ Imagine we are starting in the small, low-entropy box (which represents, for example, an unburned gallon of gasoline). The current state of the system (meaning the exact velocities of all the atoms, and the stresses of all the bonds) is represented by a single point in the state space. (For those who know what this means, these state spaces are very, very high-dimensional spaces, with multiple dimensions for every atom in the system.) The point representing the current state zips around the small box at random, exploring all the possible states that the system can be in.

Now imagine that we suddenly dissolve the walls of the small cube, releasing the system to move out into the much larger cube around it. This corresponds to letting the reaction happen: burning the gallon of gasoline. Now the point representing the current state of the system can start randomly moving around in the much bigger state space. This represents time moving in a forward direction: the system has moved into a higher-entropy condition.

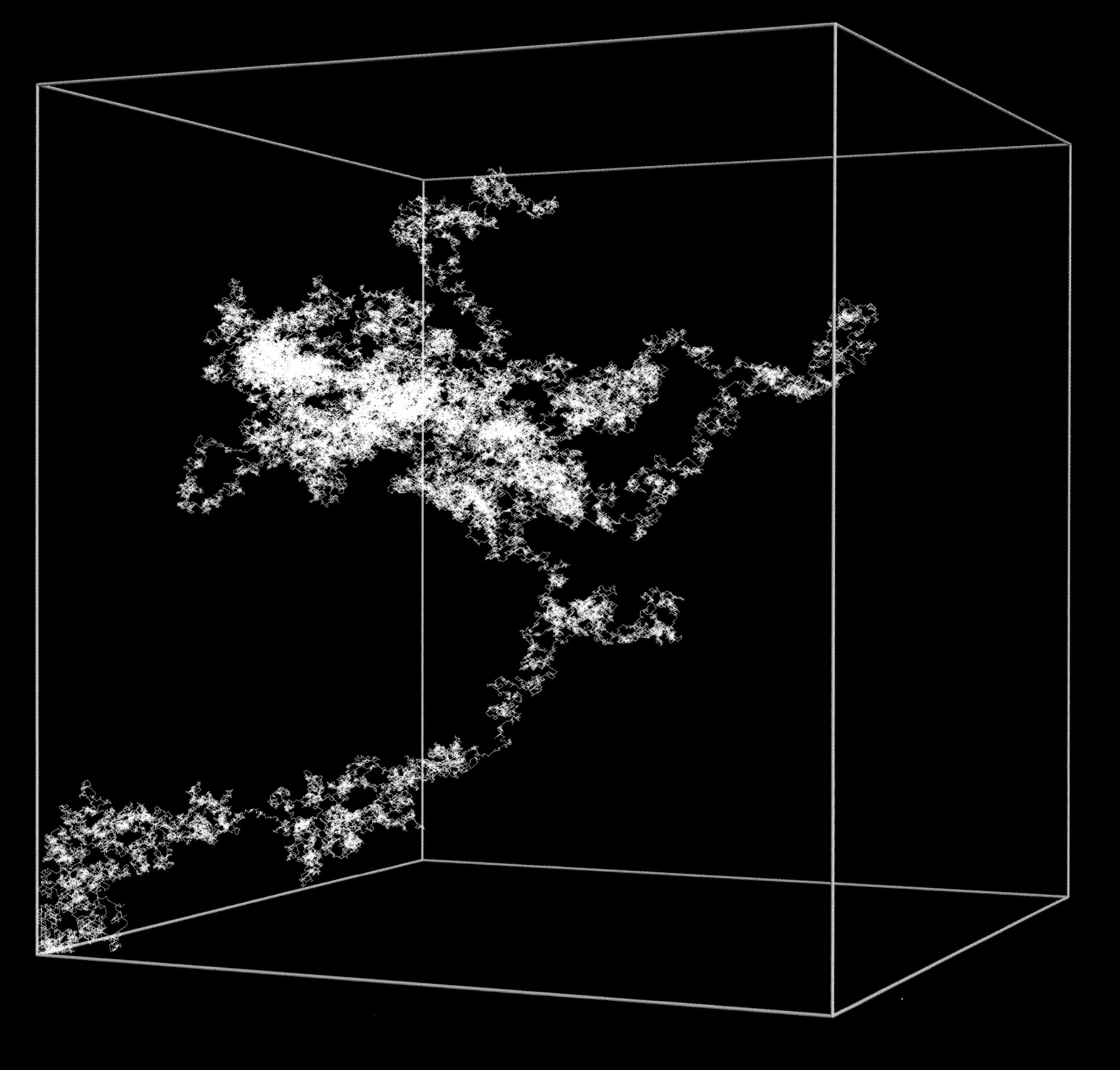

What would it take for the system to move back into its former, lower-entropy condition? The point representing the current state of the system would have to *spontaneously return to the tiny cube in the corner*. That's really unlikely, even in the example we've drawn, where the small volume is only about a thousand times smaller than the big one.

Remember that these volumes represent abstract state spaces, not physical volumes. In real-world entropy situations we can calculate the virtual size of the state spaces involved, and in a typical case the difference in size is so vast that it would take a *preposterously* long time for the state ever to return to the small corner. The age of the universe doesn't even begin to register compared to the how long it would likely take for a typical real-world system to spontaneously return to a lower-entropy state.

Systems always move from low-entropy to high-entropy configurations simply because that's overwhelmingly more likely than the reverse. If there are a billion billion billion billion billion billion billion billion billion billion billion billion more ways for the system to be in a high-entropy condition, then that's where you're going to find it.

Entropy is a deep and satisfying subject that takes time to sink in. Feel free to reread this section tomorrow if it didn't make sense the first time around.

In the meantime, now that we have learned a little bit about elements, molecules, reactions, and the laws of energy and entropy that control them, we are ready to go out into the world and see where we can find some interesting, beautiful, frightening, important, and useless reactions.

FACT CHECKING

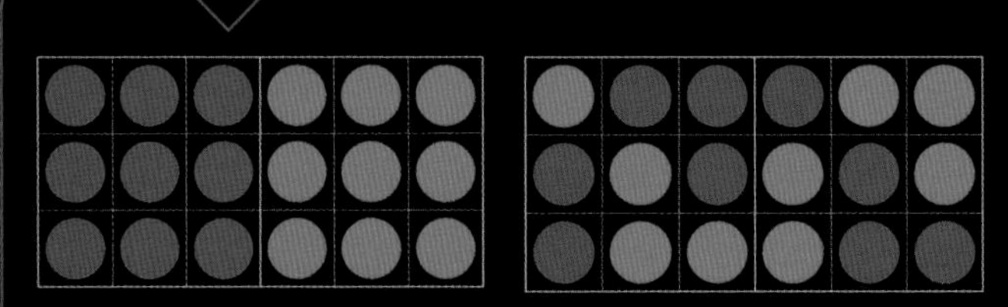

△ You will often—almost always—read that entropy is a measure of randomness. You'll hear that a highly ordered system, like this one with all the blue balls on one side, is low entropy, while a random one with the balls all mixed up is high entropy. In popular, nonscientific language, entropy means randomness or chaos. That's fine; people can use the word any way they like outside of chemistry. What's not fine is that this is also how entropy is usually described in high school chemistry textbooks. It is an absolutely false description. Entropy in chemistry does not have anything to do with randomness. Yes, the systems described by entropy operate by random fluctuations, but entropy itself is not a measurement of randomness. As we have learned in this chapter, it is a measurement of the spreading out of energy, not its organization.

△ You will sometimes hear a certain kind of poet talk about how they have learned that energy can never be destroyed. They say how wonderful this is because it means that all the energy that people create with their love will survive forever. Their soul mate may have died in a tragic banjo accident, but the eternal energy of their love will always be with them.

Except that love is not a phenomenon of energy, but rather a consequence of the distribution of energy. It comes about because of patterns and organized structures in your brain, built of matter and energy. In other words, love is a child of entropy, not energy.

Those patterns that are love do not have any conservation law backing them up the way energy does. They come and go. In fact, the law of increasing entropy says that they *will* go, sooner or later. What had been pattern will become void, like a chalk painting washed away by the rain—the chalk survives even as it is washed down the drain, but the painting is gone.

Your love and the love of everyone who has ever loved you will be washed away in the inevitable, unstoppable, universal flushing down the drain that is the fundamental law of increasing entropy. And then the sun will engulf the earth and reduce it to ash before shrinking into a white dwarf and dying in obscurity.

Sorry.

3

Fantastic Reactions and Where to Find Them

"CHEMISTRY" MAY BE a class you hated (or expect to hate, or are currently hating) in high school, but real chemistry—the kind that exists all around you—is different. The chemistry of the world outside of school is a beautiful array of spectacular colors, smells, sounds, and experiences that can be found everywhere and always. In this chapter we're going to take a romp through some interesting reactions, organized by the places in the world you can find them.

How do you recognize when a chemical reaction is happening? Pretty much any time something changes form in the world, it's due to a chemical reaction. Food turns into poop? That's a chemical reaction. Car drives across town? That's a chemical reaction (plus some machinery). There are so many different kinds of reactions that we can't possibly begin to discuss them all in this book.

Since it's impossible to be comprehensive, I will pick and choose reactions to illustrate what I think are the most interesting and the most powerful ideas behind a chemical view of the world.

In the Classroom

THERE IS A SET of reactions that pretty much everyone who teaches chemistry shows their class. Or at least there used to be. Sadly many of these are becoming less and less common as safety worries and small budgets overtake concern for the need to educate a future generation of scientists and keep human civilization functional.

You'll often hear these reactions called "science experiments," but this is a sloppy use of the word "experiment." The whole point of an experiment is that you don't know what the result is going to be. An experiment is supposed to lead to new knowledge.

These reactions are more properly called "demonstrations" because they should be the complete opposite of experiments. If you're not absolutely sure of *exactly* what the reaction is going to do, you probably shouldn't be doing it in front of a class full of students!

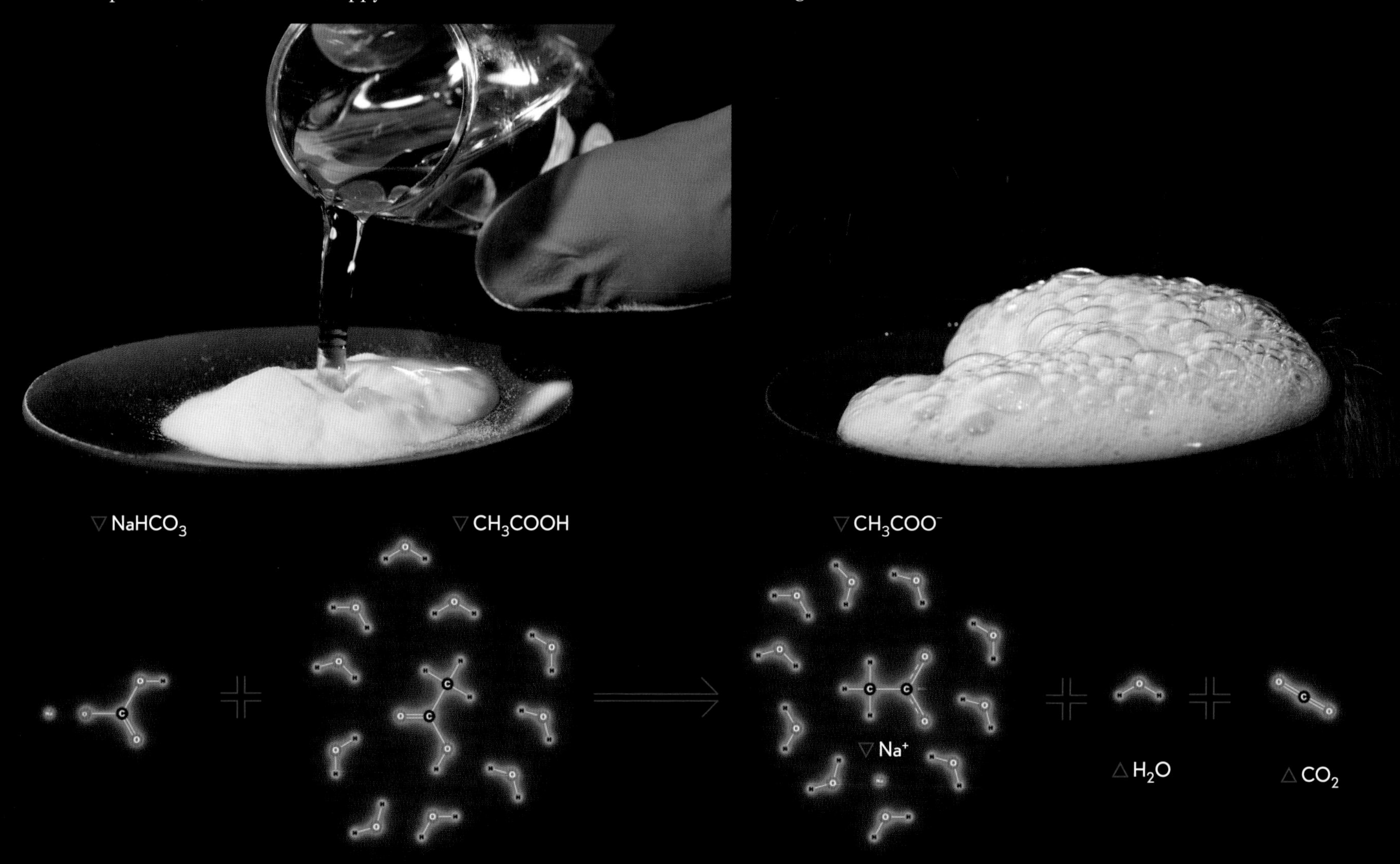

△ This is *the* definitive "Funtastic Chemistry Science You Can Do at Home" reaction. In my ten years of writing about science demonstrations for *Popular Science* magazine I made a point of *never* talking about this reaction. For me, it defines boring chemistry. It's what you do when you're doing what your parents think you should be doing.

But there is a certain charm to it, and the ingredients are usually lying around the kitchen. Baking soda is the common name for the chemical sodium bicarbonate ($NaHCO_3$). Vinegar's main ingredient (other than water) is acetic acid (CH_3COOH). When you combine one molecule of each, you get four things: a molecule of water (H_2O), a dissolved sodium ion (Na^+), a dissolved acetate ion (CH_3COO^-), and a molecule of carbon dioxide (CO_2). Because CO_2 is a gas, you get bubbles. Yay!

▽ There are lots of kits that use the baking soda and vinegar reaction to create a "volcano." I felt personally lied to by these kits when I was a kid. Volcano? I don't think so. A volcano spews out red hot molten lava that incinerates everything in its path. The kit I got did absolutely nothing of the sort. Didn't even scorch the kitchen table. So disappointing.

Later in life (about two minutes ago when I looked it up), I learned the full extent of the lie. This reaction is actually what's called "endothermic," which means it absorbs energy (like the cold packs we learned about in the last chapter, but not as extreme). Not only does it fail to create hot lava, it actually makes a bubbly water mix that is slightly *colder* than the ingredients you started with. Sheesh.

It's a total scam, and to add insult to injury, not once in the instructions do they tell you that if you just substituted iron oxide and aluminum powder for baking soda and vinegar, you could actually create red hot molten lava, which would likely burn a hole right through your kitchen table. Come on, how hard would it be to give a kid a hint how to get the thing to work as advertised?

OK, I admit, maybe it's a good thing that, at the age when I was most disappointed with baking soda volcanoes, I was not yet old enough to know about (or get the ingredients for) the chemistry demonstration that *does* create red hot liquid rock. (Technically it's red hot liquid iron, but that's actually even better. Can you imagine a full-size volcano that erupts with liquid iron? That would be so cool.)

▽ This reaction can be done in a classroom setting, but it's pushing the limits. Respect to any teacher who performs this demonstration.

▷ Liquid iron

The actually-like-a-volcano demonstration is done with a mixture called thermite, which contains aluminum powder (the element Al) and iron oxide (both forms work: red rust, Fe_2O_3, or magnetite, Fe_3O_4). When the mixture is lit, the oxygen atoms move from the iron to the aluminum, resulting in iron metal (the element Fe) and aluminum oxide (Al_2O_3). The bonds between aluminum and oxygen in Al_2O_3 are *much* stronger than those between iron and oxygen in Fe_2O_3 or Fe_3O_4. As we learned in the last chapter, that means the electrons in the aluminum-oxygen bonds are in much deeper potential energy holes, which means a lot of energy got released when they were formed. That energy release is why you don't just get iron; you get white hot, lavalike liquid iron.

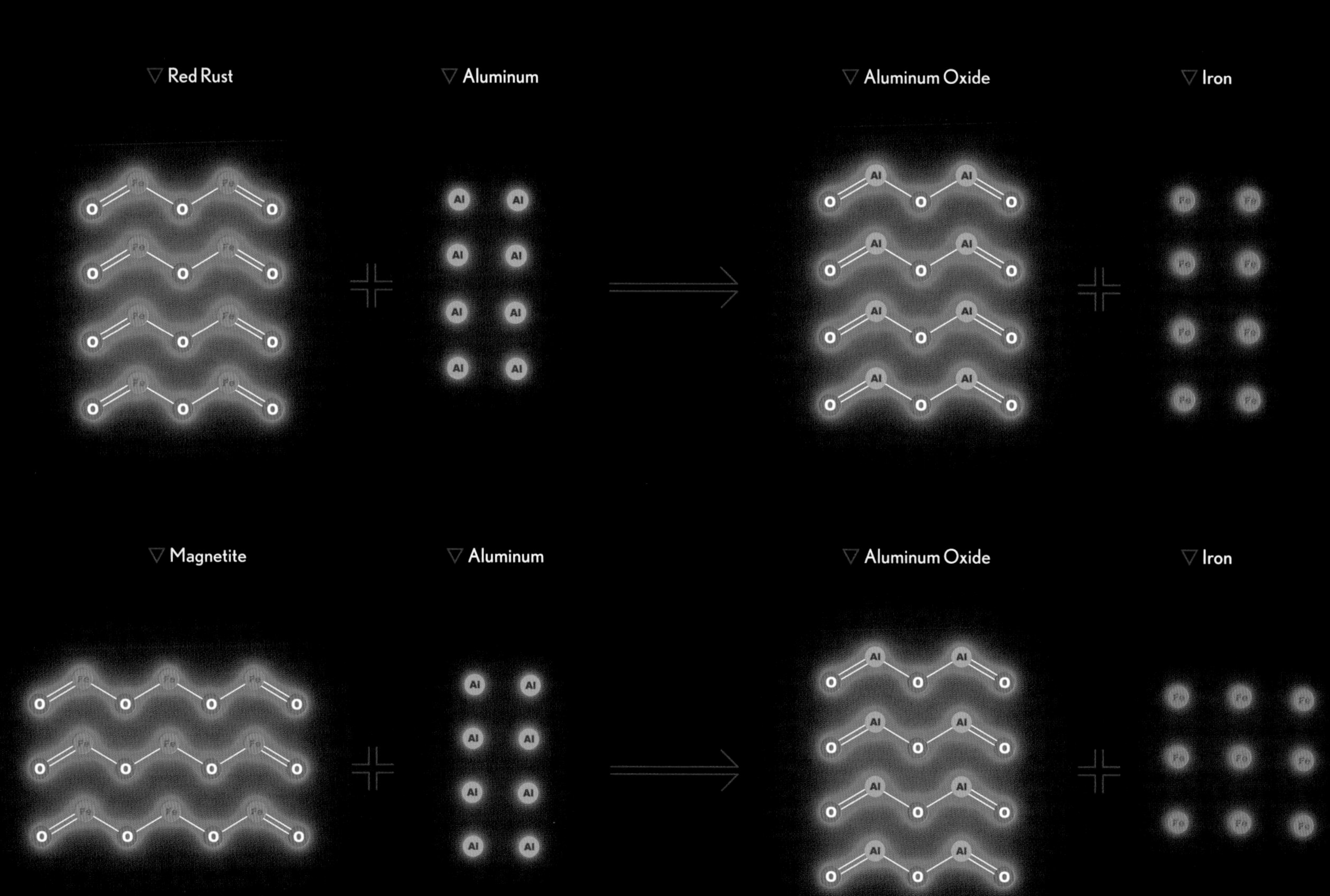

If you want your demonstration to be even more like a real volcano, you can let the iron fall into a bucket of water, creating a steam explosion that flings the molten iron high into the air, much like an erupting volcano. This is highly *not recommended*, especially in a classroom setting. I'm not going to name names, but a highly respected and experienced college professor once told me the story of the last time he made this mistake. The phrase "coeds having their bottoms irrigated by the fire department" stuck in my mind. Ah, for the simpler days of old when you could nearly incinerate the front row of students, make an inappropriate joke about it, and still keep your job. (But seriously, folks, this was a very dangerous incident, and it was only by sheer luck that no one was seriously injured. Thermite must be treated with respect or it will bite you.)

▷ By the way, when doing the thermite demonstration, it's probably best not to use the very cheapest clay pot.

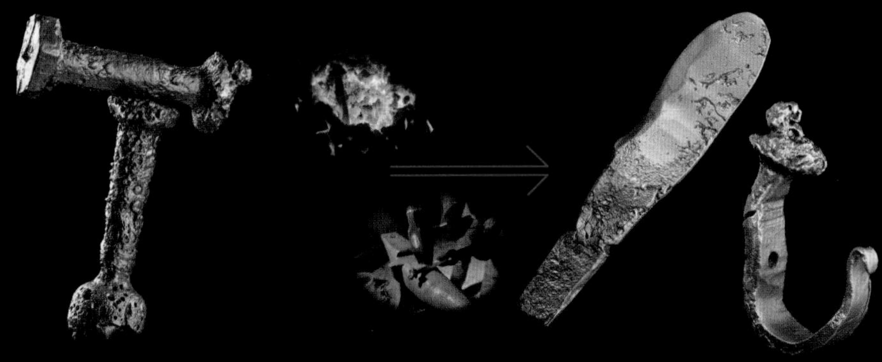

△ This is the kind of rough metal you get from a crude thermite demonstration in which the pot breaks. But it is real metal, rings like a bell, and can be forged into useful things. This crude knife and coat hook were hammered out of the raw castings in the traditional blacksmithing way, with coal, fire, hammer, and anvil. Pretty amazing that they came entirely from sand I collected with a sandal-shaped bottle opener while lying out on the beach. (See page 92 for that story, and page 96 for what it looks like when thermite is done professionally.)

▽ The thermite reaction is quite slow. As you can see in this cutaway view, it takes about five seconds for the reaction to burn its way from the top of the pot down to the bottom (which is quite a long time when you're watching liquid iron form before your eyes). When the reaction reaches the bottom, it melts through an aluminum plug covering the hole in the bottom of the pot, releasing the liquid iron.

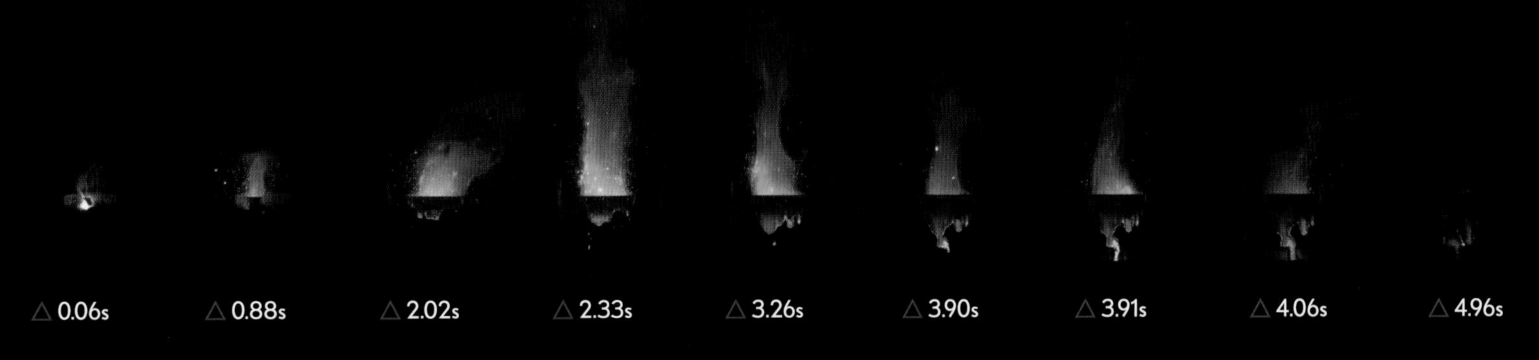

△ 0.06s △ 0.88s △ 2.02s △ 2.33s △ 3.26s △ 3.90s △ 3.91s △ 4.06s △ 4.96s

The reaction on this page, together with thermite, are just about the upper limit of what a typical chemistry teacher at the high school or first-year college level will attempt in front of a classroom. The idea is to take soapy water and blow bubbles with hydrogen gas (H_2). Hydrogen is lighter than air, so the bubbles form clumps that float up and away.

Hydrogen is also flammable: It reacts with oxygen to form water. Using a candle tied to the end of a long stick, you can run after the bubbles and try to light them before they reach the ceiling.

▷ If the bubbles are blown with pure hydrogen gas, they burn gently with nothing more than a sort of whoosh, because the rate of the reaction is limited by the rate at which hydrogen gas can mix with oxygen from the surrounding air. The resulting fire jellyfish are very attractive, and last for a good fraction of a second.

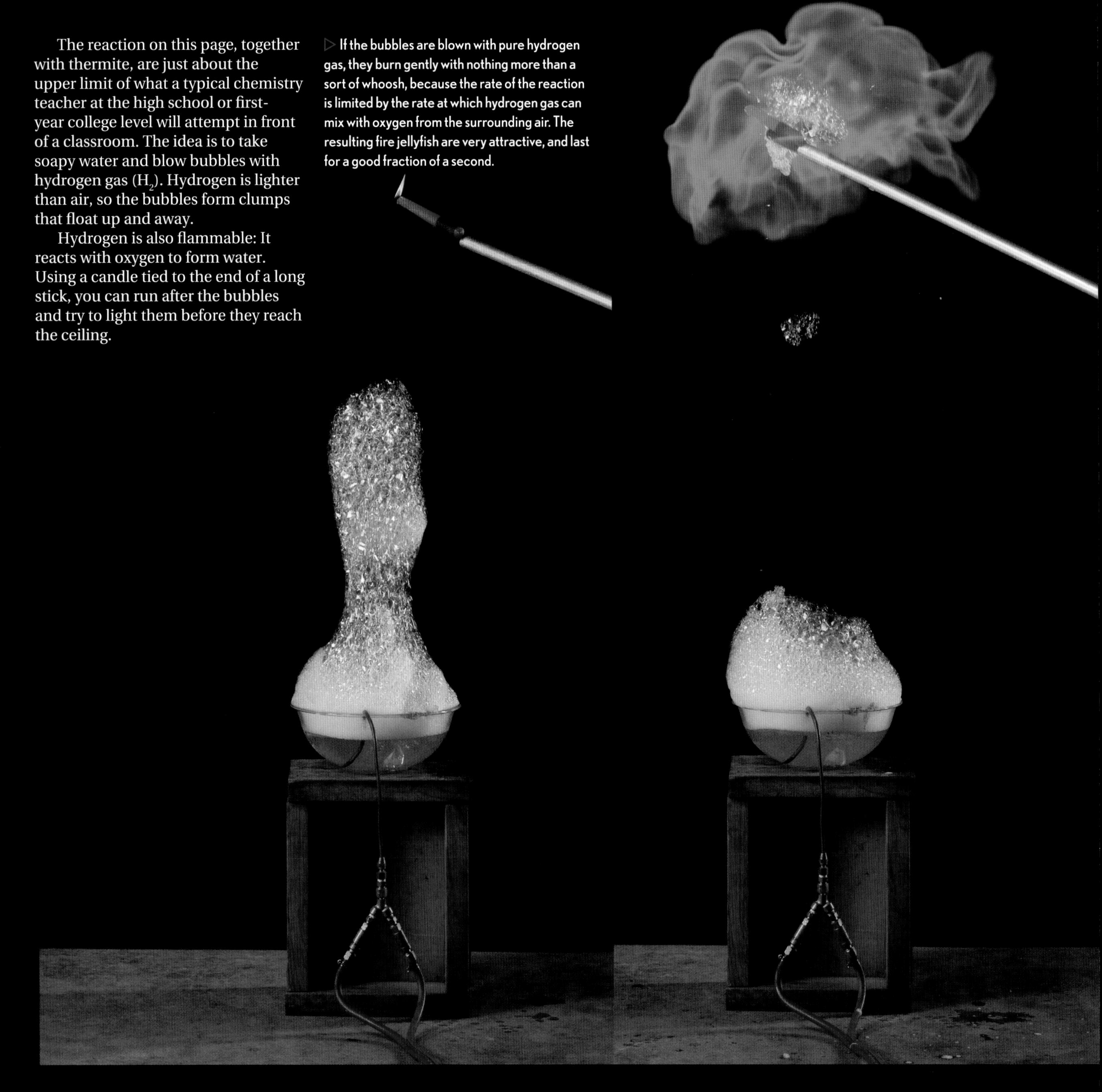

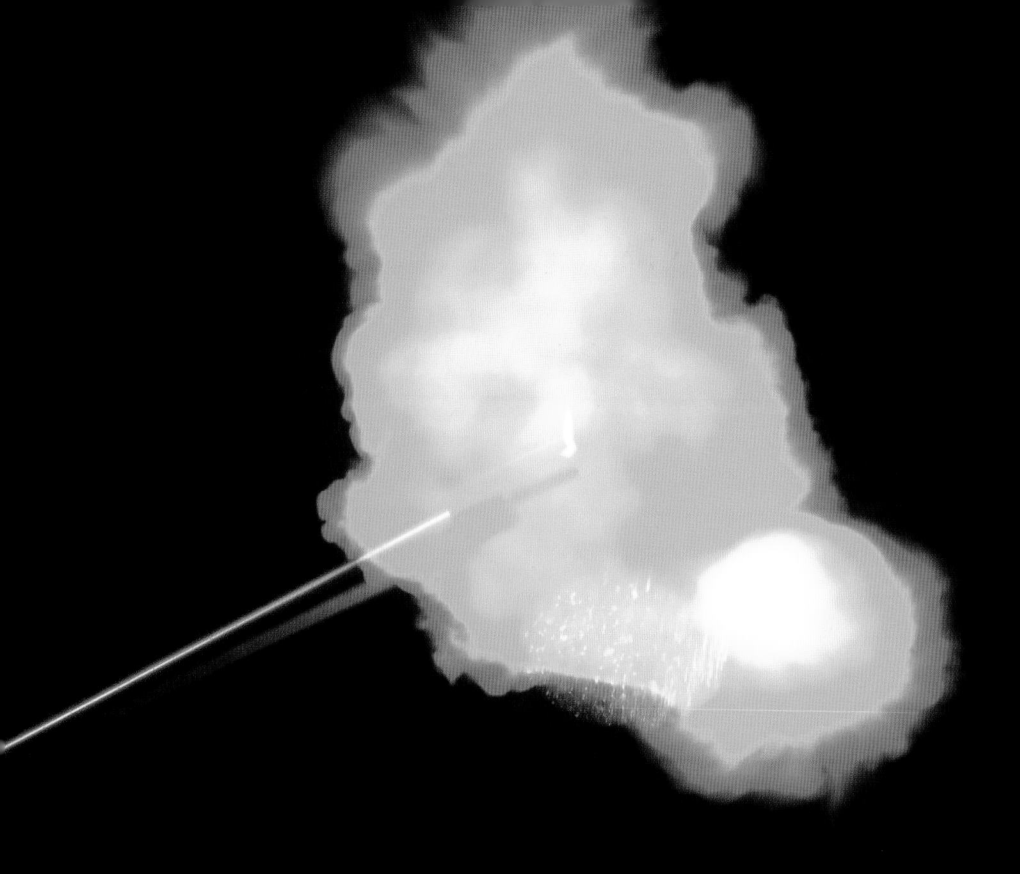

Let me tell you a bit about where I live. The nearest neighbor is about two-thirds of a mile away as the crow flies. They know me and nothing surprises them. This is rural Illinois; people have guns and sometimes enjoy shooting them. A lot. It's farm country. Nobody cares. The police came when we filmed hydrogen soap bubbles.

I first saw the hydrogen/oxygen soap bubble demonstration done by the same professor who told the story of messing up a thermite demonstration. At that time chalkboards were still actually boards that you wrote on with chalk. I'm sure he made a point of having a lot of writing on the board before starting the demonstration, just so he could make a joke about chemistry teachers suffering "white lung disease" from all the chalk dust literally knocked off the board by the impact of the explosions.

▷ If you blow the bubbles with a mixture of hydrogen and oxygen, you get something quite different. With the two gases premixed, there's nothing to slow down the reaction. The more oxygen you mix in, the faster and more powerful the reaction gets, until the point where you're mixing in exactly half as much oxygen (by volume) as hydrogen. At this "stoichiometric" ratio, you get what are basically gunshot bangs—very loud—of a wake-up-the-neighbors type.

▷ This hydrogen bubble ended badly. The famous *Hindenburg* disaster happened when the hydrogen gas in this otherwise fine airship caught fire. Many people died, but from falling, not from the fire. Hydrogen is so light that the flames went upward very quickly and largely stayed away from the unfortunate people.

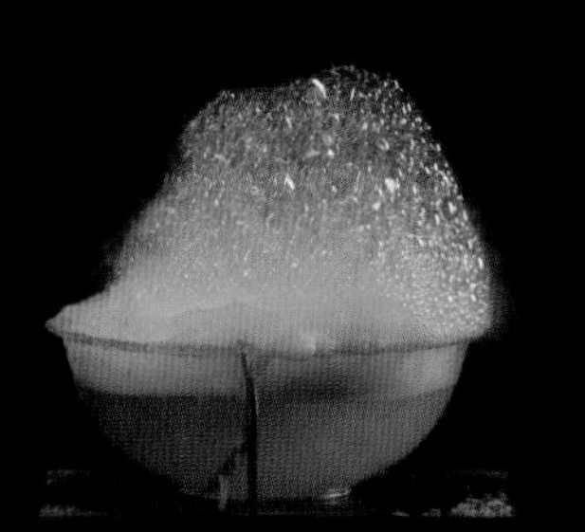

It's rarely discussed, but it's a really good thing that the *Hindenburg* was filled with pure hydrogen gas, and not a mixture of hydrogen and air. The Hindenburg *burned slowly* just like our pure-hydrogen soap bubbles. Had it been filled with a mixture of fuel and air, it would have been the equivalent of a very large fuel-air bomb, and would have leveled buildings for a considerable distance in all directions.

Fuel-air bombs work by first dispersing a flammable gas out into a large volume of air, and then a fraction of a second later igniting the mixture. If the gas were lit first, it would just burn gently, but if it's lit after mixing, the result is an explosion so powerful that the blast wave can be used to clear land mines over a radius of hundreds of yards (meters).

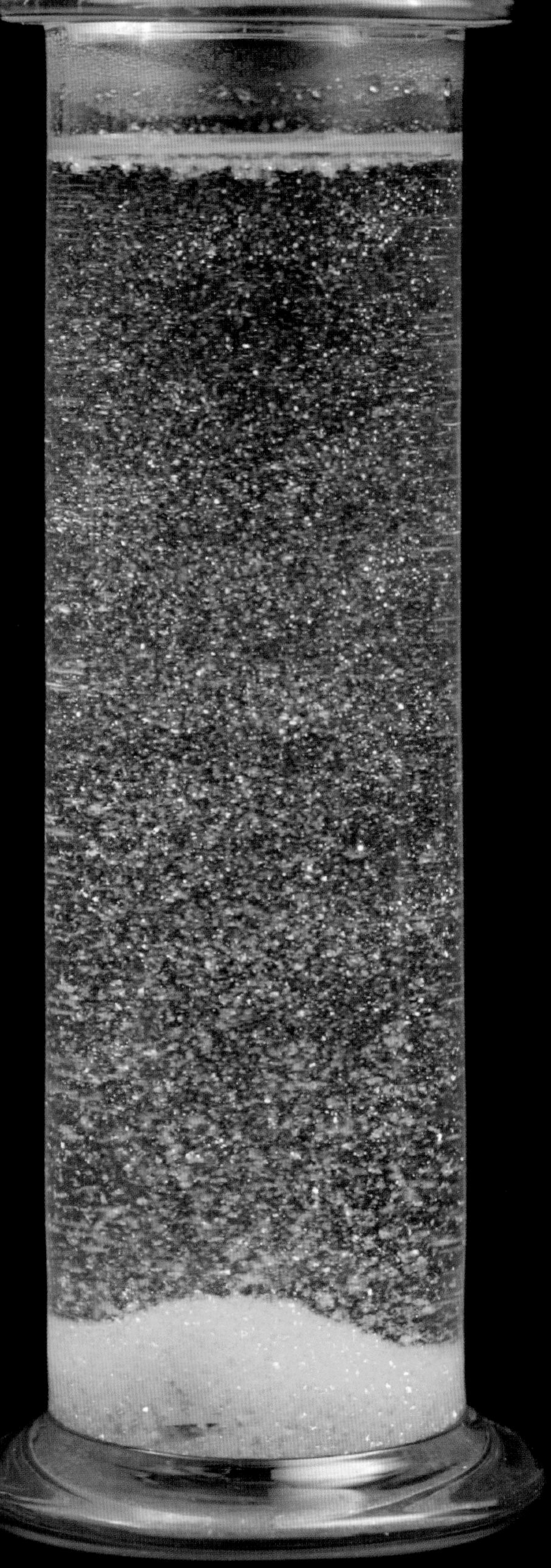

There are a lot of reactions used for classroom demonstrations that fall in the category of "pour this into that and something happens." The something is almost always that the color changes, a solid forms, or there are bubbles. Most of these are very boring, if you ask me. I mean, you learn something interesting from each one, but, well, whoop-de-doo, it changed color.

This unusually beautiful member of the genre is called the Golden Rain demonstration. It might more appropriately be called the Beautiful Death Rain demonstration, because the "rain" is lead iodide, which would poison your farm fields for a generation if it fell on them.

Golden Rain is produced by combining a solution of potassium iodide (KI, which is highly soluble in water) with a solution of lead nitrate ($Pb(NO_3)_2$, also highly soluble). Atoms of iodine quickly find matching atoms of lead and combine to form tiny crystals of lead iodide (PbI_2), which is *not* highly soluble in water. The golden-colored lead iodide crystals precipitate out of the solution, forming the "rain" you see here.

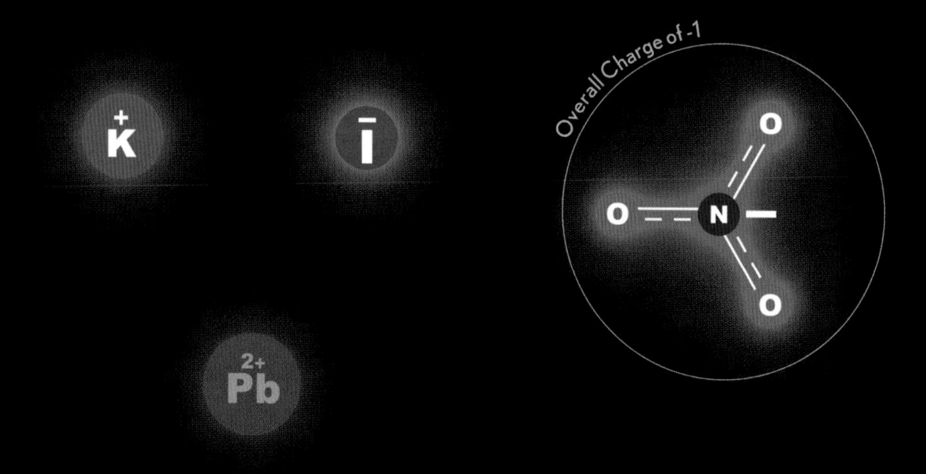

The potassium, lead, and iodine in this demonstration are present in the starting solutions in the form of "ions." Back in the last chapter we discussed how atoms normally have exactly the same number of negatively charged electrons as they have positively charged protons in their nucleus. So atoms normally have zero overall electric charge. They are "neutral." But atoms can also exist with too many or too few electrons, giving them an overall negative or positive electric charge. In this condition they are called ions. Lead and potassium atoms, for example, can fairly easily lose electrons and run around as Pb^{2+} or K^+ ions—atoms with a positive electric charge. Iodine atoms, on the other hand, prefer to hang out with one too many electrons in the form of I^- ions—iodine atoms with a negative electric charge.

Multi-atom molecule fragments can also exist as ions. The nitrate ions in this demonstration are bonded groups of four atoms (one nitrogen and three oxygen) that, overall, have a -1 electric charge. The bonds in a nitrate ion cannot be classified as either single bonds (one pair of shared electrons) or double bonds (two pairs). Instead they are something in between, with the extra electron shared symmetrically between all three oxygen atoms. They are thus best drawn with one solid and one dashed line, to indicate "it's complicated."

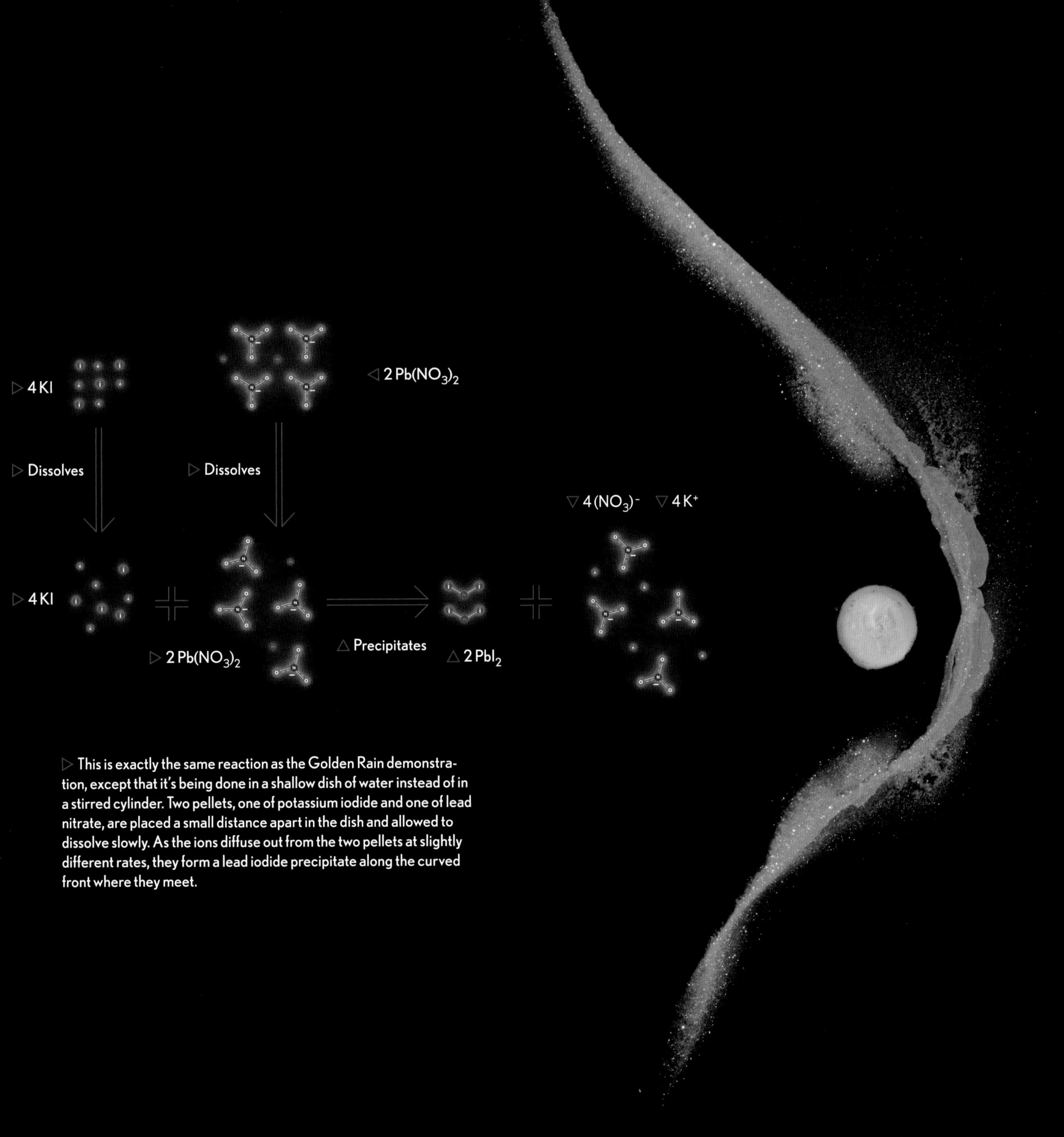

This is exactly the same reaction as the Golden Rain demonstration, except that it's being done in a shallow dish of water instead of in a stirred cylinder. Two pellets, one of potassium iodide and one of lead nitrate, are placed a small distance apart in the dish and allowed to dissolve slowly. As the ions diffuse out from the two pellets at slightly different rates, they form a lead iodide precipitate along the curved front where they meet.

▷ These images, and the video they came from, are perhaps the most beautiful chemical demonstration I've ever been involved with filming. To create this lovely constellation of smoke and sparks, we simply dropped a small wad of folded-up aluminum foil into a flask with a shallow pool of liquid bromine in the bottom. After several seconds of not seeing much, the reaction gets going and builds rapidly into an intense burning as the materials heat up.

Bromine atoms are stealing electrons from aluminum atoms, forming aluminum bromide (Al_2Br_6). Aluminum will do the same thing with any element from the second-to-right-most column of the periodic table (the halogens), but bromine is the only halogen that's a liquid at room temperature, which makes it particularly convenient for use in a demonstration.

Less convenient is the fact that bromine is both highly volatile (evaporates rapidly) and highly toxic. Tiny amounts smell a bit like a swimming pool, which is no surprise since the smell we associate with swimming pools is the smell of chlorine, bromine's neighbor in the halogens column of the periodic table. (Larger amounts of either bromine or chlorine smell like a blowtorch pointed up your nose.)

△ 2 Al

△ 6 Br_2

△ Al_2Br_6

▷ As we learned in the last chapter, chemical reactions happen when electrons move from one molecule to another, or from one chemical bond to another. There's another name for electrons in motion: an electric current. So it should be no surprise that you can use electricity to drive chemical reactions.

This demonstration shows how an electric current can move chromium atoms (in the form of dissolved Cr^{6+} ions) onto the surface of a set of monkeys, where the atoms form a thin layer of chromium metal. Just a few volts is all it takes to get chromium to "plate out" of the solution and onto the monkeys (which are connected to the negative terminal of a battery).

This kind of "electroplating" can be done as a classroom demonstration, but it's also done on an industrial scale to plate everything from cheap jewelry to shiny chrome car bumpers.

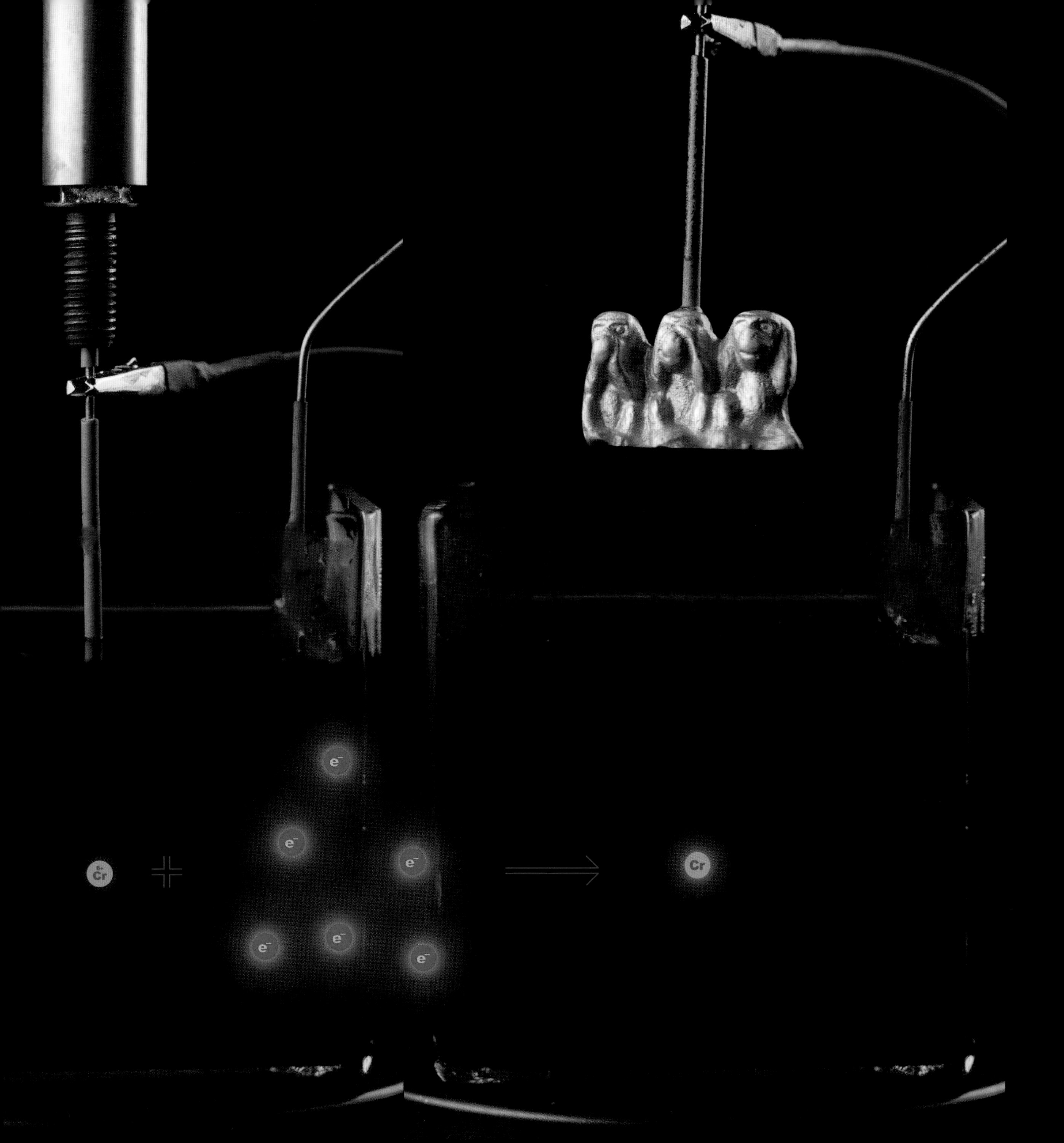
Cr^{6+}
+
e^-
e^-
e^-
e^-
e^-
e^-
→
Cr

In the Kitchen

KITCHENS ARE JUST *packed* with chemicals and chemical reactions. The everyday acts of cooking even match the formal procedures of laboratory chemistry. You take a bunch of different chemicals, combine them according to a list of weights and volumes, dissolve some of them in solvents like water or alcohol, and then you initiate a series of chemical reactions by mixing, heating, or cooling the reaction vessel. In other words, you follow a recipe.

What makes this so funny is how many people do all this chemistry while talking about how they use only natural and chemical-free ingredients—every one of which is, in fact, a chemical. Look, people, if you're eating it, it's a chemical, period. Get over it, and let's talk about the chemistry that you can do with these delicious chemicals.

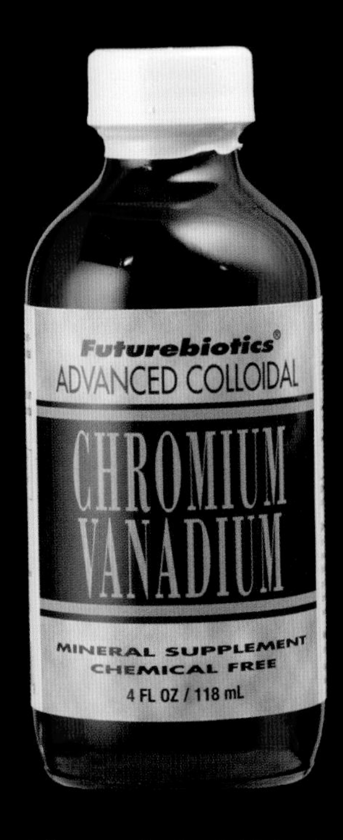

△ In my earlier book *Molecules* I included a picture of some indigo dye that proudly declares itself to be chemical-free. The problem is, indigo, the very name of the product, the reason it exists, the thing you're trying to buy, is a chemical ($C_{16}H_{10}N_2O_2$). I didn't want to repeat myself in this book, but it didn't take long to find more examples. First, here is a package of proudly chemical-free chromium and vanadium dietary supplements. *Chromium and vanadium are chemicals!* (Coincidentally, both are used to alloy with iron to make "chrome-vanadium steel." Do they really not see the iron-y?)

◁ Slightly less insane, but still a bit sad to see, is this "chemical free" polenta. Yes, like all other foods, polenta is made of chemicals (starches, sugars, cellulose, and no doubt a hundred or more other minor constituents).

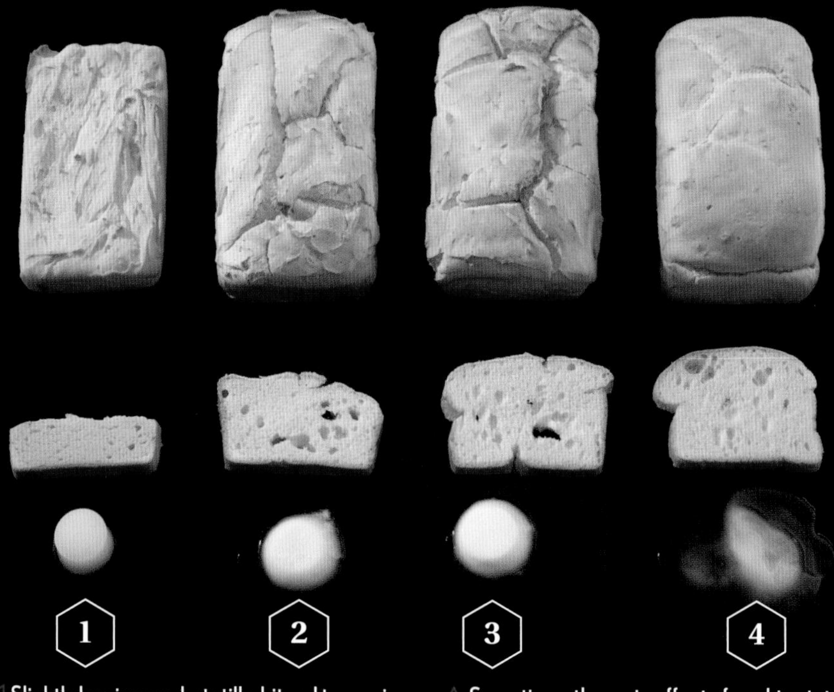

△ Sometimes the main effect of cooking is to make things harder. Soft, squishy bread dough becomes stiffer bread when gluten proteins expand and link up with each other, forming a connected network tying the whole bread together.

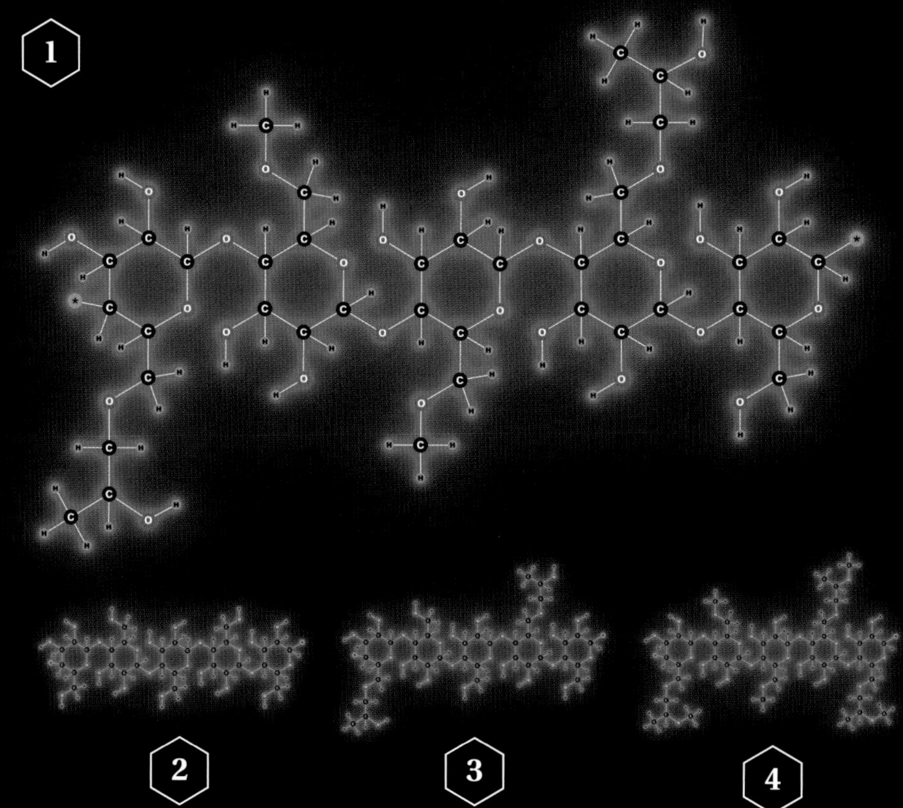

◁ Some people either are, or think they are, allergic to gluten, but still want to eat bread. There is no other natural substance known that can do what gluten does, so if such a person is also not wanting to eat synthetic chemicals, they are out of luck when it comes to bread. A more open-minded person can use synthetic gluten substitutes, which are cleverly engineered molecules based on plant cellulose modified with side-groups that link up to each other. The side-groups allow the long cellulose molecules to form networks just like those formed by gluten.

As more side-groups are added to the molecules, more cross-linking becomes possible, making for a harder end result. The sweet spot is somewhere in the middle, where the bread rises just the right amount.

▷ This is an example of my preferred method of cooking: with a blowtorch. The final step in the preparation of crème brûlée involves using an intense flame to "caramelize" a layer of sugar on top of an egg custard. The high temperature burns the sugar, turning it brown and creating the caramel flavor and smell. "Burn," of course, is a word that describes a set of chemical reactions. Starting with sucrose (table sugar, $C_{12}H_{22}O_{11}$), the high temperature creates a rich mixture of reaction products, including polymer chains (long molecules combining multiple sugar molecules), and entirely new molecules created by breaking up and rearranging the sugar molecules. The combination of these chemicals creates the caramel flavor.

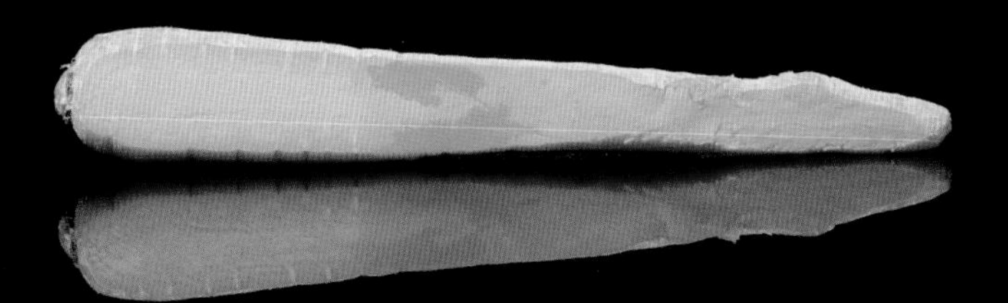

△ Sometimes you cook things to make them harder, other times it's to make them softer. A hard, crunchy carrot becomes soft enough to cut with a fork when heat and water combine to convert some of the insoluble (not able to be dissolved by water) chemicals (pectins) in the carrot into soluble ones that are carried away by the boiling water.

There was a delightful era, now fading, in which it was fashionable to use actual laboratory equipment to cook. A rotary vacuum evaporator (pictured on page 90) for example, is used to pull a volatile liquid (typically water or alcohol) out of a mixture—while continuously stirring it and keeping it at a precisely controlled temperature. Common and useful in the synthesis of laboratory chemicals, less common in the synthesis of food chemicals.

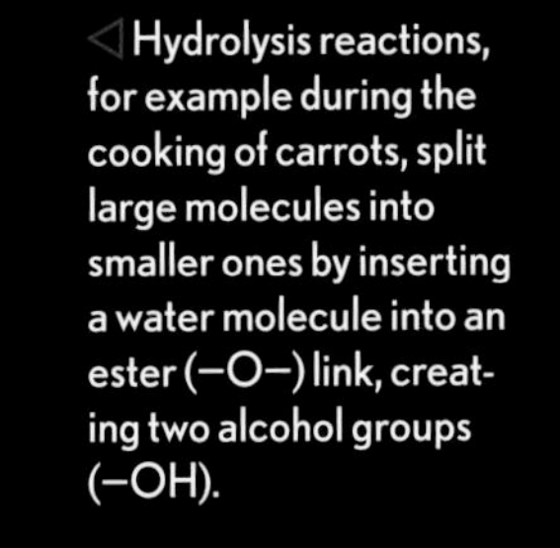

◁ Hydrolysis reactions, for example during the cooking of carrots, split large molecules into smaller ones by inserting a water molecule into an ester (–O–) link, creating two alcohol groups (–OH).

▽ There are some excellent books that have been written on "kitchen chemistry," and this is by far the most extreme example. It arrives in a large box with a "Two Man Lift" sticker on it, because it weighs more than regulations permit one warehouse worker to lift!

There's no way any book under fifty pounds can compete, so let's move on to the other place people mix chemicals according to a recipe.

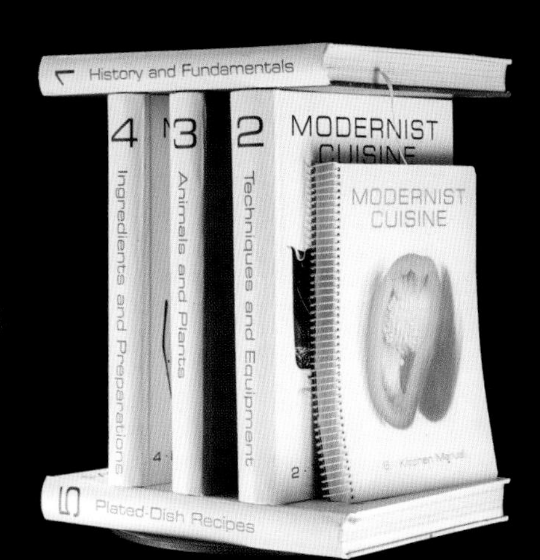

In the Lab

A LABORATORY FILLED with boiling flasks and elaborate glassware is, I'm pretty sure, what the majority of people think of when asked to imagine a chemical reaction. Which is fair enough: This is what chemical reactions typically look like when they are being carried out in a traditional laboratory setting.

◁ This is the stereotype of where chemistry happens: in a boiling flask. Which begs two questions: Why is chemistry so often done in a liquid, and why is that liquid so often boiling?

The great majority of chemicals (in the sense of things you find in bottles in a chemical lab) are normally solids, and they are nearly always stored at room temperature. Wouldn't it make more sense to do chemistry with room-temperature solids instead of boiling liquids?

The problem with solids is that they don't mix. Sure, it looks like these two powders (iron oxide and aluminum) are mixed, but even under an ordinary microscope you can see that they are still separate light and dark particles. In order for chemical reactions to happen, the substances intended to react with each other have to come together at the molecular level, molecule against molecule. Each of these particles, despite being nearly too small to see with the naked eye, still contains countless billions of atoms, and the few on the surface are not nearly enough to get any sort of reaction going between them.

△ This is thermite powder, which you can see reacting in a couple of places in this chapter. Doesn't that prove I'm wrong about solids not reacting? No, because thermite only starts reacting once it's turned into a liquid. Thermite is very difficult to light. A match will just go out if you push it into a thermite mixture. Even a propane torch won't set it off. The reason is that you have to get the mixture hot enough to *melt* a substantial number of the particles and keep them molten long enough for them to start reacting with each other. Only then will the reaction take off and start supplying enough heat to melt more and more of itself.

So, despite appearances, the thermite reaction actually also takes place in the liquid state. Solids really are a lousy way to do chemistry. What about gases?

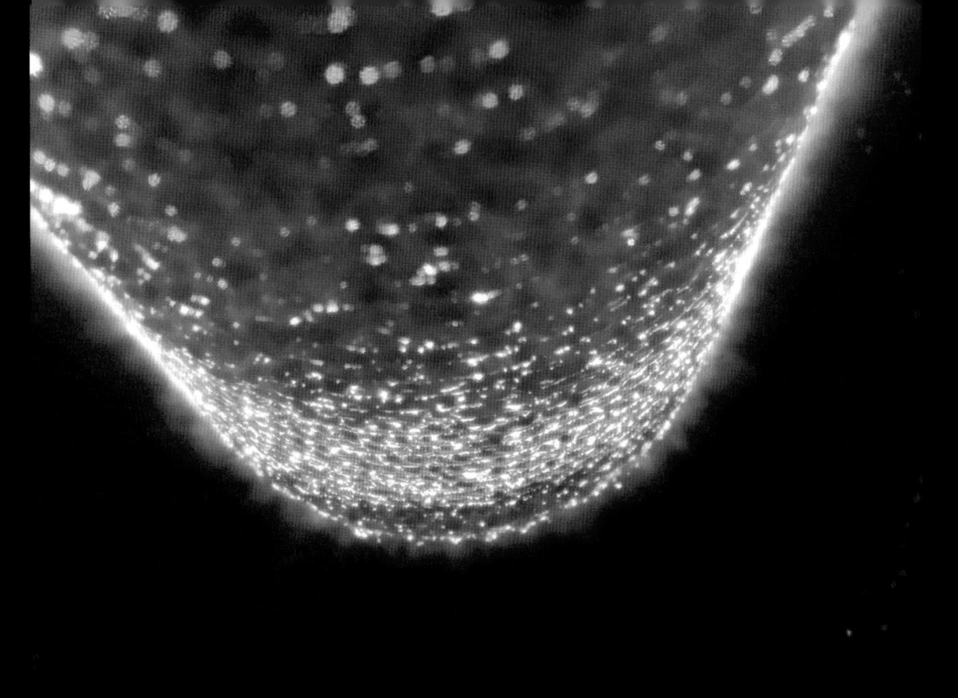

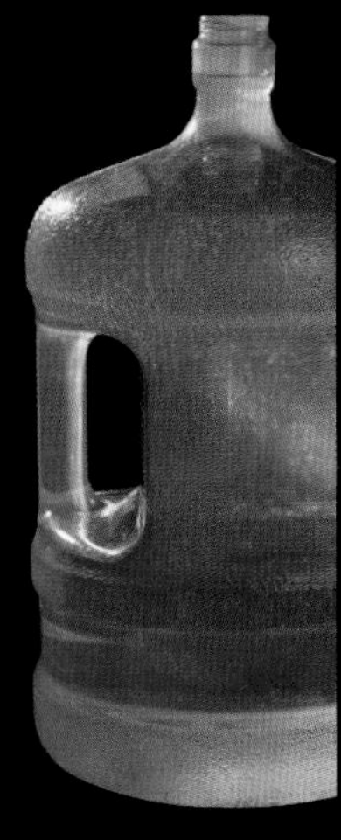

◁ I set up a five-gallon jug to capture the freshly synthesized water created by the furnace in my house. I really like this jug. We think of water as such an "elemental" thing, a force of nature, something that just *is*, not something you create. But water is not an element, it's a compound, a molecule made of three atoms of two different elements. This water didn't used to be water. Just a few weeks before this picture was taken it was methane pumped out of the earth, and oxygen floating around in the air. This water didn't come into my house, it was *made in my basement.* Yet it is absolutely identical in every possible way to all other water. Yes, I drank a bit of it, and it tastes fine. Because I'm not sure of the purity of the gas it was made from, I'm not going to drink any more, but with a bit of filtering it would be as wholesome as any water.

This is an example of a reaction that happens entirely in the gas phase. Natural gas (methane, CH_4) combines with oxygen from the air (O_2) to form carbon dioxide (CO_2) and water (H_2O). Many buildings are heated this way (for the time being, until the gas runs out and we all switch to solar or wind power).

In older, less efficient furnaces, the water exits the area in the form of steam. In other words, it stays a gas until it's far away from where the reaction happened. But more efficient furnaces—like the shiny new downdraft boiler in my basement—are able to extract so much heat from the escaping steam that it condenses back into liquid water, which must then be pumped away from the furnace.

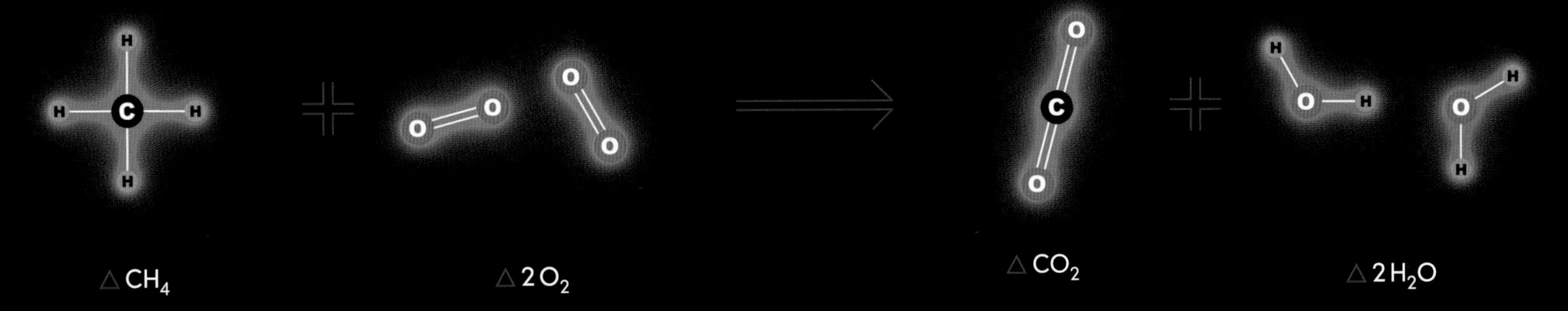

△ CH_4 △ $2O_2$ △ CO_2 △ $2H_2O$

The reaction of methane and oxygen is a rare example of a widely practiced chemical reaction that happens entirely in the gas phase. There are others, but compared to the number of important reactions that happen in liquids, gas-phase reactions are rare.

This is not hard to explain—gases are really impractical to work with! They are constantly trying to escape, need to be kept under pressure, and if they escape they immediately fill the whole room, potentially poisoning everyone in it. Furthermore, a great many interesting molecules cannot be turned into a gas because they decompose (breaking down into smaller molecules or individual atoms) at a temperature lower than their boiling point.

Think of, for example, cookies. You can't vaporize cookie chemicals because they char and burn and decompose long before they melt, let alone evaporate. If you want to do chemistry with cookies, you can't do it in the solid phase because the chemicals won't mix, and you can't do it in the gas phase because you can't create gaseous cookies. The only option left is to make the cookies liquid.

But you can't melt cookies any more than you can evaporate them. This is a very general problem. Chemistry is most conveniently done in liquid form, but many important chemicals are impossible or impractical to melt. What's to be done about this unfortunate situation?

Thus we arrive at by far the most common way chemistry is done in labs, factories, and living bodies—by dissolving otherwise solid chemicals in a solvent, such as water, alcohol, hexane, or some other convenient, naturally liquid chemical solvent.

Solutions (the name for a liquid solvent with something dissolved in it) are *perfect* for chemistry. The molecules in a liquid are constantly moving around, bringing together new combinations. If you dissolve two different substances in a solvent, their molecules will constantly be encountering each other, providing countless opportunities for reactions to happen.

Solutions are often even more useful than the pure chemicals would be in liquid form because you can vary the *concentration* of each molecule independently. If a reaction works best with twice the concentration of one over the other reactant, you can easily set up the solutions that way.

Next question: Why do we so often heat up chemicals to get them to react (not only in the lab, but also in the kitchen)? This has to do with the *rate* at which reactions happen. As a rule of thumb, nearly any reaction will run about twice as fast for every 10°C (20°F) warmer you make it.

It takes luck and energy to get two molecules to react with each other, and heat supplies both.

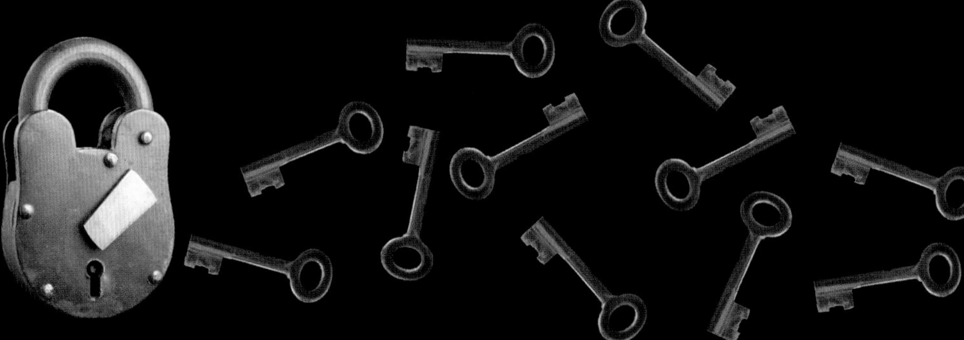

△ You can't just throw a key at a lock and expect it to open. The key has to approach the lock in exactly the right orientation or it won't go in. And once it's in place, it takes energy to turn the lock. If you're planning to open a lock by throwing keys at it, you'll need to throw a *lot* of keys in the hope that one of them will randomly happen to arrive at just the right angle. And you should probably also throw a lot of other things at the lock so that eventually one of them will bang into the key just right to get it to turn.

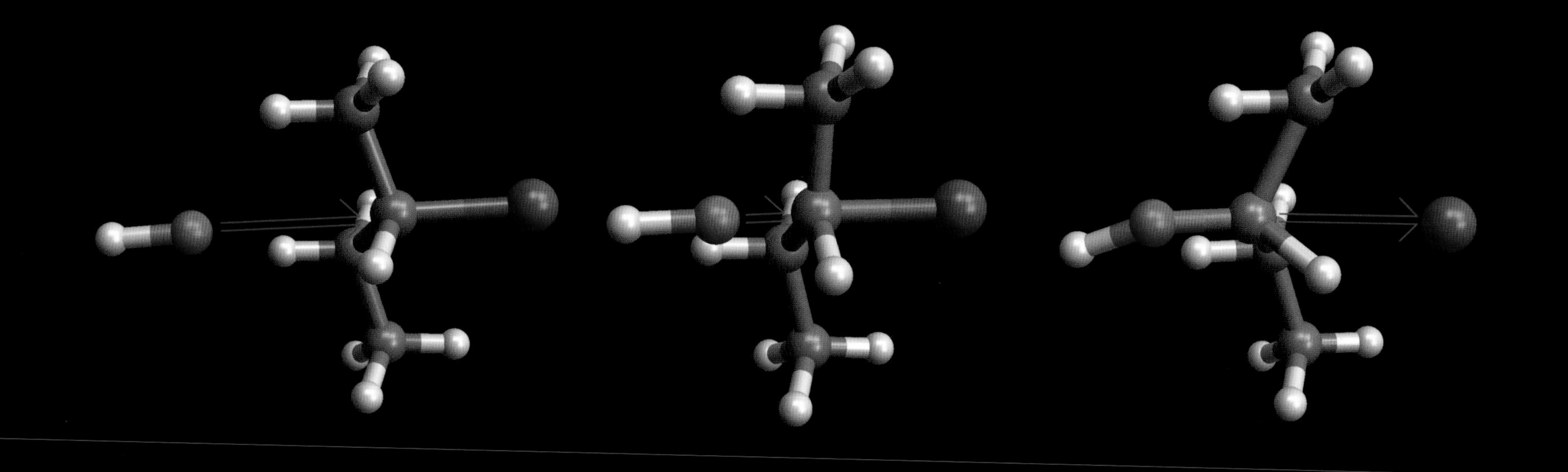

Throwing keys is a very inefficient way to open locks, but this is pretty much how chemistry works much of the time. Getting two molecules to react with each other generally involves those molecules bumping into each other many, many times until eventually they happen to come together in just the right orientation, and with enough force to overcome the strength of the existing bonds and allow new ones to form. In this reaction, for example, a hydroxide ion (OH^-), playing the part of a key, has to come at the larger molecule with the (red) oxygen atom facing in, exactly opposite the (darker red) bromine molecule on the right. If it comes from any other angle or in any other orientation, or not fast enough, it will just bounce off. It's only because random collisions like this happen trillions of trillions of times a second that anything ever gets done in a chemical reaction.

Heat helps in two ways. Remember from the last chapter that heat is expressed as the random movement of the atoms and molecules in a substance. The hotter something is, the faster its molecules are moving. Faster movement means more frequent collisions, which means a greater chance of getting it right. And faster movement also means each of those hits is harder, and thus more likely to have enough energy to trigger a reaction.

The result of these two factors is that we often try to get chemicals as hot as possible, because who wants to wait around for a slow reaction? In a lab that's a waste of some graduate student's time, and in a factory it's a waste of money. Of course there's always some upper limit: Beyond a certain temperature the molecules involved will decompose, or more likely the solvent will boil away.

In many cases the practical upper limit of temperature is simply the boiling point of the solvent. So there you have the answer. Chemical labs are full of boiling flasks not just because it looks cool, but because that's often the practical compromise at which the reactions will go fast enough, without having to use expensive pressure vessels to keep the solvent from boiling. (In commercial settings where reactions are being done on a large scale, and every bit of efficiency translates into profit, it is quite common to see high-temperature, high-pressure reaction vessels. It's worth it in those cases because the same reaction is going to be carried out many, many times.)

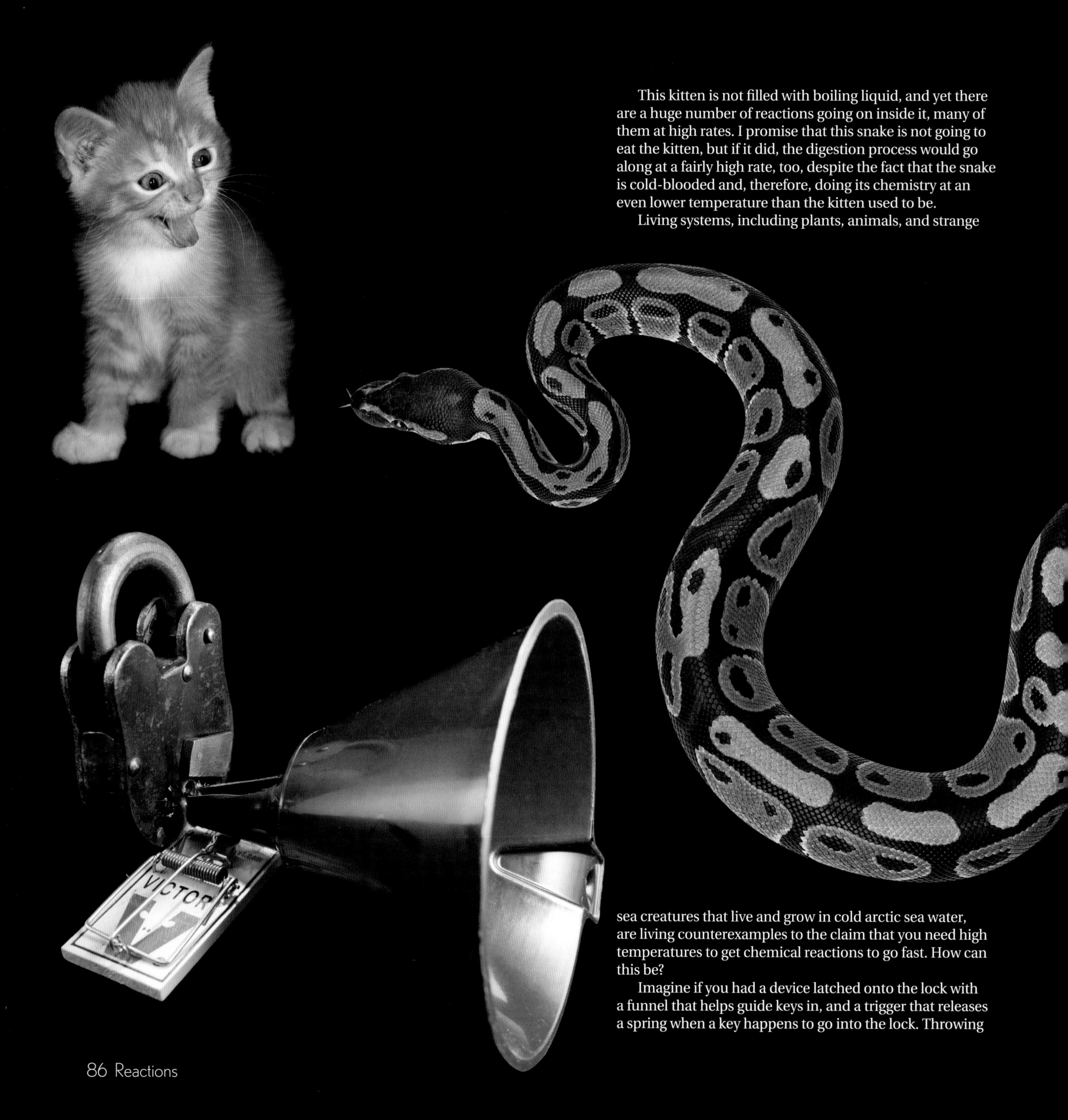

This kitten is not filled with boiling liquid, and yet there are a huge number of reactions going on inside it, many of them at high rates. I promise that this snake is not going to eat the kitten, but if it did, the digestion process would go along at a fairly high rate, too, despite the fact that the snake is cold-blooded and, therefore, doing its chemistry at an even lower temperature than the kitten used to be.

Living systems, including plants, animals, and strange sea creatures that live and grow in cold arctic sea water, are living counterexamples to the claim that you need high temperatures to get chemical reactions to go fast. How can this be?

Imagine if you had a device latched onto the lock with a funnel that helps guide keys in, and a trigger that releases a spring when a key happens to go into the lock. Throwing keys at *that* lock would be a whole lot more successful than throwing keys at the bare lock. It might still take a lot of tries, but many fewer than without the mechanism. (Don't look at my model mechanism too closely. It's a metaphor, not something that's actually supposed to work!)

This is exactly how living systems handle the problem of near-room-temperature chemistry. They are filled to the brim with thousands of specialized protein molecules called enzymes, the function of which is to align molecules in the necessary orientation to get them to react. Some enzymes just align the molecules, others actually have the molecular equivalent of springs that push them in the right direction once triggered. We'll see an example of this when we talk about watching grass grow in The Boring Chapter (Chapter 5).

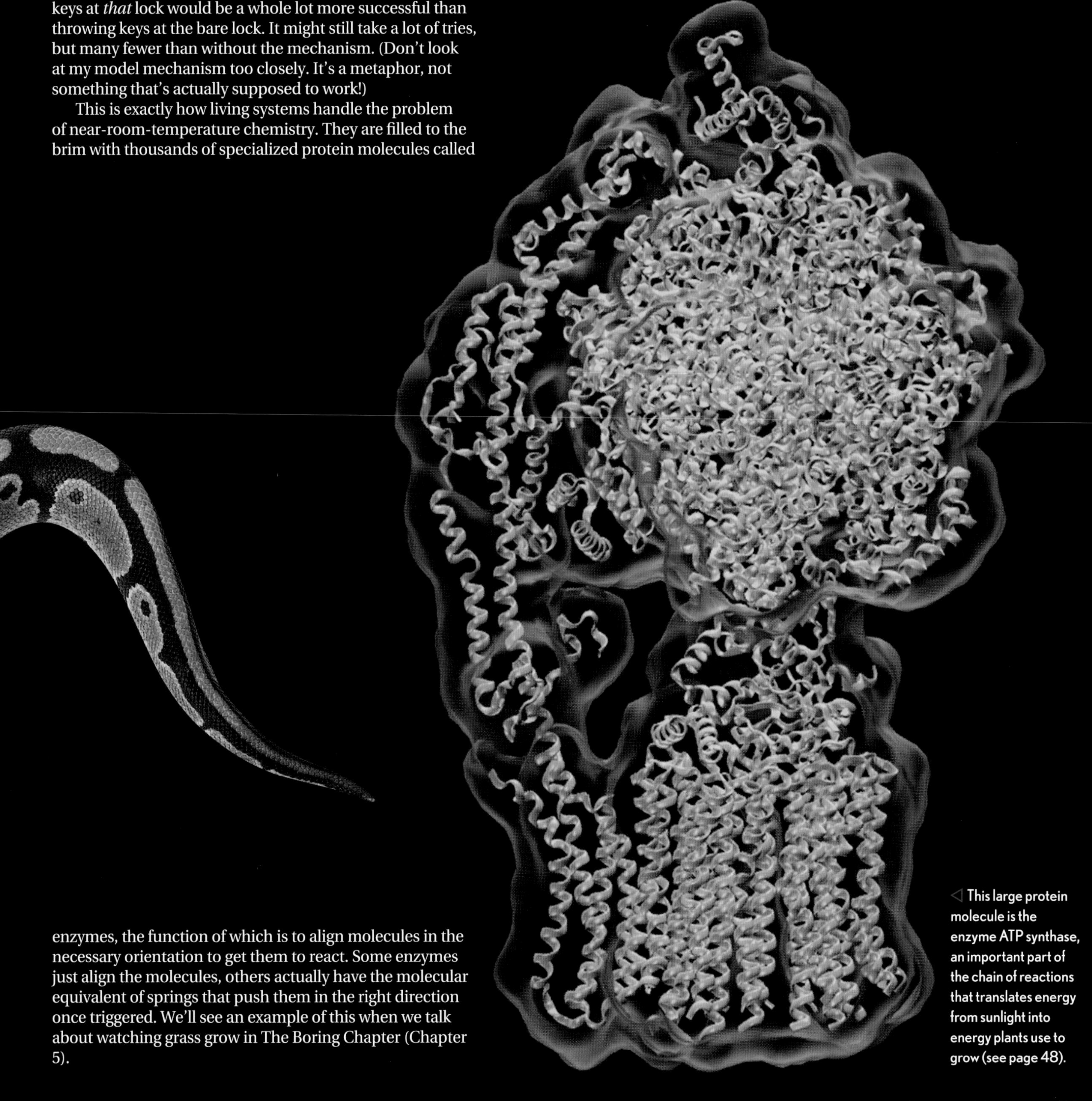

◁ This large protein molecule is the enzyme ATP synthase, an important part of the chain of reactions that translates energy from sunlight into energy plants use to grow (see page 48).

In a Factory

GIANT CHEMICAL FACTORIES look tremendously complicated, but if you pick apart what you're looking at, you can usually figure out that each component is really just a puffed-up, oversized version of something you can find in a lab, kitchen, or backwoods shed.

For example, any time you see a tall thin column, optionally with smaller pipes coming off the side at various levels, you are almost certainly looking at a distillation column, no matter how big or small it is.

All distillation setups have the same basic parts. At the bottom is a boiling flask (pot) where a mixture of liquids is heated (or sometimes a vessel in which some reaction is going on). Vapors rise from this vessel up into a distillation column, where some of them recondense and fall back into the pot, while others continue on until they encounter a condensing coil (often cooled by water flowing through it) off to the side. There the remaining vapors condense back into a liquid and drip into a receiving vessel.

The net effect is to transport the substance from the boiling flask to the receiving flask. Why would you want to do that in such a complicated way? Why not just pour it? Because only *some* of what's in the boiling flask evaporates and makes it past the column. Distillation is a way of separating substances based how easily they evaporate—those with lower boiling points come out first.

Condensing Coil (Often water-cooled)

(Optional) reflux baffles encourage condensation and re-evaporation, improving separation by boiling point ("fractionation").

Receiving Flask

Column

Boiling or Reaction Flask

This much larger commercial alcohol still works exactly the same way, and has exactly the same major parts, as the moonshine still.

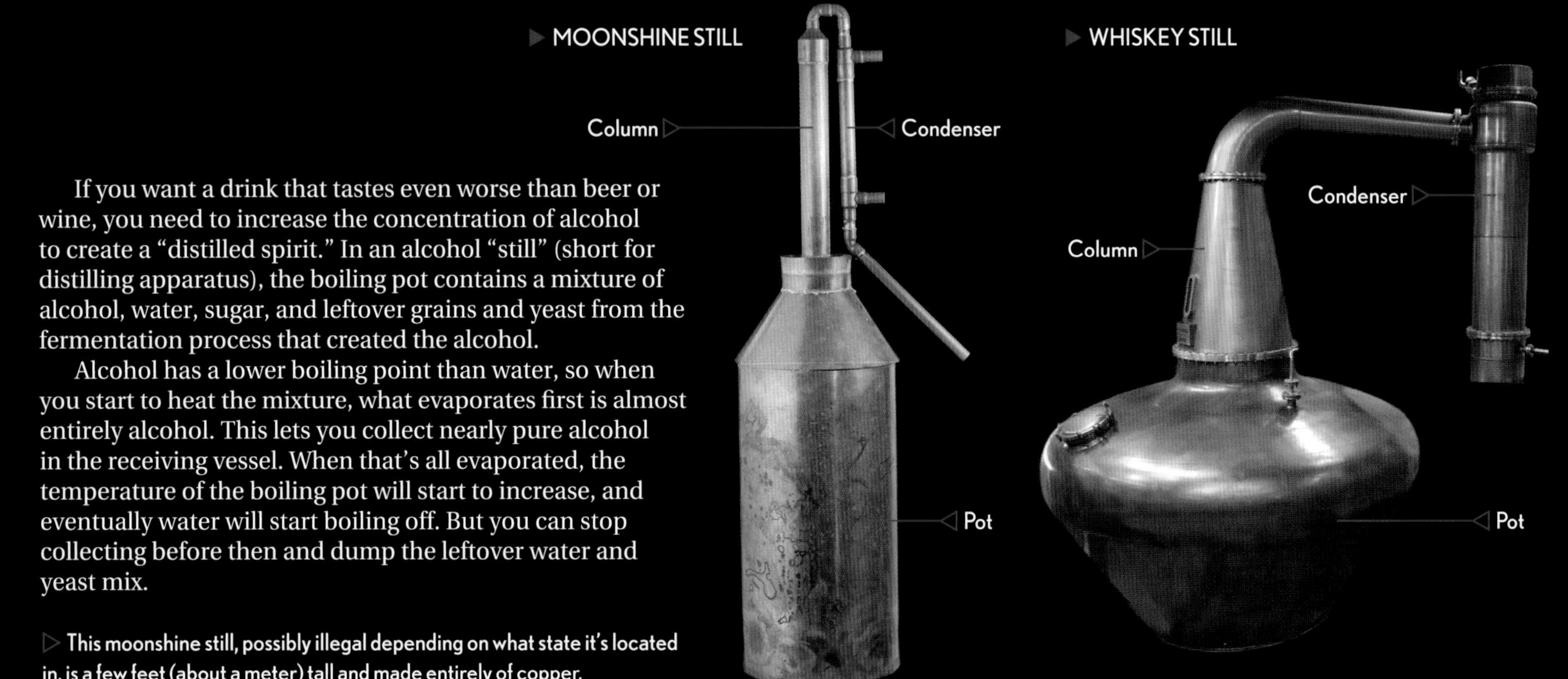

If you want a drink that tastes even worse than beer or wine, you need to increase the concentration of alcohol to create a "distilled spirit." In an alcohol "still" (short for distilling apparatus), the boiling pot contains a mixture of alcohol, water, sugar, and leftover grains and yeast from the fermentation process that created the alcohol.

Alcohol has a lower boiling point than water, so when you start to heat the mixture, what evaporates first is almost entirely alcohol. This lets you collect nearly pure alcohol in the receiving vessel. When that's all evaporated, the temperature of the boiling pot will start to increase, and eventually water will start boiling off. But you can stop collecting before then and dump the leftover water and yeast mix.

This moonshine still, possibly illegal depending on what state it's located in, is a few feet (about a meter) tall and made entirely of copper.

△ When alcohol distillation is done on a (fancy) commercial scale, it looks like this, with 20-foot-tall copper pot stills (seen here in an Irish whiskey distillery in Dublin). These stills are much bigger than the moonshine still we just saw, but the basic parts, the pot, column, and condenser, are identical in their function.

◁ According to the bubbly Canadian tour guide at one Irish distillery, to be legally called "Irish whiskey," distilled alcohol must be aged in a barrel for at least three years and a day (because Scottish whisky requires only three years and they wanted to be one day better). This product, on the other hand, is the collected distillate straight from the condensing coil, as pure and colorless as a properly run distillation operation can make it. (Truly straight from the still it would be about 95 percent ethyl alcohol, but that's not legal to sell for human consumption, so this is diluted with water down to the legal limit.)

◁ A large commercial alcohol still operates at such a high volume that the condensing alcohol forms a mini-river in this viewing cabinet (designed to allow the master distiller to view the product in real time).

△ Laboratory distillation setups are the most "visible" because they are made of clear glass (which does not contaminate the chemicals almost no matter what is being distilled) and because they are designed to be flexible and reconfigurable, so all the parts are separate with joints between them.

There is a delightful variety of distillation columns, with the most elaborate being used for "fractional" distillation (see page 21 for more pictures). This is when you are trying to separate multiple different volatile (able to be evaporated) substances that may differ in boiling point by only a few degrees. By carefully controlling the rate of boiling and the temperature of the column, you can create a situation where the substance is being continually evaporated and recondensed as it rises through the column, teasing apart the similar compounds and bringing them off the top one at a time in separate "fractions."

This lovely copper still in the south of France is used to distill lavender fragrance from crushed lavender flowers. A common problem with industrial chemical equipment is that it's often made of metal instead of glass, so you can't see what's inside. Because it's always used to process exactly the same chemicals over and over again (unlike laboratory equipment) it doesn't have to be impervious to everything, only to the specific things it's going to be used for.

Copper is a common choice for commercial stills because it lasts nearly forever in contact with water and alcohol. Although copper is a relatively expensive metal, it's easy to work with and solder, so the total cost of producing a still like this in copper is likely lower than if it were made out of stainless steel or aluminum. On a side note, yes, we really do crush beautiful flowers and distill their essence to spray in our toilets. Fortunately there are also synthetic lavender fragrances, so maybe at least the cheapest stuff is free of cruelty to pretty flowers.

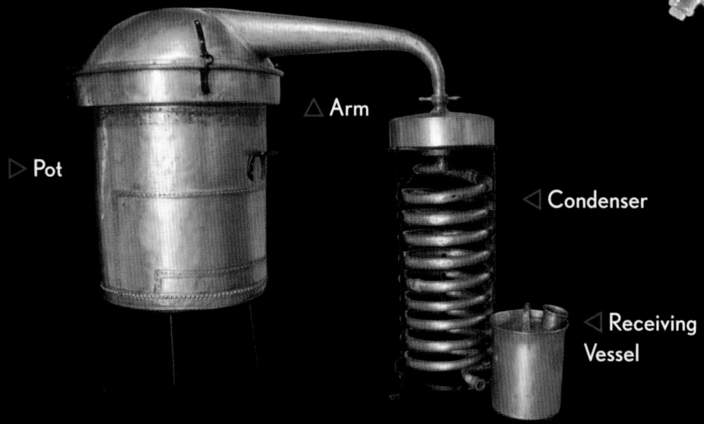

△ Arm

▷ Pot

◁ Condenser

◁ Receiving Vessel

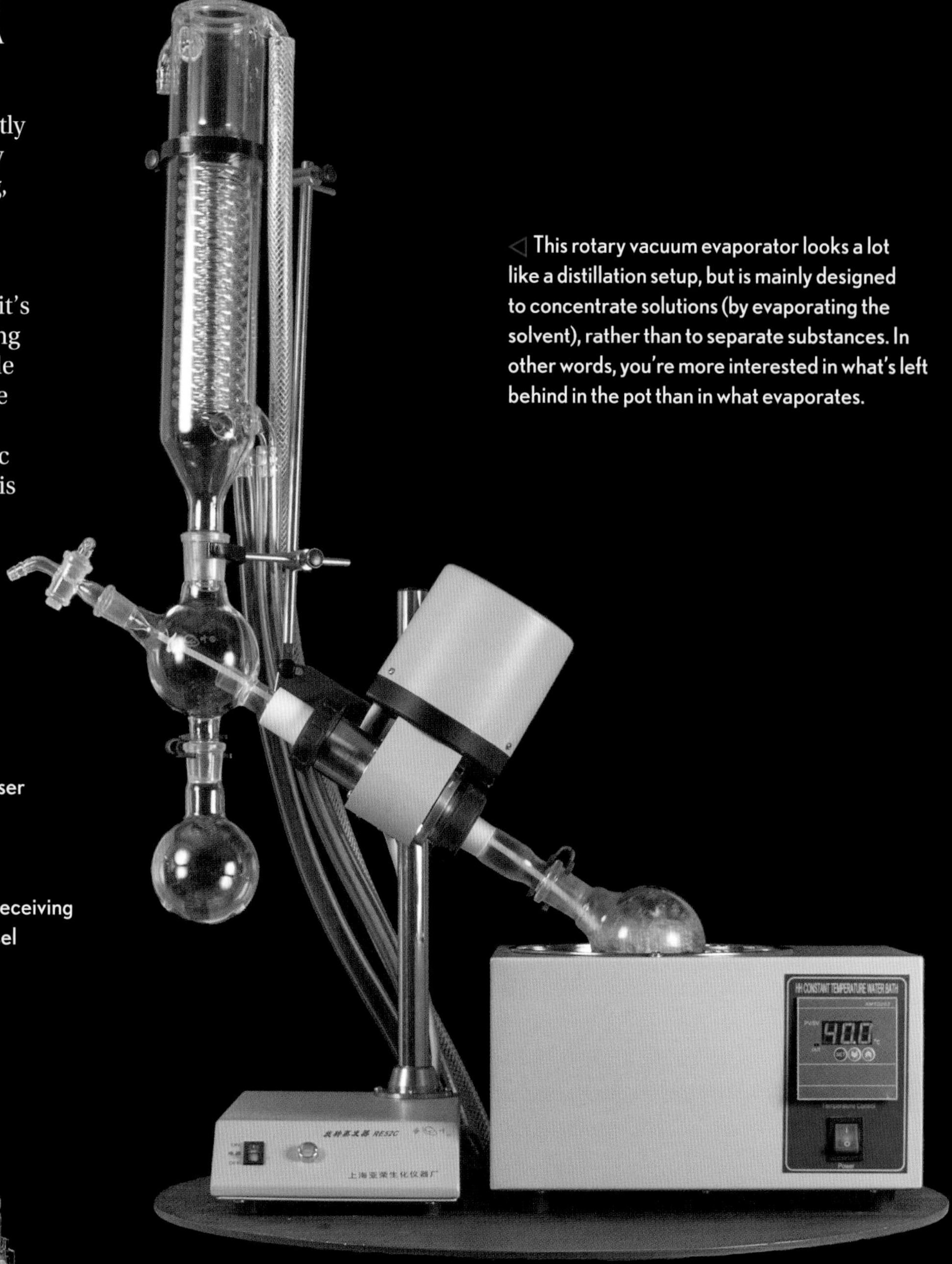

◁ This rotary vacuum evaporator looks a lot like a distillation setup, but is mainly designed to concentrate solutions (by evaporating the solvent), rather than to separate substances. In other words, you're more interested in what's left behind in the pot than in what evaporates.

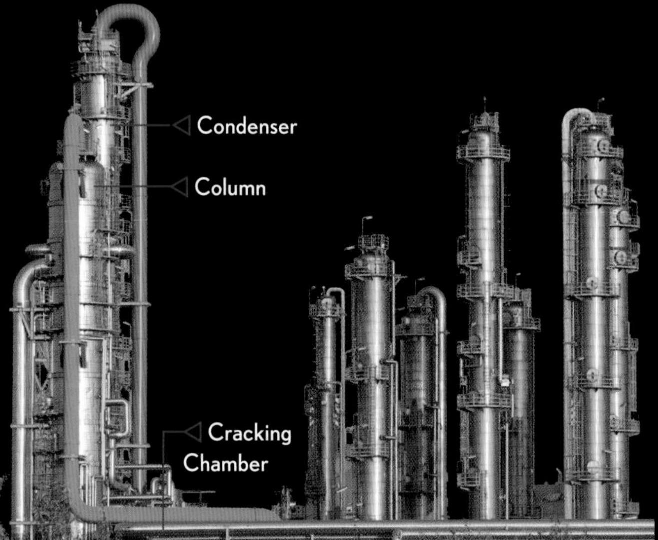

◁ Condenser

◁ Column

◁ Cracking Chamber

◁ These are simply distillation columns, nothing special. The fact that they are a hundred feet (30 meters) tall doesn't change how it works. On the bottom, crude oil is heated to the point where most of it evaporates. The vapors rise up in the column and become progressively cooler the higher they get. Each "fraction" condenses out in order according to its boiling point.

Near the bottom of the column, the first vapors to condense are heavy oils. Next up are lighter oils, then kerosene, then compounds found in gasoline, and finally "naphtha," the lightest fraction that condenses at all. At each different level there is a take-off that draws the products off to be further processed. What doesn't condense even at the very top consists mainly of natural gas (methane and ethane). These gases are drawn off the top, purified, and either used to power the oil refinery, piped off for processing into various useful chemical products, or collected to be sold for heating homes.

This miniature perfume still is less than a foot (30 cm) tall.

Column

Condenser

Pot

Receiving Vessel

Iron is very, very common in the world. You can dig it up practically anywhere, but you always find it in the form of iron oxide, either red rust called hematite (Fe_2O_3) or black magnetite (Fe_3O_4). Any exposed iron metal (elemental Fe) quickly rusts away to iron oxide.

To get useful iron, you need to "reduce" the iron oxide into iron metal. We've seen how to do that using the thermite reaction. This is great if you need a few pounds of white hot liquid iron in a hurry, but it's no good as a way to refine iron ore on a large scale. Why? Because thermite requires aluminum metal, and aluminum also can only be found as its oxide (Al_2O_3). Turning aluminum oxide into aluminum metal is even harder (and more expensive) than turning iron oxide into iron metal, so on an industrial scale it really doesn't help that you can use aluminum metal as a way of getting iron metal.

Fortunately you can reduce iron oxide to iron metal simply by heating the iron oxide in the presence of carbon. Simple in principle, tricky in practice! The reason the Stone Age and Bronze Age came well before the Iron Age is because of how hard iron is. The main problem is the high temperature required. You have to build a furnace and a fire able to achieve a much higher temperature than any ordinary fire, and sustain it for hours while the reduction takes place.

How common is iron ore? You can't even escape it at the beach! Literally while writing this chapter in the dead of winter I was forced to spend time at a beach resort in Panama, and what should I find everywhere, even between my toes? Black magnetite sand! I bought a souvenir bottle opener and used the magnet on the back of it to collect a bag of the stuff. (You never know when you might need to make some thermite.) Magnetite, as the name implies, is magnetic, and if you run a magnet over the surface of the sand, the black magnetite particles leap up and stick to the magnet, allowing you to separate and collect them.

You can see pictures of the iron objects created from this sand on page 71.

△ Black magnetite sand in place on Playa Blanca, Panama. Playa Blanca means "white beach."

△ The author in place on Playa Blanca, Panama, posed in front of black magnetite boulders littered along the sea shore. The black sand on the beach comes from the erosion of these rocks by wind and waves. (At least I think it does: Full disclosure, I am not a geologist and might be quite wrong about this. But the beach was nice, and the sand was definitely magnetite.)

◁ Iron ore comes in many forms. This pretty magnetic rock is from an abandoned iron mine near the aptly named Ironton, Minnesota.

The first attempts to make iron from ore never actually resulted in liquid iron. Instead, layers of charcoal or wood (which burn and supply carbon) were alternated with layers of iron ore in a clay oven and set on fire. Bellows were used to pump air into the furnace for many hours, resulting in a "bloom" of impure, soft-but-not-molten iron. This bloom was then hammered, folded, heated, and hammered again in an incredibly labor-intensive process that eventually resulted in a small, precious lump of usable iron.

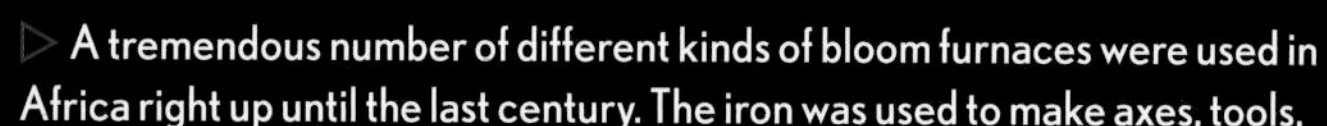

▷ A tremendous number of different kinds of bloom furnaces were used in Africa right up until the last century. The iron was used to make axes, tools,

▽ The Japanese version of the bloom furnace looks different but works the same way. The most famous use for Japanese bloom iron was to make Samurai swords. Coincidentally, the layered, inconsistent nature of the iron you get with this process actually makes for particularly good swords.

◁ When iron is very high in carbon, it's hard and keeps a sharp edge, but is brittle. When iron is low in carbon, it is strong and shock resistant, but softer. By alternating thin layers with different carbon content, you can get a sword that is both strong and sharp. Ancient Samurai swords first achieved this by accident. Today some steel is intentionally made with some of the same characteristics, but since we now understand in detail how to control the properties of steel, ours is far superior to anything created by the Samurai, or any of the other ancient steel makers.

▷ Modern "Damascus" steel is made with alternating layers of high and low carbon steel folded in an ancient tradition similar to that of Japanese sword making. Etching with acid reveals the patterns, and makes for particularly beautiful knives. While ancient and traditional, this material is no match for modern tool steel alloys, or tungsten carbide bits, which cut it like butter.

▷ The modern blast furnace is similar in basic design to the ancient bloom furnaces, in that it too has a stack of iron ore alternating with coke (carbon fuel derived from coal). But instead of a hand-powered bellows, the blast furnace uses high pressure preheated air to stoke the fire to a much higher temperature, turning the iron completely liquid. These huge furnaces are operated continuously. Liquid iron flows out the bottom and new ore and coke is loaded in the top, for months on end without interruption.

When one of these immense beasts of industry is allowed to go cold, it is enormously expensive and time-consuming to start it up again. When a steel mill closes for good, the last heat leaking from a dying furnace is a source of great sorrow for those who tended it and saw the life it gave to their communities. Those who ran it know better than anyone that there is a finality, a kind of death, that comes when the cold seeps in.

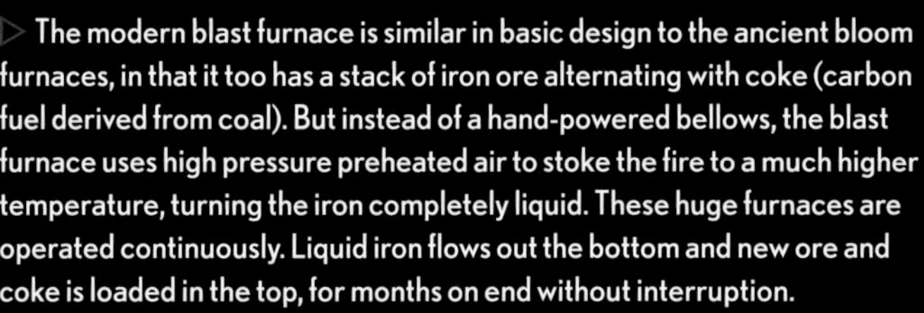

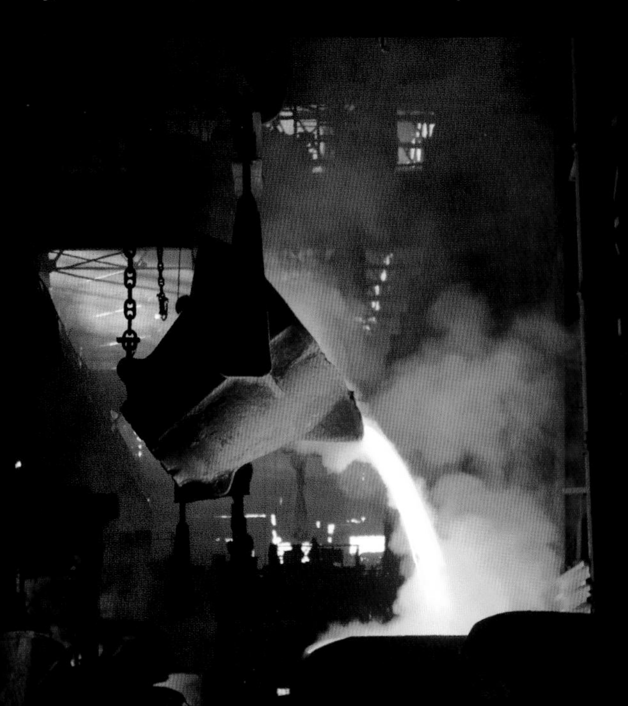

◁ With the iron in a completely liquid state, it can be separated from impurities much more thoroughly. They either float or sink and can be skimmed off. Blasts of oxygen can burn out excess carbon in the iron, and alloying agents, including vanadium, molybdenum, and many others, can be dissolved into the molten iron. This kind of chemical processing allows us to create an incredible range of different kinds of iron-based alloys, including super-hard tool steels, stainless steels that never rust, spring steel that can be bent countless times without losing its shape, and so on.

Iron smelting was first achieved so long ago that there's an age named after it (about 3,000 years ago). It can be done as a purely chemical process, a reaction between iron oxide and carbon. Aluminum smelting, on the other hand, is a far more recent development—the first aluminum metal was only created in 1825, and commercially practical smelting started in the 1880s. Practical aluminum had to wait for the invention of electricity, because massive electric currents are the only way to extract aluminum from its ore on an industrial scale.

Each cell, of which there are hundreds in a typical aluminum smelting factory, produces about a ton of aluminum a day. That's about 20,000,000,000,000,000,000,000,000,000 (20 billion billion billion) atoms of aluminum, each of which has to be individually reduced using three electrons (as shown in the reaction expression below, it takes three electrons, each with a -1 charge, to reduce an aluminum ion with a +3 charge down to a neutral aluminum atom with zero charge). If you work out how much electric current (the rate at which electrons are flowing) is required to get enough electrons to produce this amount of aluminum metal in a day, it comes out to about 100,000 amps (and taking into account inefficiencies in the process, typical aluminum refining cells operate at twice that current or more).

▷ These aluminum refining cells are overgrown versions of the tabletop demonstration we saw earlier in this chapter, where chromium was plated onto small monkeys for decorative purposes (see page 78). Instead of chromium, these cells are plating aluminum metal out of a molten bath of its ore. And instead of plating a microscopically thin layer just to make a shiny surface, the cells keep working, plating more and more aluminum, until literally tons of it have built up.

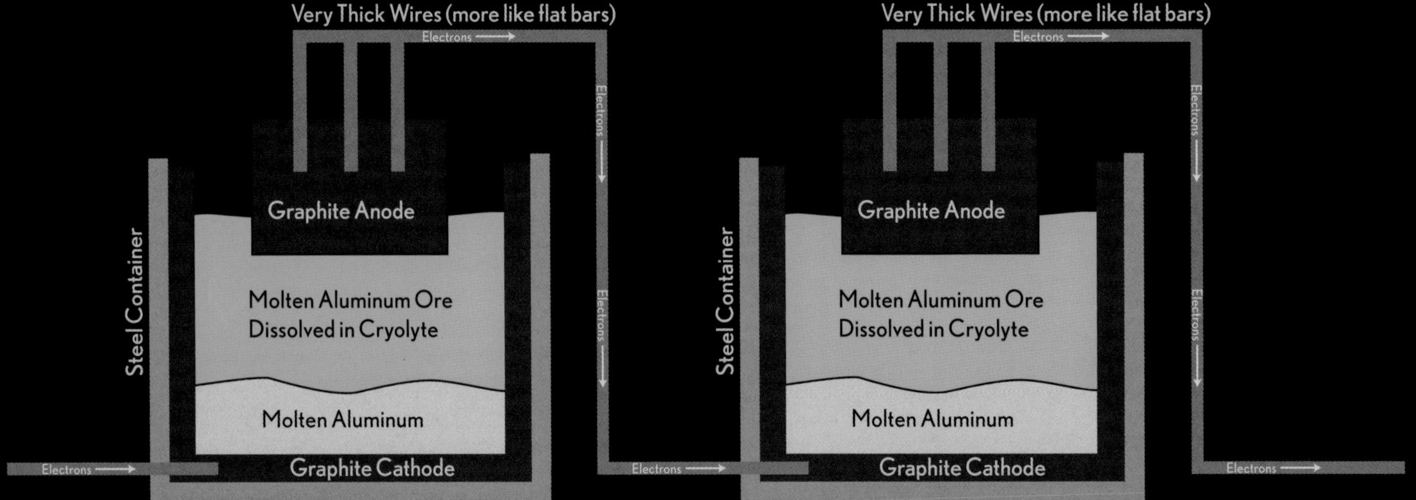

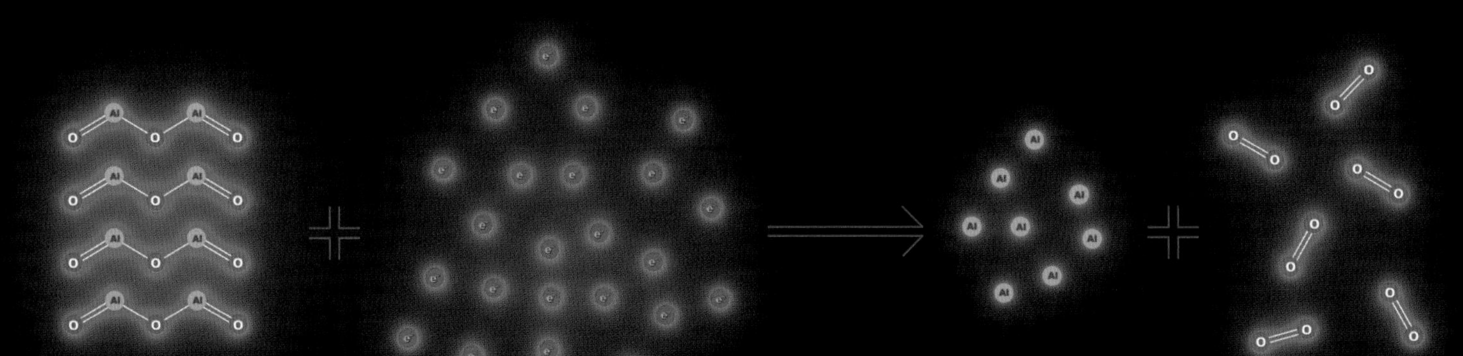

△ How much is 100,000 amps? Two amps (at a similar voltage to what's used in aluminum refining) will charge your phone, so one of these cells is using enough current to charge every single cell phone being held up by every single fan taking a video of every single thing the latest pop star does on stage in a 50,000-seat arena. Just take a look at the size of the cables it takes to route that much current to the cell!

△ Geothermal power plant in Iceland.

◁ It's common to locate iron smelting plants near the source of iron ore. Makes sense, right? Why transport the ore over large distances when you can process it onsite near the mine and ship only the higher-value iron? But driving the chemical reaction to produce aluminum metal requires so much electricity that it turns this logic on its head. Aluminum smelting plants are usually located near a source of inexpensive electricity: hydroelectric plants in Canada, geothermal electricity in Iceland, or nuclear plants in various countries. The aluminum ore is shipped to where the electricity is rather than the other way around.

On the Street

ROAD FLARES ARE USED to warn passing cars of an emergency, like a car off the side of the road or a fallen tree. A flare is a cardboard tube filled with a mixture of strontium nitrate, potassium nitrate, sawdust, charcoal, and/or sulfur. Strontium nitrate and potassium nitrate are both "oxidizers" which supply oxygen to support burning of the sawdust, charcoal, and sulfur. Some flares will even burn underwater because they do not need oxygen from the air.

Flares are good for when there's a breakdown on a road. A breakdown on a railroad requires an incendiary mix of a different sort.

Welding together lengths of iron train track is difficult. In fact, welding isn't really the right word for what needs to be done. To stand up to continuous pounding by train wheels, both ends of the track must be made fully liquid at the same time, so they blend together into a single, continuous piece of track. Train tracks are very thick and if you try to heat up one end, the heat will be drawn away into the rest of the track pretty quickly. There's no practical way to make this kind of joint with an ordinary welding torch or arc welder. These tools are nowhere near powerful enough.

What's needed is a way to fill up a gap between the rails—about the width of a finger—with liquid iron that's hot enough to melt a bit of the neighboring rails before it cools. (The rails are also typically preheated before the gap is filled, to be sure there's no chance of an unmelted spot.)

When a new section of track is being laid from scratch, you can bring along a train car–sized piece of equipment to heat and join the sections. But when field repairs need to be made, you need some kind of very portable setup that can deliver a tremendous amount of heat—and about ten pounds of liquid iron. You couldn't really ask for anything better than thermite.

△ Flare mixture is a lot like gunpowder (potassium nitrate, charcoal, and sulfur) but it burns far more slowly because the ingredients are not mixed as finely, and strontium nitrate is not as powerful an oxidizer as potassium nitrate. (See page 193 for the gunpowder reaction expression. The reaction you see here is basically the same, but substituting strontium for potassium.) Strontium nitrate is chosen not just because it's an oxidizer, but also because the strontium atoms in it give rise to the intense red color of the flare's flame. (In Chapter 4 we will learn about how different atoms create light of different colors.)

▽ Earlier in this chapter we saw thermite as a particularly high-end classroom demonstration. Iron oxide (rust that's been collected and ground into a powder) and aluminum metal in powder form react with each other to create very hot liquid iron. This is a cool classroom demonstration, but think about it: liquid iron on tap from an easily portable reaction? That could be really useful!

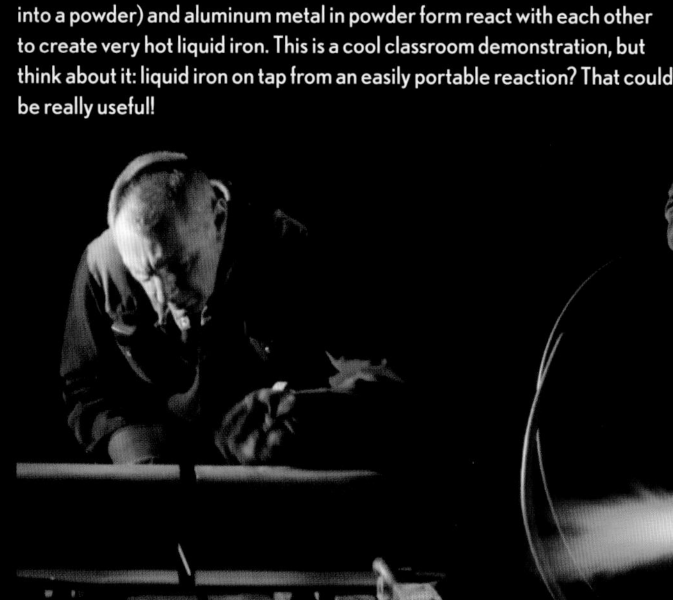

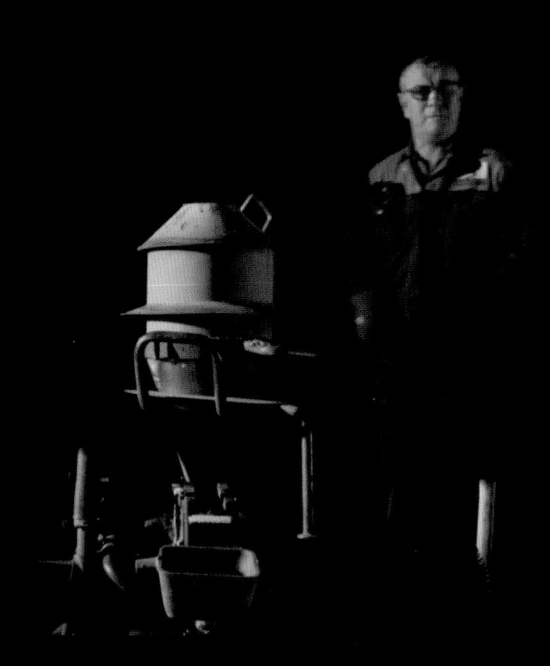

△ Lighting thermite is surprisingly difficult. Various methods can be used, but by far the simplest is to use an ordinary fireworks sparkler (the kind with a wire inside and a silvery coating). They burn very hot, and will reliably start any kind of thermite. (Professional setups use custom-made sparklers, but there's very little difference between them and the just-for-fun kind.)

▷ After the thermite is lit, it huffs and puffs for a good thirty seconds before anything further happens. During this time the reaction is progressing from the top down, and a pool of liquid iron is forming inside the concrete pot (exactly as we saw in the cutaway view on page 71). At the bottom of the pot there is a hole stopped up with an aluminum plug of carefully chosen thickness. At just the right moment, when the liquid iron is fully formed and consolidated into a clean pool, the aluminum plug melts through, releasing the iron.

▽ Iron is much heavier (denser) than any of the other substances present in a thermite reaction (iron oxide, aluminum, and aluminum oxide). So the first thing to come out when the aluminum plug melts is fairly clean liquid iron, which flows straight down into a clay mold that has been constructed around the gap in the rails.

Once the mold is full, the excess spills out into catch basins on each side. Some of what's spilling out is excess iron, but a lot of it is the other reaction product, aluminum oxide (which looks pretty similar when it's a white-hot liquid). Cooled down and solidified, aluminum oxide is known as corundum, and it's what one kind of sandpaper is made of (because it forms very hard, sharp-edged crystals). You know you've got a hot reaction when it creates not only liquid iron, but also liquid sandpaper.

▽ After the iron cools, the two sections of track have become one. It remains only to grind the top and side surfaces down to form a smooth joint. If you're ever on a high-speed train going a couple hundred miles per hour, the bumps you don't feel are not there because of the perfection of thousands of these joints flying under your wheels.

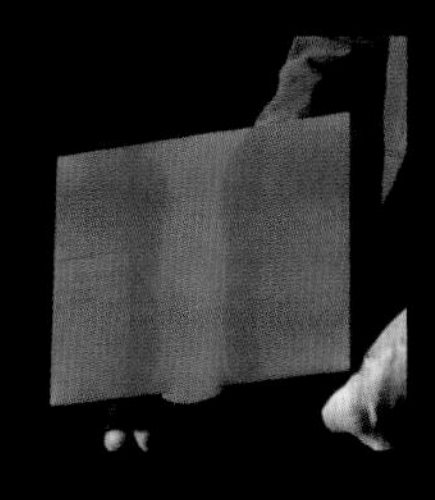

◁ A cross section sawn through the joint shows that what had been two pieces of iron are now one.

High explosives are widely used when you have something and you would like to not have it anymore. Things like bridges, enemy tanks, large boulders in the way of a road, or entire buildings that have become inconvenient—all can be "taken care of" with a few high explosive charges.

This, however, is not an example of high explosives at work (we won't get to those until Chapter 6, The Need for Speed). Instead, this concrete block has been broken up *slowly*, over the course of several hours, by nonexplosive expanding demolition powder. When this powder, which is mostly simple calcium oxide and calcium hydroxide lime, is poured into bore holes, a slow transformation happens that causes the powder to expand. With nowhere to go, it exerts more and more pressure on the concrete until it cracks wide open. For people without the necessary license to possess high explosives, but with inconvenient concrete, this is a great concept.

A very similar thing happens when water seeps down into cracks in roadways or other concrete structures, and then freezes. Because water expands when it freezes (a very unusual property discussed in more detail on page 203), ice can exert tremendous pressure on the concrete, cracking it and ruining millions of dollars worth of road work. This is just one of the many ways ice can be annoying.

Fortunately, we have salt to deal with ice!

When you spread salt on ice, it lowers the melting point of the ice, turning it back into water (as long as the new melting point is lower than the current ridiculously cold temperature that no human being should ever have to endure). You can use ordinary sodium chloride table salt down to about 20°F (–7°C), but other kinds of salt work at lower temperatures—the champion is calcium chloride, which works down to –25°F (–30°C).

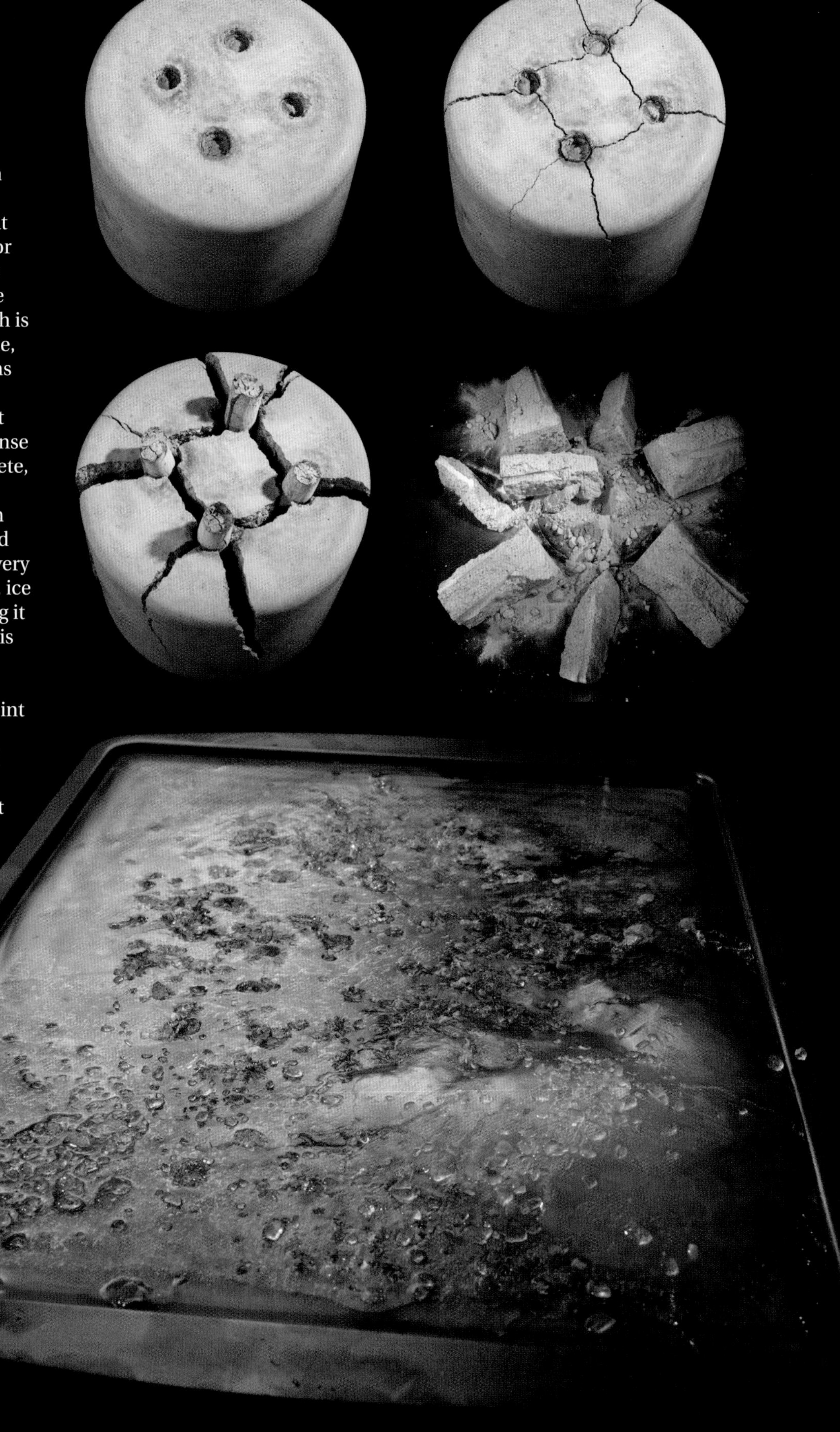

▷ Those of us who have the privilege of living in places where the roads regularly freeze over in sheets of solid ice know all about this reaction. Or is it a reaction?

◁ Solid grains of salt are crystals made of alternating electrically charged atoms (ions). In this example of table salt the ions are Na^+ (sodium atoms that have a +1 charge because they are missing one negatively charged electron) and Cl^- (chlorine atoms with a −1 charge because they have one extra electron).

When a salt grain is dissolved in water, its ions get split up and surrounded by water molecules. The solid crystal disintegrates into individual atoms and vanishes into the water.

Salt water has a lower freezing point than pure water because the dissolved ions interfere with the freezing process. As liquid water gets colder and colder, its molecules start aligning with each other, forming networks and eventually whole three-dimensional grids of connected water molecules. These are ice crystals.

▷ Dissolving

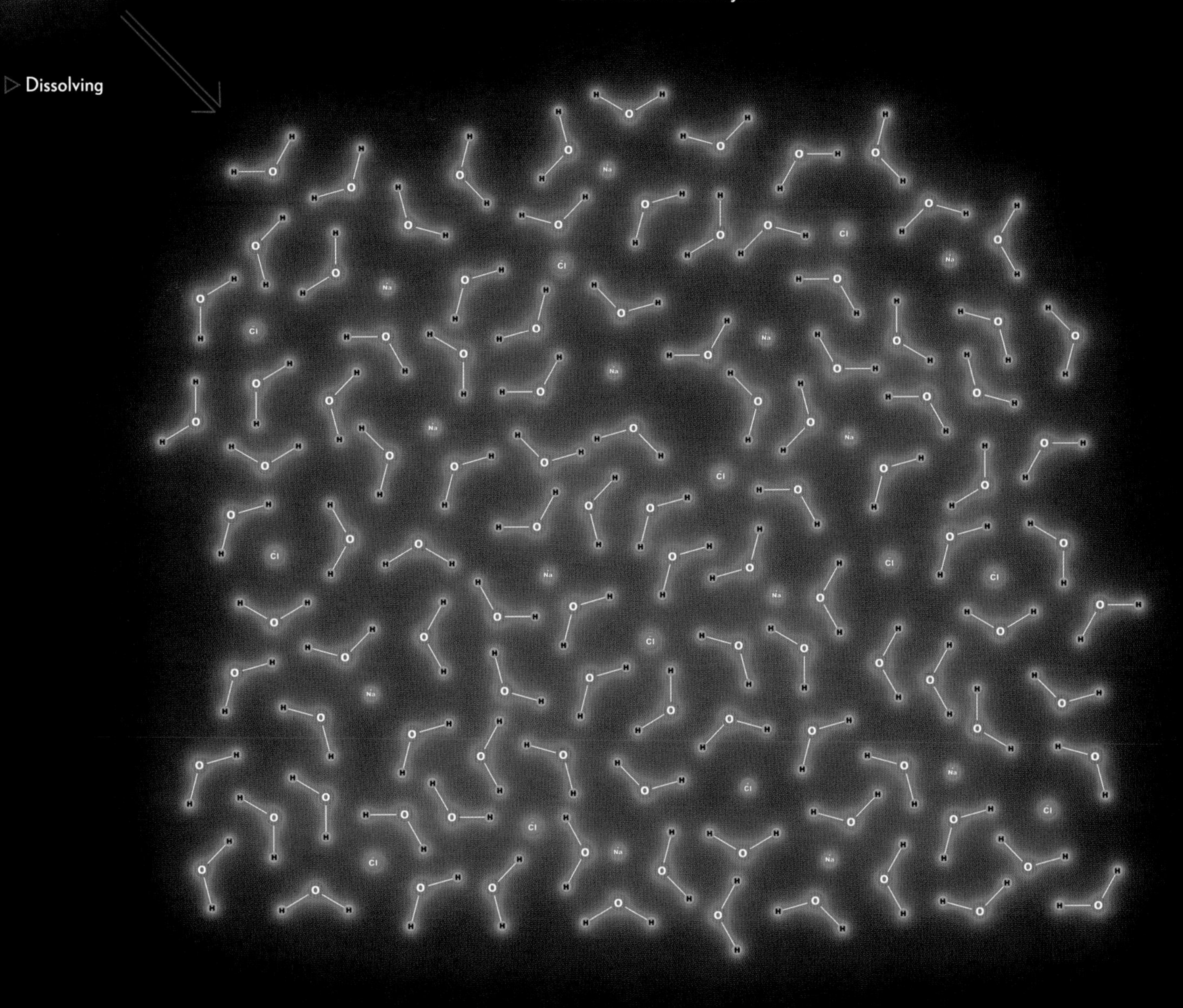

▽ When there is salt dissolved in the water, the ions block water molecules trying to join the crystals as they are forming, tipping the balance in favor of remaining liquid down to a lower temperature.

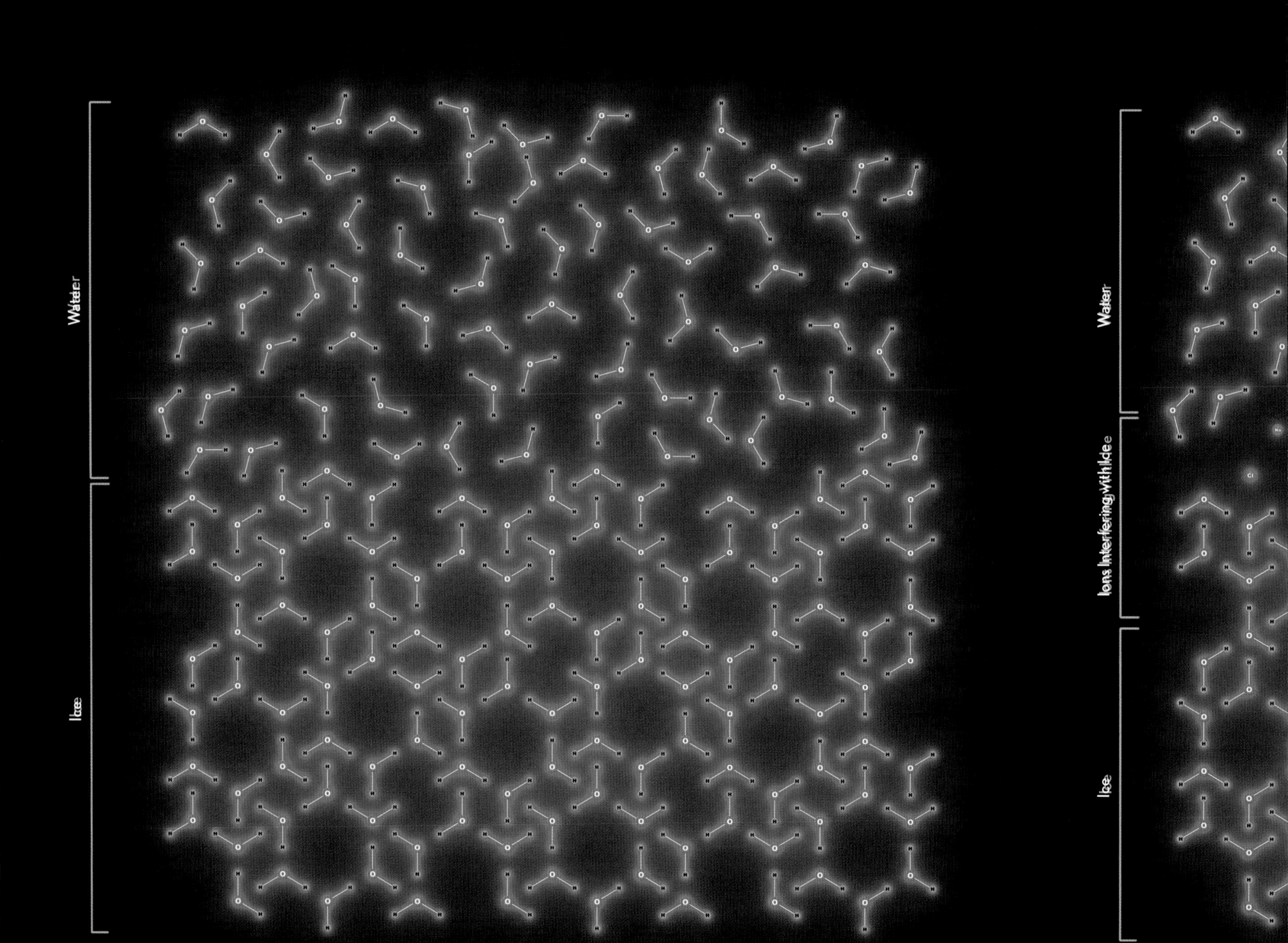

Is Dissolving a Reaction?

I'VE ASSERTED SEVERAL TIMES in this book that nearly everything you see going on around you is a chemical reaction. But how far exactly does this claim extend? Is it a chemical reaction when a substance is dissolved in water?

Textbooks often make a big deal about the difference between chemical changes (reactions) and physical changes that are not reactions. Examples given for nonreactions typically include melting, boiling, and . . . dissolving. But, as is usually the case if you make a big deal about the definition of a word, edge-cases can trip you up.

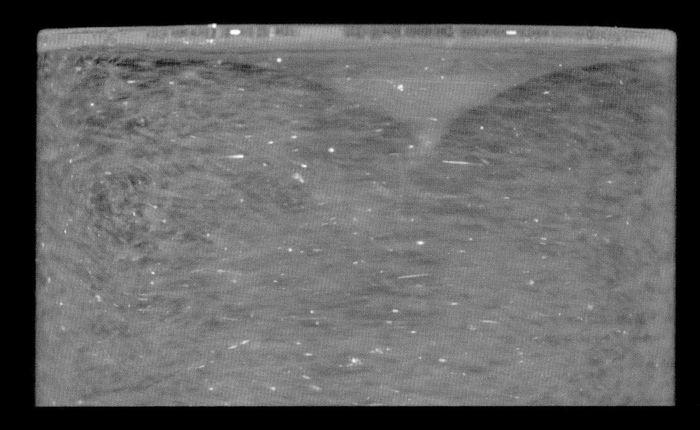

It is pretty easy to find a problem with the claim that a salt dissolving is not a chemical reaction. You just have to look at the reverse process. Remember the Golden Rain demonstration from page 74? It's a demonstration of a *chemical reaction*, right? Well, what's happening in that demonstration is precisely the opposite of what happens when a salt dissolves. Three ions, one positive (Pb^{2+}) and two negative ($2\ I^{-}$) combine to form a solid crystal (PbI_2), which we see as the golden precipitate.

You can't really have it both ways. If a salt precipitating is a chemical reaction, then the reverse process, a salt dissolving, must surely also be a reaction. And it fits the standard definition of a reaction: Chemical bonds between ions are being broken, and making or breaking bonds is what reactions are all about.

Some texts classify a salt dissolving as a chemical change—more like a reaction than just a physical change—but even they draw the line at sugar dissolving.

▲ SALTY WATER

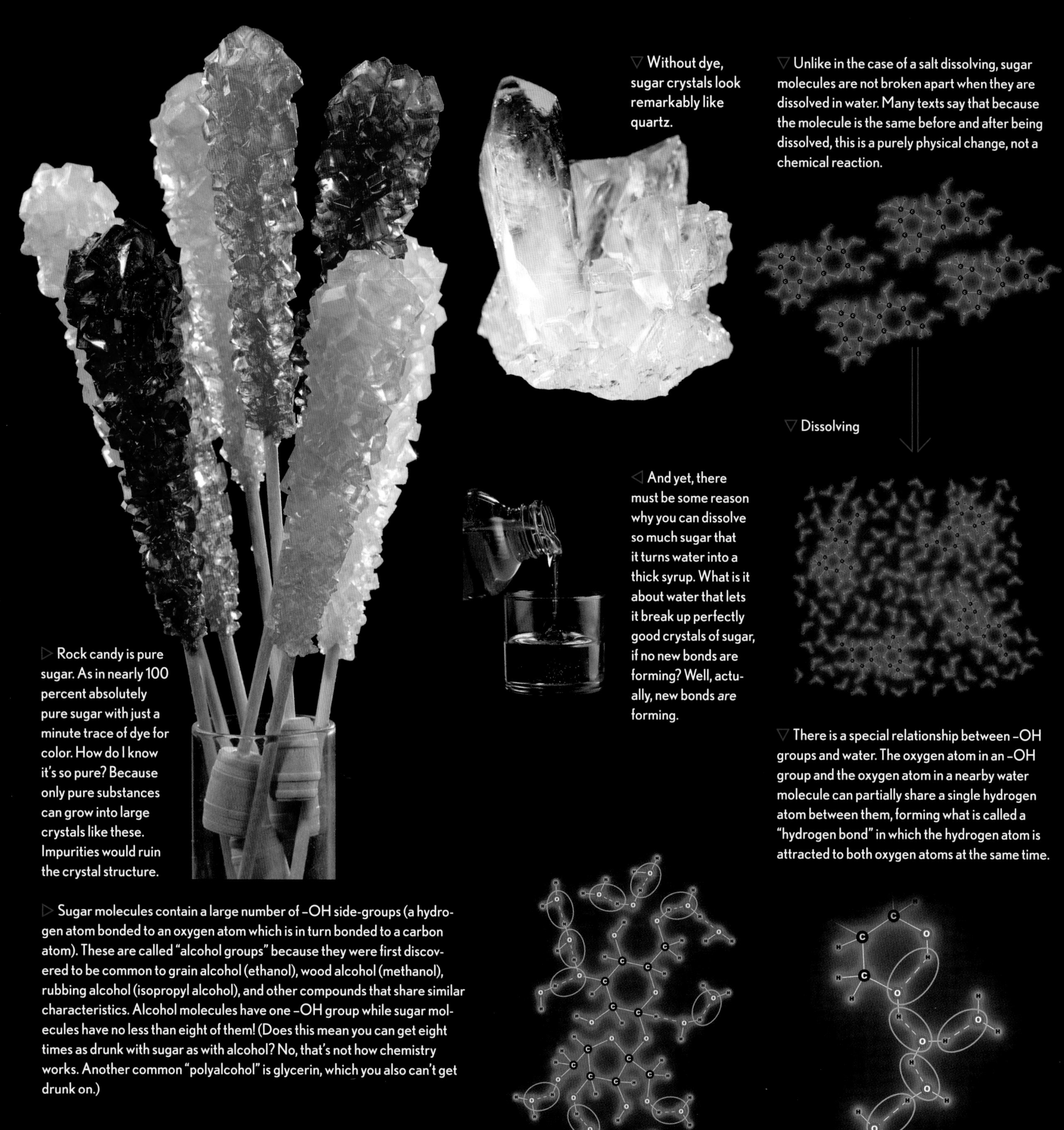

▽ Without dye, sugar crystals look remarkably like quartz.

▽ Unlike in the case of a salt dissolving, sugar molecules are not broken apart when they are dissolved in water. Many texts say that because the molecule is the same before and after being dissolved, this is a purely physical change, not a chemical reaction.

▽ Dissolving

◁ And yet, there must be some reason why you can dissolve so much sugar that it turns water into a thick syrup. What is it about water that lets it break up perfectly good crystals of sugar, if no new bonds are forming? Well, actually, new bonds *are* forming.

▷ Rock candy is pure sugar. As in nearly 100 percent absolutely pure sugar with just a minute trace of dye for color. How do I know it's so pure? Because only pure substances can grow into large crystals like these. Impurities would ruin the crystal structure.

▽ There is a special relationship between –OH groups and water. The oxygen atom in an –OH group and the oxygen atom in a nearby water molecule can partially share a single hydrogen atom between them, forming what is called a "hydrogen bond" in which the hydrogen atom is attracted to both oxygen atoms at the same time.

▷ Sugar molecules contain a large number of –OH side-groups (a hydrogen atom bonded to an oxygen atom which is in turn bonded to a carbon atom). These are called "alcohol groups" because they were first discovered to be common to grain alcohol (ethanol), wood alcohol (methanol), rubbing alcohol (isopropyl alcohol), and other compounds that share similar characteristics. Alcohol molecules have one –OH group while sugar molecules have no less than eight of them! (Does this mean you can get eight times as drunk with sugar as with alcohol? No, that's not how chemistry works. Another common "polyalcohol" is glycerin, which you also can't get drunk on.)

Hydrogen bonds are not very strong, but there are a *lot* of them in a volume of water because water can form hydrogen bonds with *itself* as well as with –OH groups on other molecules. Hydrogen bonds are why sugar is so willing to dissolve in water.

In other words, it's the formation of bonds—specifically hydrogen bonds—that drives the dissolving of sugar. That means sugar dissolving is actually very much a chemical reaction. Which goes directly against what nearly every textbook is going to tell you about sugar dissolving. (But, by the way, if they ever ask you on a test, say no, it's not a reaction. Anyone who asks is almost certainly expecting the answer no.)

Does it matter whether we call sugar dissolving a reaction or not? No, that's just arguing about words, and that's never very important or interesting. The interesting reality is that there are subtle things going on that defy rigid definition, even in something as simple as sugar dissolving.

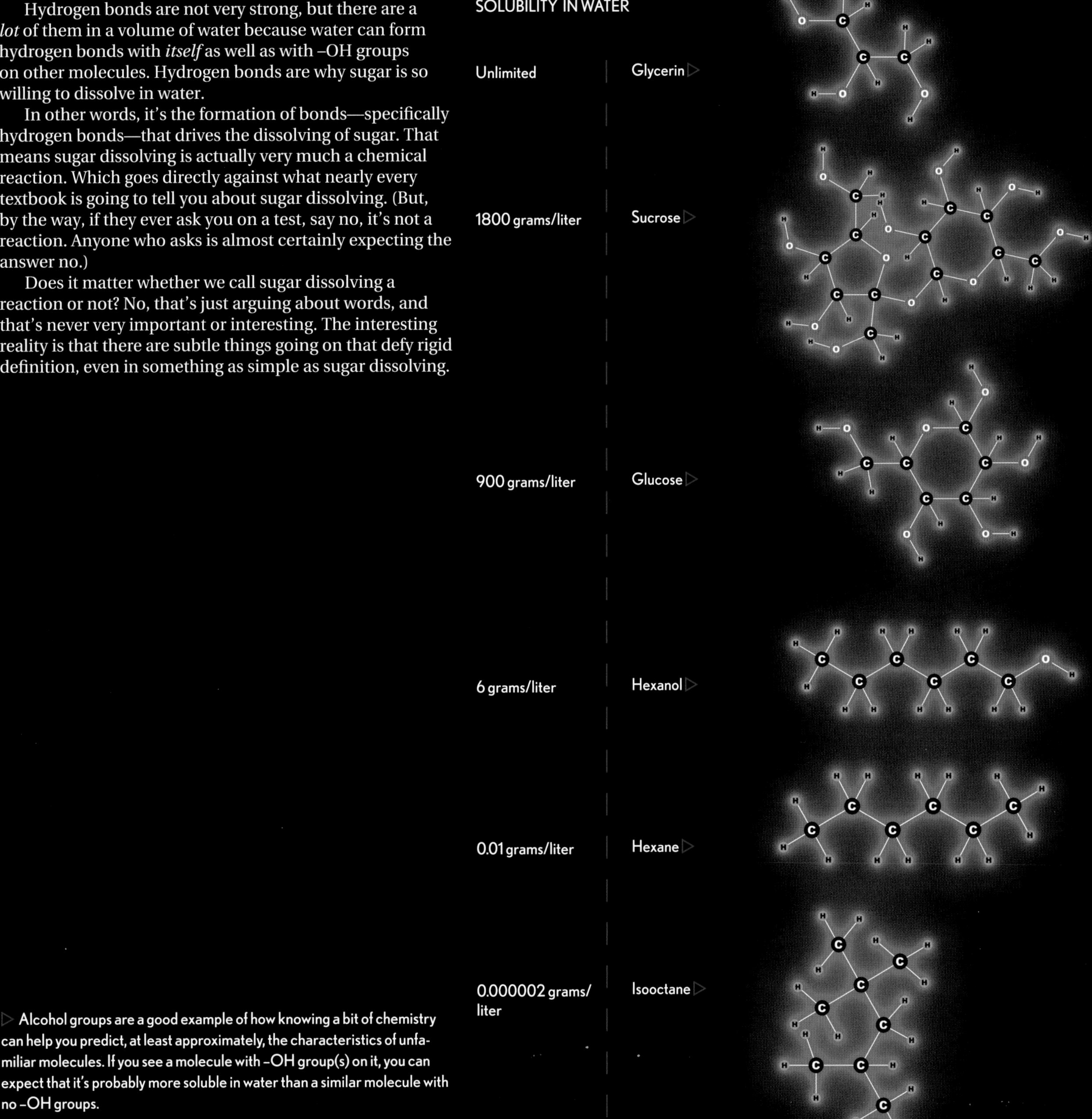

▷ Alcohol groups are a good example of how knowing a bit of chemistry can help you predict, at least approximately, the characteristics of unfamiliar molecules. If you see a molecule with –OH group(s) on it, you can expect that it's probably more soluble in water than a similar molecule with no –OH groups.

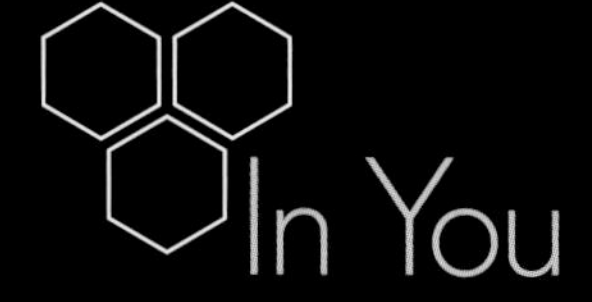

In You

YOU ARE MADE OF chemical reactions. From digestion to death, it's all chemistry. But is that really a useful way of thinking about it? I could just as well say that you are made of elements, or that you are made of protons, neutrons, and electrons. Or that you are made of quarks and gluons, or whatever those things are made of. All these things are true. So to understand life, should we study physics, chemistry, biochemistry, medicine, or what?

It comes down to this: What kind of language is the most helpful for understanding the particular phenomenon you're interested in? If you want to figure out why a certain sport is popular, it's probably not useful to talk about the chemical reactions that move the athlete's muscles. You'd want to use the language of psychology or sociology or politics to understand how that sport got popular and how its managers use the media to trick people into thinking they care about a certain team and giving them money.

Every field of study is useful at its own level, and there is a definite order in which these fields build on each other.

Political science and sociology are useful for understanding the actions of people acting (often badly) in large groups. To do that, you need to know something about how individual people behave, which is covered by the field of psychology.

Psychology is useful for understanding the thinking of one or two people at a time. People's thoughts are influenced, but not fully explained, by the mechanics of what's going on in their brains, so to understand thoughts you want to know something about medicine.

Medicine is the study of how people's bodies work as whole systems, with all their interacting parts influencing each other. To do that, you need to know how those individual parts work, which is explained mostly by biochemistry.

Biochemistry is about chemical reactions—often involving proteins, DNA, and other very large molecules—that go on in living systems. To understand these very complicated reactions, you need to know how reactions work in general.

Chemistry is the study of how atoms and molecules interact with each other at the atomic level. Individual bonds being made or broken, one atom knocking out another, and so on. To study that, you first need to know how atoms and subatomic particles work, which is physics.

Physics is the study of the fundamental forces of nature. It used to be about big things like planets and gravity (and it still is) but much of it is now about things happening at scales far smaller even than an atom. This is the world of quantum mechanics.

▽ This candle and lovely slice of apple pie are both made with the same kind of beef fat known as tallow. The candle is pure tallow while the apple pie has a few other ingredients. When the candle burns, the tallow reacts with oxygen from the air and produces carbon dioxide (CO_2) and water (H_2O) as the only significant reaction products. The reaction also releases energy—you see light and feel heat from the candle.

Quantum mechanics is a set of theories, expressed in mathematical formulas, that describe all known phenomena in the world to an extraordinary degree of precision. But only if the phenomenon is incredibly small, or carefully set up to magnify quantum phenomenon into the macroscopic world. Fundamentally, quantum mechanics is all mathematics.

Mathematics is the end of the line. It is the field that transcends all individual, specific issues and speaks only in universal and absolute truths. As such, mathematics is about everything and nothing. It is the root answer to all questions, but the actual answer to hardly any of them.

These fields build on each other, and each level adds new ideas. What I find fascinating is that sometimes you can stab a needle through the stack, connecting a phenomenon at a very high level with something happening at a much lower level.

That's how we ended up with dancers wearing gas masks.

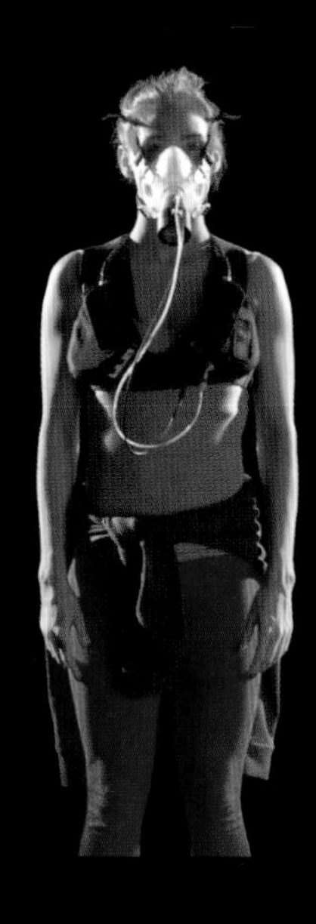

▷ The reaction when you eat the apple pie is much more complicated, but the end result is the same: pie in through your mouth, CO_2 and water out through your lungs and elsewhere. In chemical language, the *pathway* of the reaction is different, but the reactants (inputs) and products (outputs) are the same.

It's a general law in chemistry that if the reactants and products of a reaction are the same, then the total amount of energy released by the reaction must also be the same, regardless of the pathway. So, as long as we digest and metabolize it all, an equal amount of energy must be released when we consume an ounce of tallow as when that same ounce is burned in a candle.

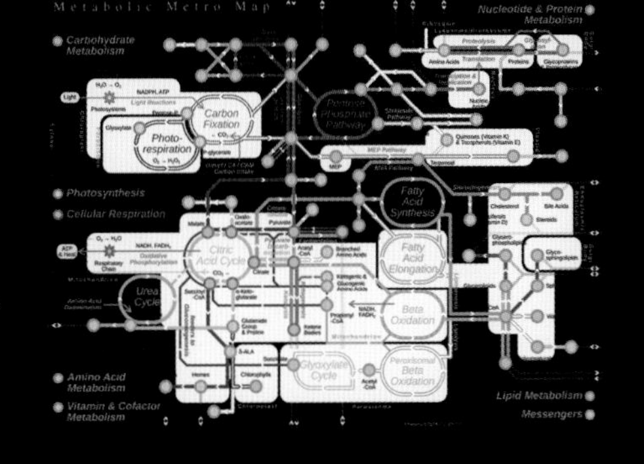

▷ A candle glows from the energy released by its reactions, and so does a human who's eaten a slice of apple pie. People just glow at a lower temperature. Instead of emitting visible light, they emit invisible infrared (IR) light. An IR camera lets us see this heat-light. A dancer's body glows brighter in the infrared as it heats up from internal combustion (metabolism) of food during vigorous activity.

Apple pie is made with more than just tallow fat. There's also sugar and other carbohydrates in there. Can we tell whether this dancer is getting her energy from the fat or the sugar?

△ Fat △ Sugar △ Starch △ Protein

Our bodies can run either on carbohydrates (sugar, starch, etc.) or on fat and proteins (everything that tastes good but isn't sugar). Both these fuels circulate in the blood stream, and the body can choose which one to burn at any given time.

Regardless of which kind of fuel is being burned, the process is complicated. Hundreds of individual chemical reactions, some of them the same, some of them different depending on the fuel source, are required to turn food into the movement of a muscle. How can we possibly untangle what's going on and figure out whether the body is burning fat or carbs? Surely this requires a detailed study of biology and biochemistry?

No, actually this is a situation where we can bypass a whole lot of complicated, messy biology and get right down to the most elementary rules of chemistry: balancing simple chemical reactions.

▽ Here's what happens if you let sugar (in this case finely powdered confectioner's sugar) react with oxygen from the air. A lot of energy is released in a lovely fireball. The products of the reaction are the same as for any combustion of organic matter: carbon dioxide (CO_2) and water (H_2O). Because atoms can never be created or destroyed in a chemical reaction, when a reaction expression is "balanced" it must have exactly the same number of each kind of atom on both sides of the reaction.

I've drawn the reaction with more than the minimum number of each kind of atom, but what matters is the ratio between the number of carbon, hydrogen, and oxygen atoms.

The balanced reaction shows that we need one molecule of oxygen (O_2) for each molecule of CO_2 produced. Remember this 1:1 ratio, it's going to be important!

If you count up the atoms of each kind in the fat molecules shown on the next page, you will find that there are approximately two hydrogen atoms for every carbon atom, and only an insignificant number of any other kind of atom. If you look at the sugar molecule on this page, you'll find the same 2:1 ratio of hydrogen to carbon, *plus* about as many oxygen atoms as carbon atoms.

As a crude approximation, we can say that the overall, average chemical formula for the fat/oil molecules is CH_2, and for the sugar/carbohydrate molecule it's CH_2O. (The actual molecules are much bigger and have many more atoms in them, but this simplified form has about the same *ratio* of each type of atom, which is all that matters right now.)

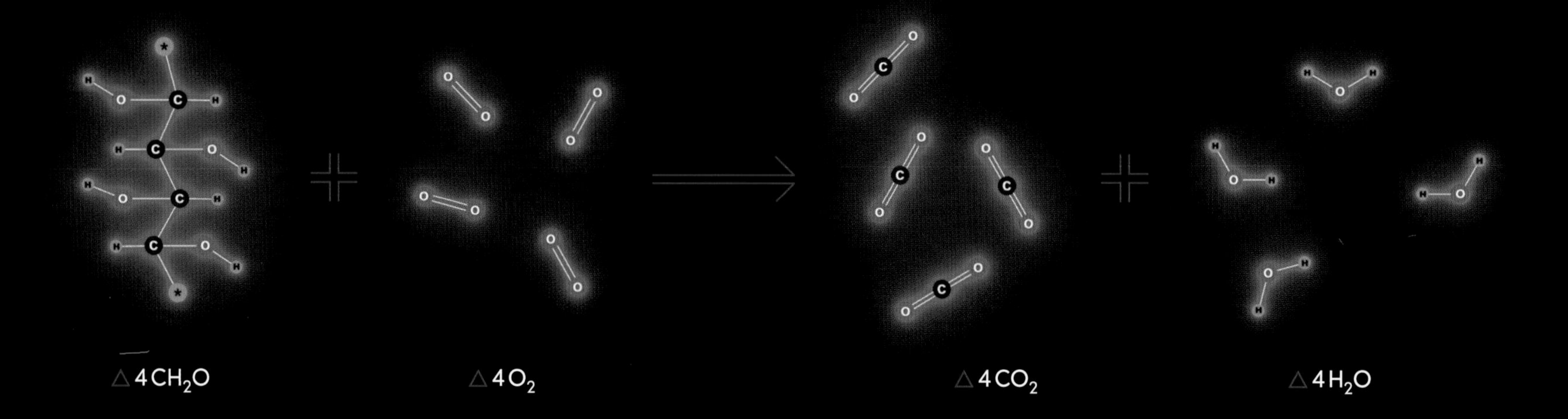

△ $4\,CH_2O$ △ $4\,O_2$ △ $4\,CO_2$ △ $4\,H_2O$

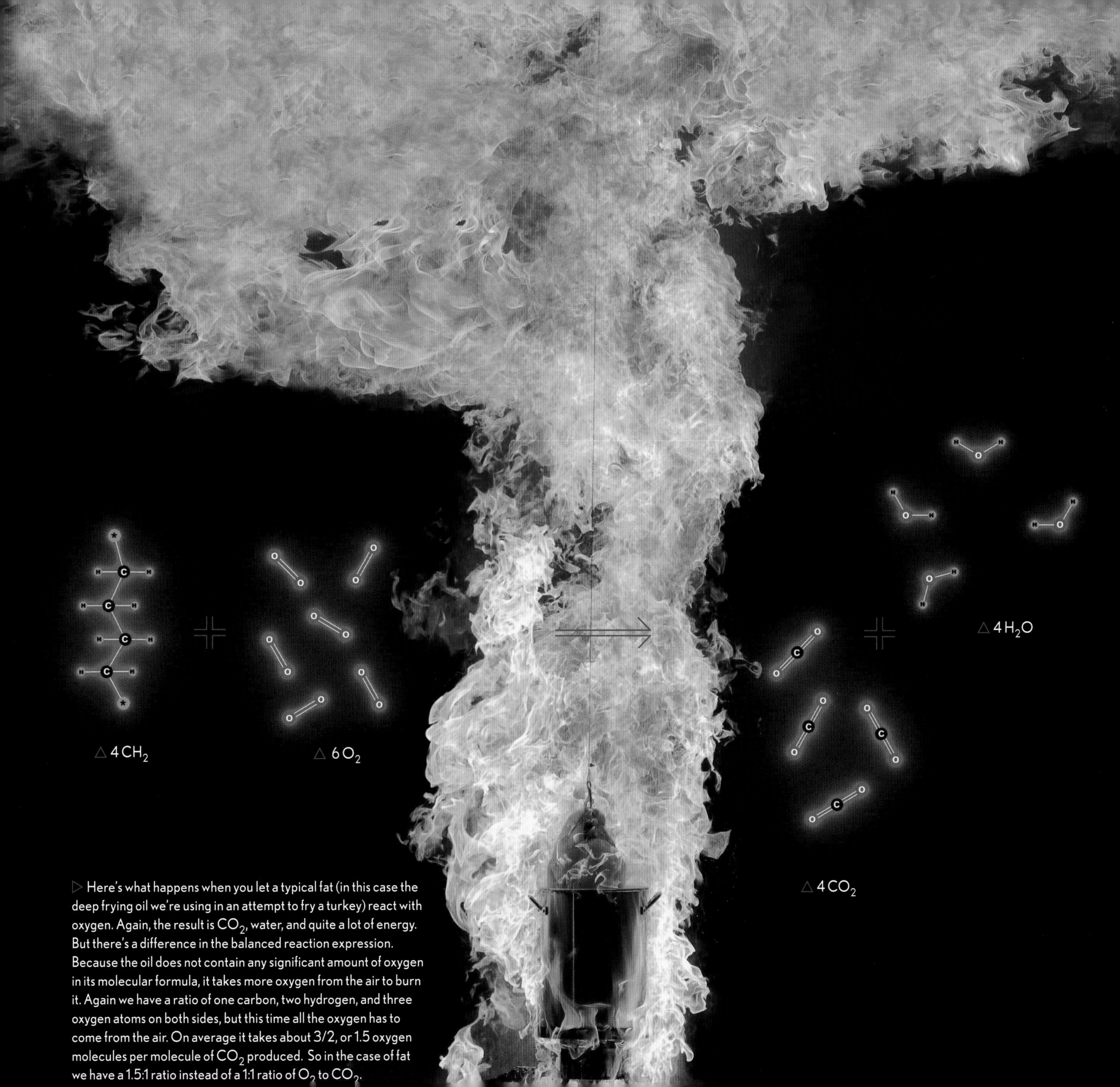

▷ Here's what happens when you let a typical fat (in this case the deep frying oil we're using in an attempt to fry a turkey) react with oxygen. Again, the result is CO_2, water, and quite a lot of energy. But there's a difference in the balanced reaction expression. Because the oil does not contain any significant amount of oxygen in its molecular formula, it takes more oxygen from the air to burn it. Again we have a ratio of one carbon, two hydrogen, and three oxygen atoms on both sides, but this time all the oxygen has to come from the air. On average it takes about 3/2, or 1.5 oxygen molecules per molecule of CO_2 produced. So in the case of fat we have a 1.5:1 ratio instead of a 1:1 ratio of O_2 to CO_2.

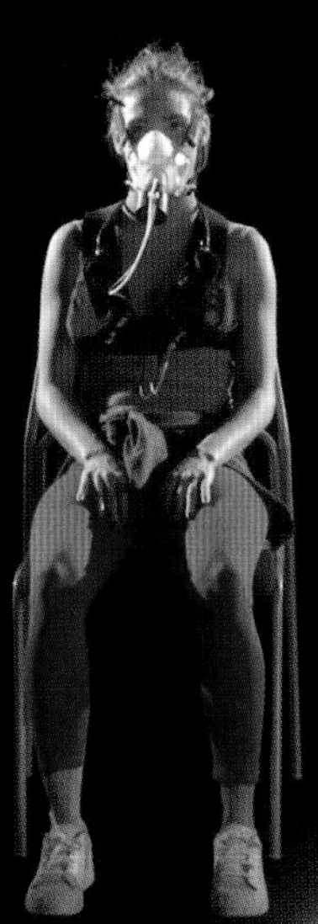

Both O_2 and CO_2 go in and out of a dancer's lungs, and we can measure how much of each she is breathing in or out using a gas analysis mask and backpack. By looking at the ratio between the oxygen she consumes and CO_2 she exhales, we should be able to tell what food source her body is running on. It makes no difference how complicated the intermediate pathways are, the gas ratio doesn't lie.

What you see from the data is quite clear. Before exercise, with the dancer at rest, the O_2:CO_2 ratio is close to 1.5:1. She is burning mostly fat.

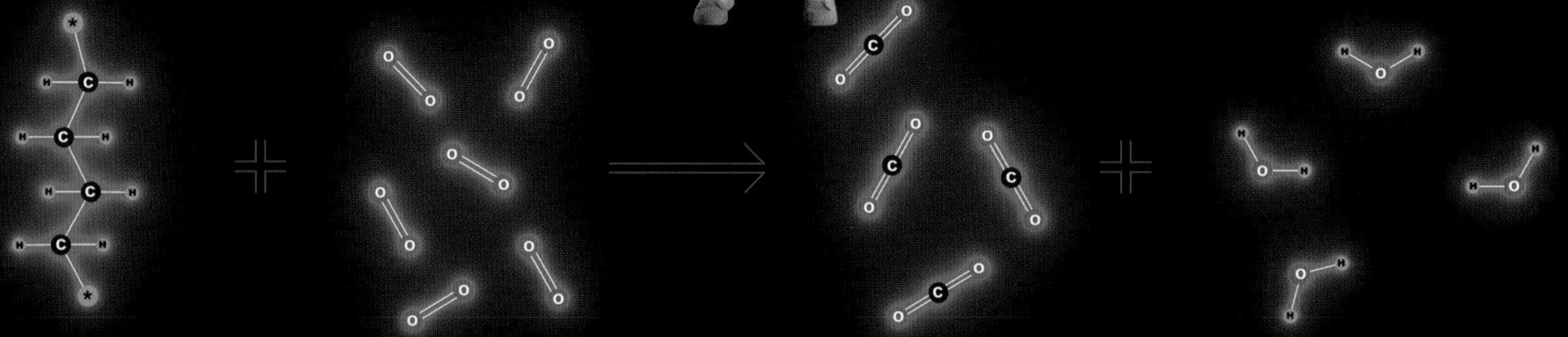

△ $4CH_2O$ △ $6O_2$ △ $4CO_2$ △ H_2O

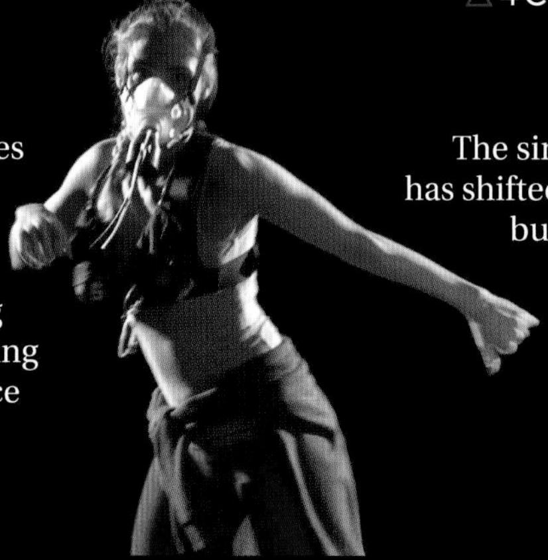

After a period of vigorous activity, the gas ratio moves closer to 1:1. She's now burning mostly sugars. Why would the body shift consumption toward sugars? One advantage of burning sugars is that, as the gas ratio implies, you need less oxygen from the air. During hard exercise, the supply of oxygen can become a limiting factor, so there's an advantage to using an energy source that requires less of it. (Of course there are many other factors: metabolism is very complicated!)

The simple thing we can say for sure is that the body has shifted toward consuming energy sources with more built-in oxygen. That alone is a good starting point toward trying to understand why the body is doing what it is. I think it's pretty nifty that this can be determined in such a direct and simple way, from a couple of balanced chemical reaction expressions (and a dancer willing to wear a gas mask).

▽ $4CH_2$ ▽ $4O_2$ ▽ $4CO_2$ ▽ $4H_2O$

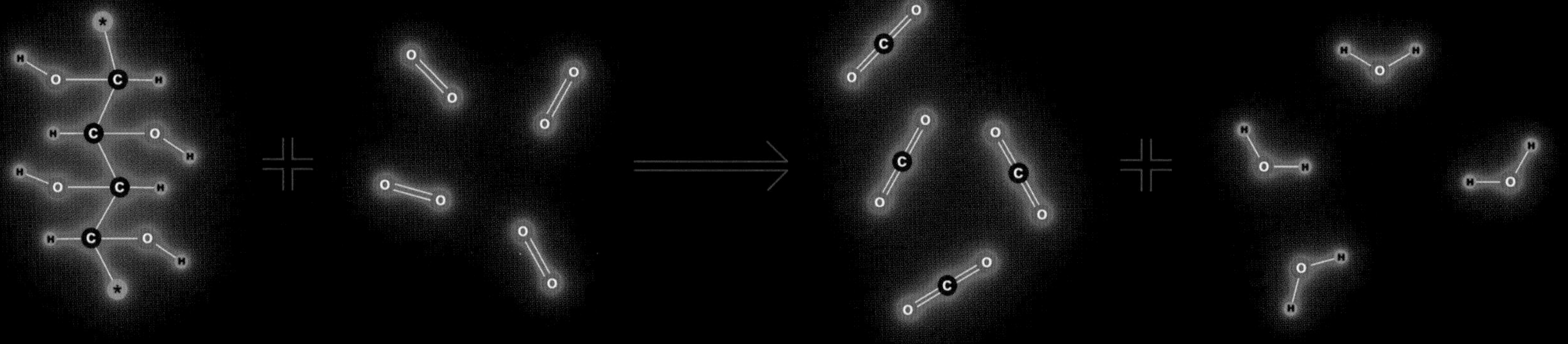

By the way, you might notice something that seems very counterintuitive. Do we really burn more fat at rest than when exercising? Could this mean that sitting still is a better way to lose weight than exercise??

This is where biochemistry gets far too complicated for this simplistic way of looking at things. Yes, technically, carbohydrates (sugars) are burned in preference to fat when exercising. But what happens if you *don't* exercise? Those carbohydrates are still there in your bloodstream, and if you don't use them up, your body will get busy converting them into fat.

What matters when it comes to weight loss or gain isn't which source of energy your body is using, but only how much energy it's burning compared to how much you're consuming in the form of food calories. If you're burning up more energy through activity than you are consuming, you will lose weight. If you digest more than you burn, you will gain weight. Period. There are no other variables in this equation. Because exercise increases the amount of energy burned, it leads to weight loss unless there's an equal or greater increase in food eaten.

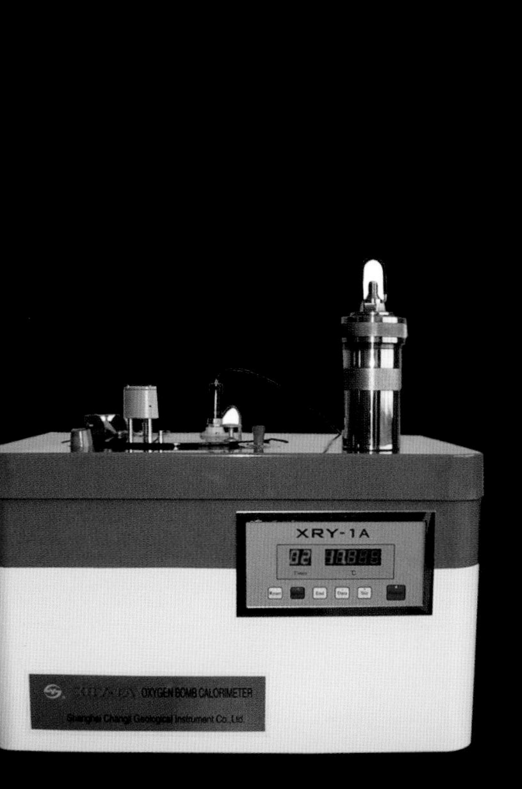

Nutrition Facts
Serving Size 2 Cakes (77g)
Servings Per Container 5

Amount Per Serving
Calories 260 Calories from Fat 80

	% Daily Value*
Total Fat 8g	**12%**
Saturated Fat 3.5g	**18%**
Trans Fat 0g	
Cholesterol 35mg	**12%**
Sodium 350mg	**15%**
Total Carbohydrate 43g	**14%**
Dietary Fiber 0g	**0%**
Sugars 29g	
Protein 2g	

Vitamin A 0% • Vitamin C 0%
Calcium 2% • Iron 4%

*Percent Daily Values are based on a 2,000 calorie diet. Your daily values may be higher or lower depending on your calorie needs:

	Calories:	2,000	2,500
Total Fat	Less than	65g	80g
Sat Fat	Less than	20g	25g
Cholesterol	Less than	300mg	300mg
Sodium	Less than	2,400mg	2,400mg
Total Carbohydrate		300g	375g
Dietary Fiber		25g	30g

You've probably seen the nutrition labels on many food products. They tell you how much energy the food contains, and thus how much of it you can eat without starting to gain weight. Have you ever wondered how they determine these numbers? Originally it was done with a bomb calorimeter. The "bomb" part means that it has a strong metal container in which the food is placed and then burned in a stream of oxygen. The "calorimeter" part means that the instrument is measuring the amount of energy released by the reaction. (These days the numbers are known for most ingredients, so manufacturers can just add up the numbers without having to measure them again.)

As we learned earlier, in theory the total energy released by a reaction is the same regardless of the pathway, so the energy measured by the calorimeter should be equal to the amount released into your body when you eat the same food. In practice, of course, things are not quite so simple. For example, if the food contains a lot of indigestible plant fibers, they will burn and contribute to the calorimeter reading, but have no nutritional energy content because we can't digest them. (In other words, you can burn wood in a fireplace and get heat from it, but unless you're a cow, you can't eat wood and get any food value from it.) This factor is compensated for in the nutrition labels.

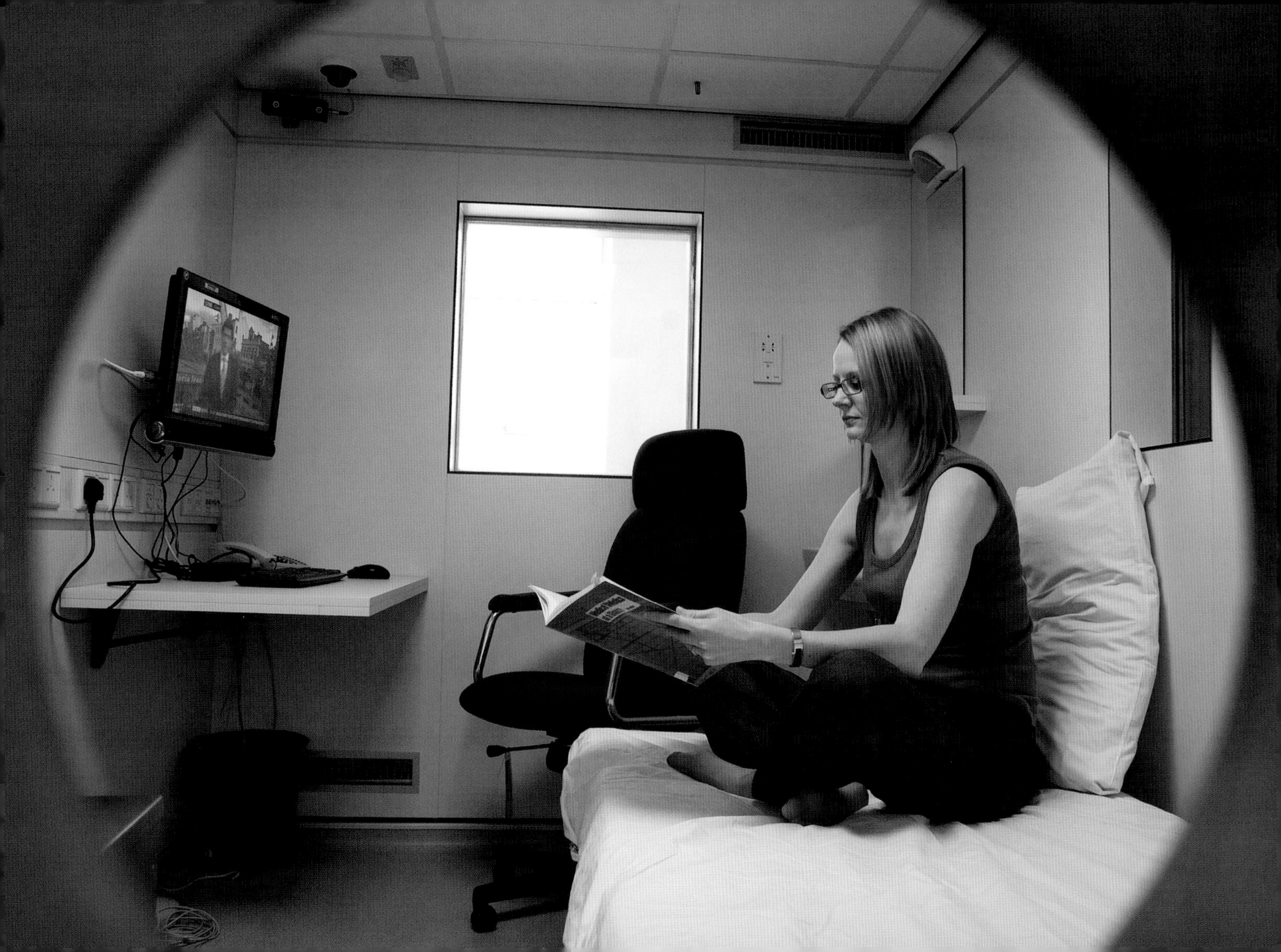

On the Origin of Light and Color

LIGHT IS EVERYWHERE and obvious. But what is light, and what is that shape of light we call color?

We learned at the very start of this book that there is an intimate relationship between light and chemical reactions. We saw in Chapter 1 how one of the molecules in an orange glow stick can use chemical energy to emit a photon, a pulse of orange light. And in Chapter 2 we learned how chlorophyll can turn a photon of light back into chemical energy. But the relationship between light and chemistry goes much deeper than just a few special molecules that are useful for luminous trinkets.

We saw in Chapter 2 that the bonds holding molecules together are defined by their energy. How much energy must be added to pull apart this bond? How much energy is released when this other bond is formed? Bonds are a way of storing potential energy, and every activity involving bonds—making, breaking, bending, stretching, rotating—has associated with it a definite amount of energy.

Light is also a way of storing energy. Every photon, every unit of light, represents a definite amount of energy. And as we will see, the amount of energy held by a photon defines its color. So for every bond, there is a color—the color of light whose photon energy equals the energy of that bond.

A molecule may have many bonds, all of which can bend or stretch in multiple ways. So for every molecule there is a palette of colors representing all the energies of these ways of moving. We call this the spectrum of the molecule.

All chemical reactions involve the trading of energy from one bond to another. So for every reaction there is another spectrum of colors corresponding to the energies of all the bonds being made and broken.

Because of the connection between light, color, and chemistry, it's worth learning about light in more depth. So, again, what is light?

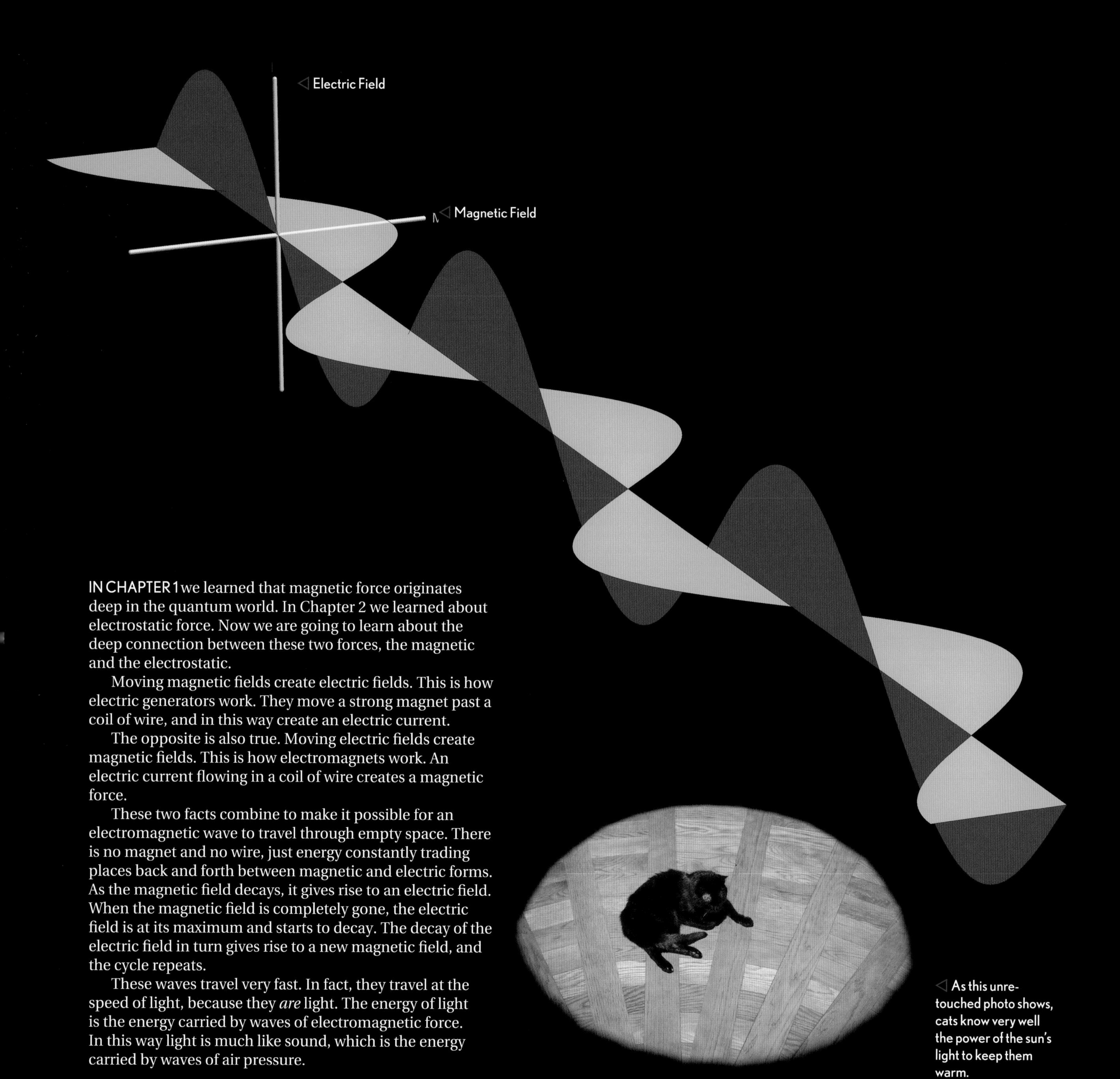

IN CHAPTER 1 we learned that magnetic force originates deep in the quantum world. In Chapter 2 we learned about electrostatic force. Now we are going to learn about the deep connection between these two forces, the magnetic and the electrostatic.

Moving magnetic fields create electric fields. This is how electric generators work. They move a strong magnet past a coil of wire, and in this way create an electric current.

The opposite is also true. Moving electric fields create magnetic fields. This is how electromagnets work. An electric current flowing in a coil of wire creates a magnetic force.

These two facts combine to make it possible for an electromagnetic wave to travel through empty space. There is no magnet and no wire, just energy constantly trading places back and forth between magnetic and electric forms. As the magnetic field decays, it gives rise to an electric field. When the magnetic field is completely gone, the electric field is at its maximum and starts to decay. The decay of the electric field in turn gives rise to a new magnetic field, and the cycle repeats.

These waves travel very fast. In fact, they travel at the speed of light, because they *are* light. The energy of light is the energy carried by waves of electromagnetic force. In this way light is much like sound, which is the energy carried by waves of air pressure.

As this unretouched photo shows, cats know very well the power of the sun's light to keep them warm.

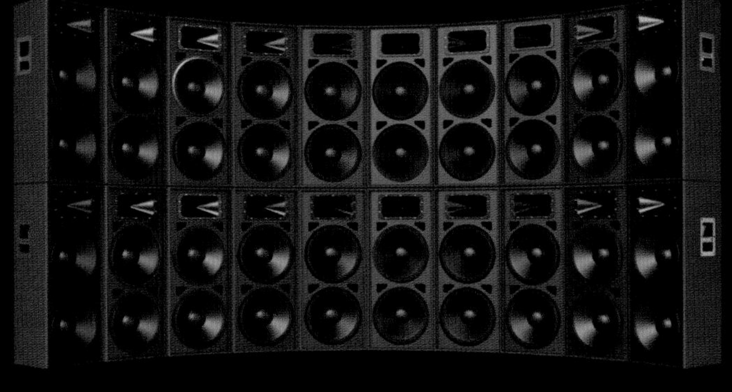

◁ The speakers at a rock concert can deliver enough sound energy to your ears to permanently damage them.

▽ The sound alone of a rocket engine at full thrust carries so much energy that its mere reflection off the ground could damage the rocket seconds after launch. That's one reason huge volumes of water are dumped all around the launch pad during launch. It's not just to cool the structure (see page 172), but also to dampen the intense sound waves. (These pictures are from a test of the space shuttle's "water deluge system," so there is no actual rocket present.)

▽ Color also has a close analogy in sound: different colors of light are like different notes in music. Light and sound don't just deliver energy: They deliver energy of a particular flavor. Just as there are high notes and low notes, so there are high colors and low colors. Blue, for example, is a high color, while red is a low color.

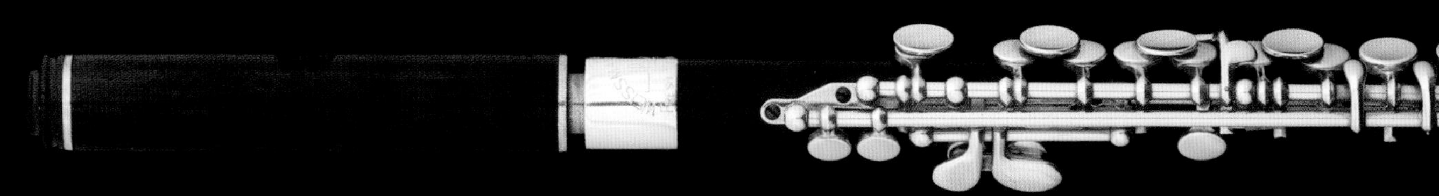

▷ And I don't mean that in some kind of wishy-washy, new age, arsty-fartsy way, I mean it in a precise and literal way.

▷ All waves, including sound and light waves, can transmit energy. This intense blue laser can easily light a match on fire.

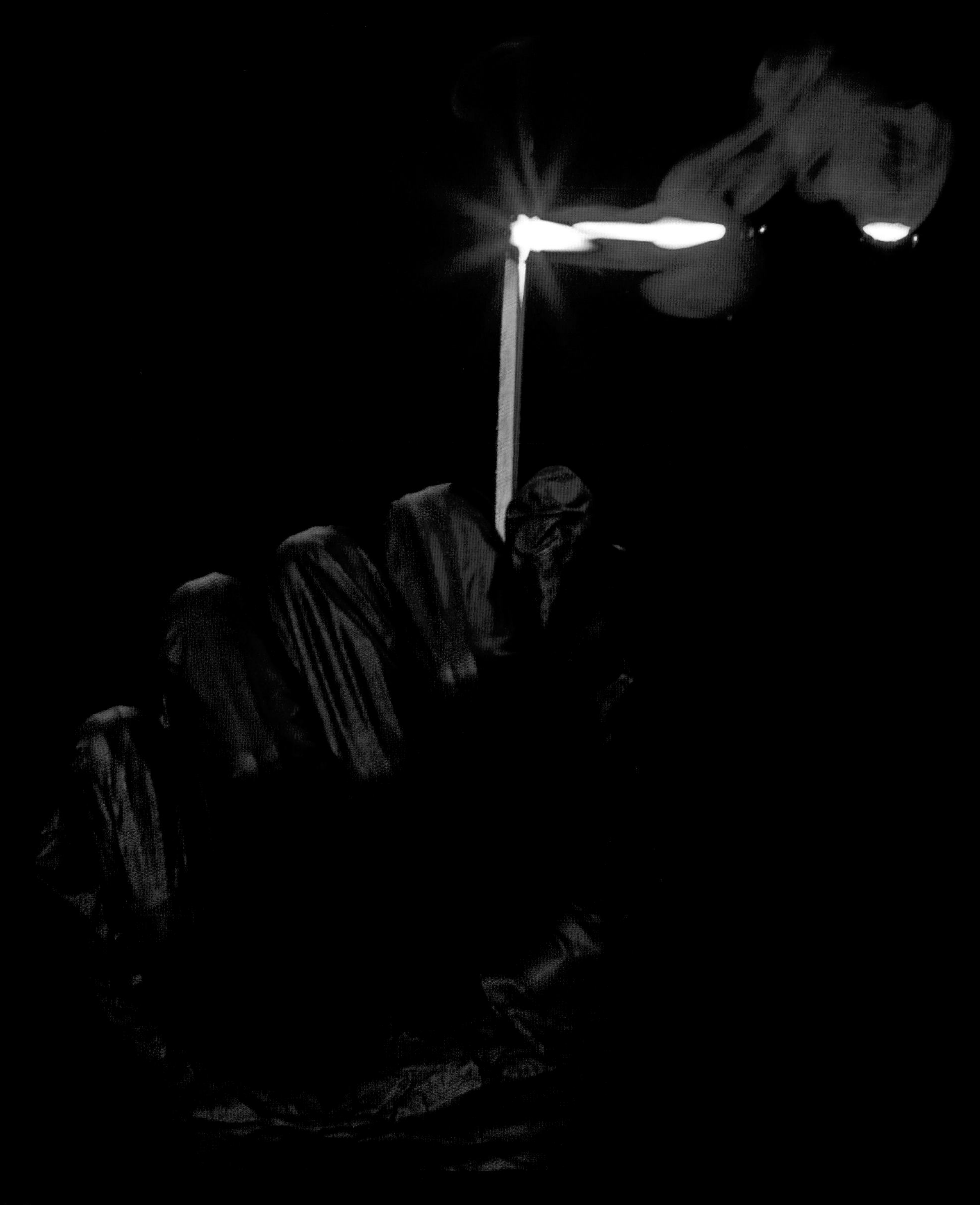

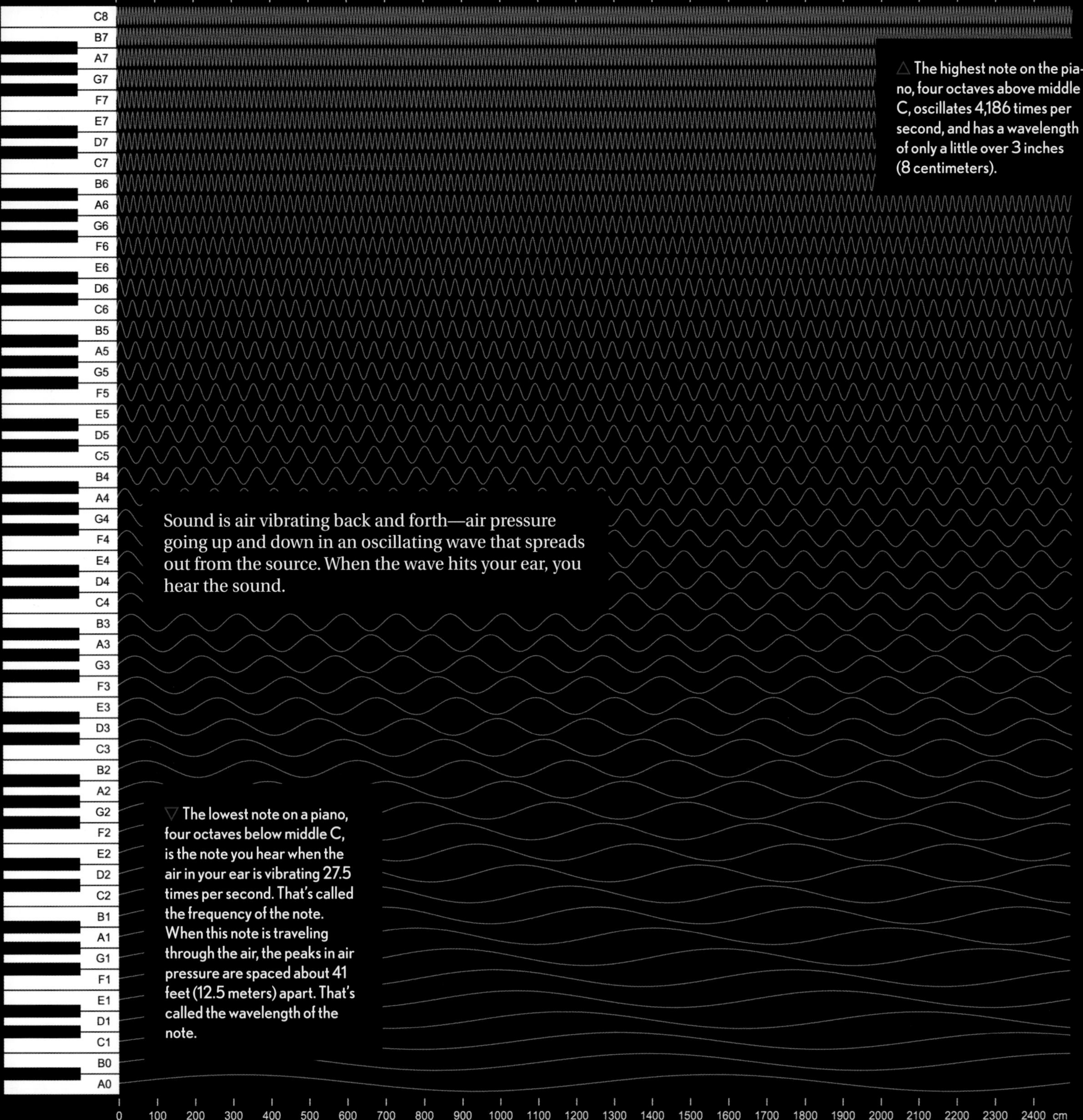

△ The highest note on the piano, four octaves above middle C, oscillates 4,186 times per second, and has a wavelength of only a little over 3 inches (8 centimeters).

Sound is air vibrating back and forth—air pressure going up and down in an oscillating wave that spreads out from the source. When the wave hits your ear, you hear the sound.

▽ The lowest note on a piano, four octaves below middle C, is the note you hear when the air in your ear is vibrating 27.5 times per second. That's called the frequency of the note. When this note is traveling through the air, the peaks in air pressure are spaced about 41 feet (12.5 meters) apart. That's called the wavelength of the note.

△ The highest note of light we can see, violet, vibrates about 750 million million times per second (750 terahertz), and has a wavelength around 400 billionths of a meter (400 nanometers). Shorter wavelengths are called ultraviolet light. Shorter still are x-rays, then gamma rays.

Just as sound is the oscillation of air, so light is the oscillation of an electromagnetic field. And just as different notes are different frequencies of oscillation, so different colors are different frequencies. Light waves vibrate much faster than sound waves. The lowest note of light we can see, deep red, vibrates about four hundred million million times per second (400 terahertz). Light waves also travel much faster: about 186,000 miles (300,000 kilometers) per second, compared to about 1,100 feet (340 meters) per second for sound.

Every frequency of light corresponds to a different unit, called a "quantum," of energy. High-frequency blue light has more energy per photon than low-frequency red light. All photons of a given color have exactly the same amount of energy.

▷ I often use the phrase "atoms are smaller than light" to explain why we can't see them in the conventional sense of the word. This is what I mean: an entire sugar molecule with 45 separate atoms in it is still less than one nanometer wide, compared to hundreds of nanometers for visible light. On the scale of these light waves, you can barely see the dot representing the whole molecule. When we use light to "see" something, we do it by bouncing many light waves off the thing and seeing which ones come back. Obviously it's not possible to use waves this big to see details inside something that is a thousand times smaller than they are.

▽ Light waves are much smaller than sound waves—about ten million times smaller! Notice that the scale on the sound waves page is in centimeters (cm, or hundredths of a meter), while the scale on the light waves page is in nanometers (nm, or billionths of a meter). The longest wavelength of light we can see is around 700 billionths of a meter (700 nanometers) long. Longer wavelengths are called infrared light. Longer still are microwaves, then radio waves.

Look at the diagrams on the last two pages. Isn't it amazing just how much wider a range of sound frequencies we can hear than light frequencies we can see? Each time you go an octave higher on the piano, the frequency is doubled. A standard piano covers a little over seven octaves, or just over 150 times higher frequency from the lowest to the highest note. Human hearing extends well beyond that, with many (young-ish) people able to hear frequencies over a range of 400 times or more (eight or nine octaves).

But we can't even see one whole octave of color! From lowest red to highest violet, the frequency of the light goes up by less than a factor of two.

We are also much better at hearing fine variations in sound frequency than we are able to see differences in light frequency. The coiled sensor in our inner ear can separately identify many hundreds of different individual pure tones. People with "perfect pitch" can even tell you exactly what each note is called, just from hearing it played by itself.

Our eyes are absolutely miserable in comparison: We take measurements at only three different light frequencies, roughly corresponding to the colors red, green, and blue. It's as if we could hear only three different kinds of sound: low notes, middle notes, and high notes, with everything in between being just a mixture of different amounts of those three ranges.

Some animals can see a few more than three colors, but only an instrument called a spectrometer (which we will meet again later) can see colors as well as we can hear notes.

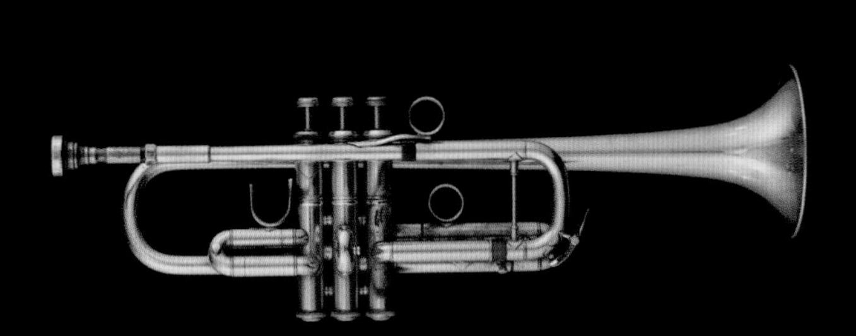

Not only can we hear hundreds of *different* frequencies, we can hear multiple frequencies at the same time. For example, these three notes, whose wavelengths are shown above, played together form a pleasant-sounding C-major chord. And we can easily tell when both a low note and a high note are being played at the same time.

For colors, we're completely unable to do any of these things. No interior decorator, no matter how pretentious, is physically capable of seeing what specific frequencies of light a particular mixed color is composed of.

The diagram on this page implies something about sound that may seem obvious, but actually isn't. Whenever we want a particular sound, say from an instrument or a loudspeaker, we create that sound by asking the device to create each of the individual frequencies we want.

If we want a C-major chord, we strike three individual strings in the piano, each of which vibrates at a different frequency. Or we pluck the right set of strings on a guitar, or blow air through an instrument of the right length. One way or another, we usually make sounds by adding together individual frequencies.

Obvious, right? I mean, how else would you do it?

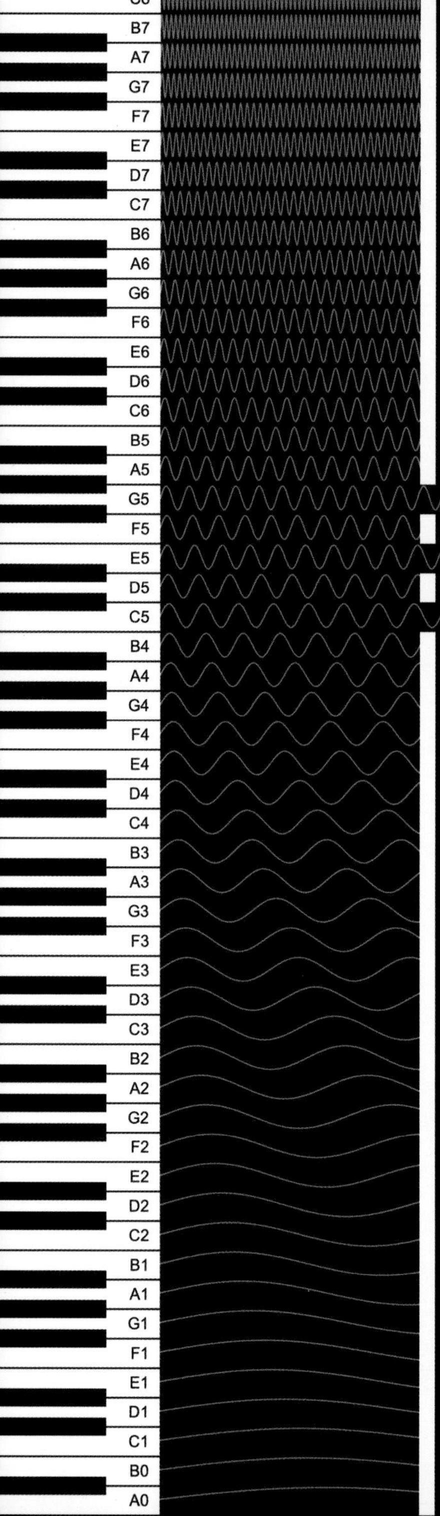

The other way you could make a C-major chord would be to play every note on the piano simultaneously, and then use some kind of special wall to block out all the frequencies except the three you need. The sound that's left would be the same as if you had created just those three frequencies by the conventional means of playing just those three notes. Crazy? Not entirely.

A piano would indeed be a silly way to create a sound of all frequencies, but such sounds (called "white noise") can easily be made with electronic circuits. White noise sounds like hissing static on the radio, but you can filter it to create "pink noise" that has a certain tonal cast to it. You can even filter it down to a small number of pure tones. (This can be done mechanically with tuned pipes, or electronically with circuits called audio filters.)

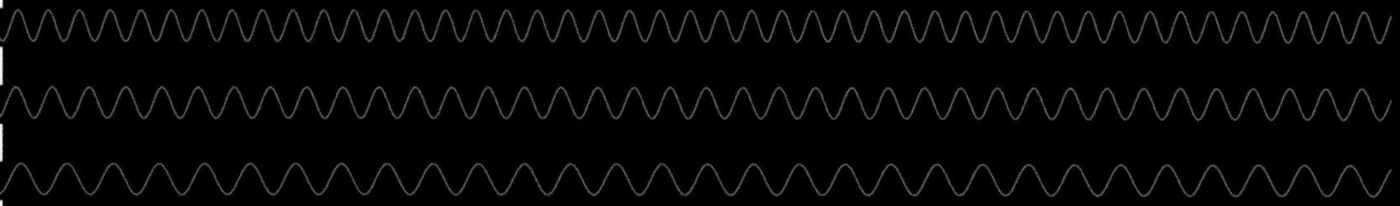

△ Many classic analog synthesizers use "subtractive synthesis" to turn white noise into odd sounds, or to take sounds that contain many overlapping frequencies and partially remove some of them to change the flavor of the sound.

You may wonder why I'm going on about the way sounds are created in a book about reactions. It's because understanding how sound works is really useful in understanding how light works, and light—color in particular—is the key to looking into the world of atoms and molecules.

Absorbing Light

JUST AS SOUND WAVES can be filtered by electronic circuits, so light waves can be filtered with special materials called . . . just about anything.

420 430 440 450 460 470 480 490 500 510 520 530 540 550 560 570 580 590 600 610 620 630 640 650 660 670 680 690

▷ The “white noise” we talked about on the last page gets its name from the more literal meaning of white—white light. Like white noise, white light contains all frequencies of light in equal amounts. Our environment is constantly bathed in white light coming from the sun and from all sorts of artificial light sources. As far as light is concerned, this picture is the equivalent of a piano with a pack of wild animals banging every key at the same time. (If you want to know what a pack of wild animals can do to a patch of perfectly nice white light, turn to page 142.)

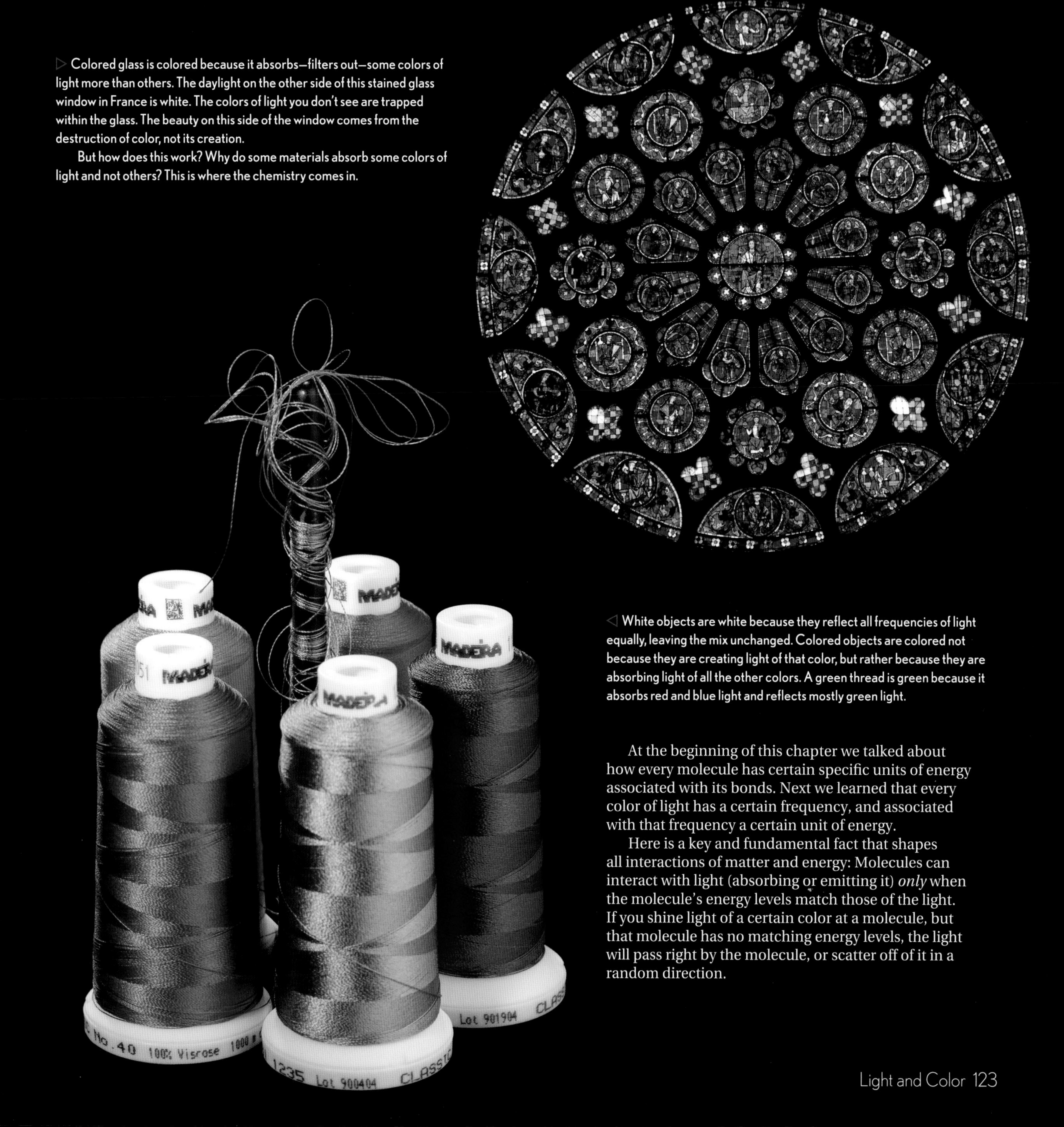

▷ Colored glass is colored because it absorbs—filters out—some colors of light more than others. The daylight on the other side of this stained glass window in France is white. The colors of light you don't see are trapped within the glass. The beauty on this side of the window comes from the destruction of color, not its creation.

But how does this work? Why do some materials absorb some colors of light and not others? This is where the chemistry comes in.

◁ White objects are white because they reflect all frequencies of light equally, leaving the mix unchanged. Colored objects are colored not because they are creating light of that color, but rather because they are absorbing light of all the other colors. A green thread is green because it absorbs red and blue light and reflects mostly green light.

At the beginning of this chapter we talked about how every molecule has certain specific units of energy associated with its bonds. Next we learned that every color of light has a certain frequency, and associated with that frequency a certain unit of energy.

Here is a key and fundamental fact that shapes all interactions of matter and energy: Molecules can interact with light (absorbing or emitting it) *only* when the molecule's energy levels match those of the light. If you shine light of a certain color at a molecule, but that molecule has no matching energy levels, the light will pass right by the molecule, or scatter off of it in a random direction.

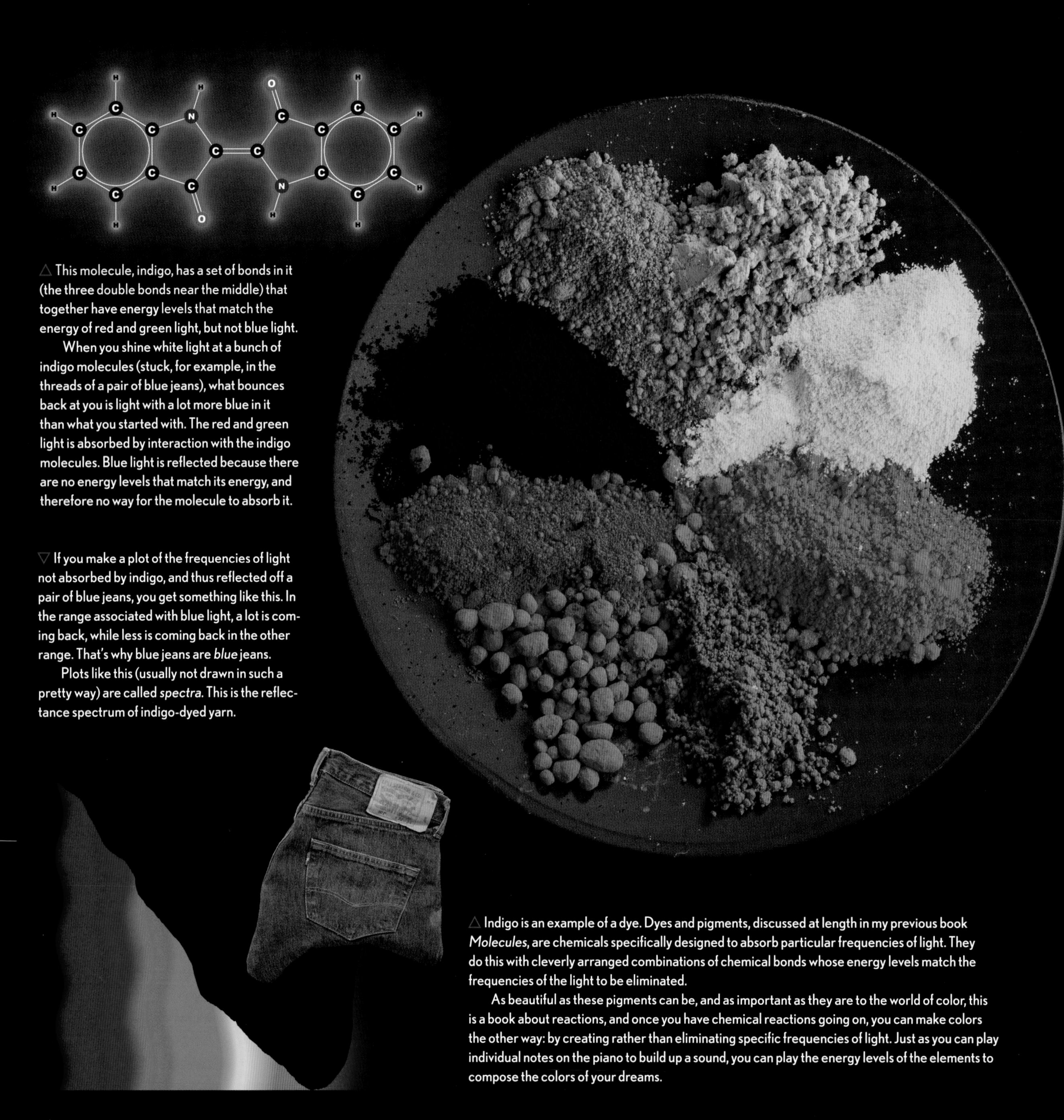

△ This molecule, indigo, has a set of bonds in it (the three double bonds near the middle) that together have energy levels that match the energy of red and green light, but not blue light.

When you shine white light at a bunch of indigo molecules (stuck, for example, in the threads of a pair of blue jeans), what bounces back at you is light with a lot more blue in it than what you started with. The red and green light is absorbed by interaction with the indigo molecules. Blue light is reflected because there are no energy levels that match its energy, and therefore no way for the molecule to absorb it.

▽ If you make a plot of the frequencies of light not absorbed by indigo, and thus reflected off a pair of blue jeans, you get something like this. In the range associated with blue light, a lot is coming back, while less is coming back in the other range. That's why blue jeans are *blue* jeans.

Plots like this (usually not drawn in such a pretty way) are called *spectra*. This is the reflectance spectrum of indigo-dyed yarn.

△ Indigo is an example of a dye. Dyes and pigments, discussed at length in my previous book *Molecules*, are chemicals specifically designed to absorb particular frequencies of light. They do this with cleverly arranged combinations of chemical bonds whose energy levels match the frequencies of the light to be eliminated.

As beautiful as these pigments can be, and as important as they are to the world of color, this is a book about reactions, and once you have chemical reactions going on, you can make colors the other way: by creating rather than eliminating specific frequencies of light. Just as you can play individual notes on the piano to build up a sound, you can play the energy levels of the elements to compose the colors of your dreams.

Emitting Light

IF YOU MAKE ANYTHING, for example this platinum rod, very hot, it will glow. If it's just a little bit very hot, it will glow a dull red. As it gets hotter and hotter, it will go through a continuous range of colors from red to orange to yellow to nothing because it melted and fell out of the flame.

This kind of light is broad-spectrum: It contains a mixture of all wavelengths of light over a wide range. The hotter the object is, the more this spectrum is shifted toward shorter-wavelength, higher frequencies of light (toward the blue end of the spectrum). When the object is hot enough (somewhere around 5,000°F [2,700°C]) it starts to appear white to us, because there's enough light from the red, green, and blue ranges to mimic the light of the sun.

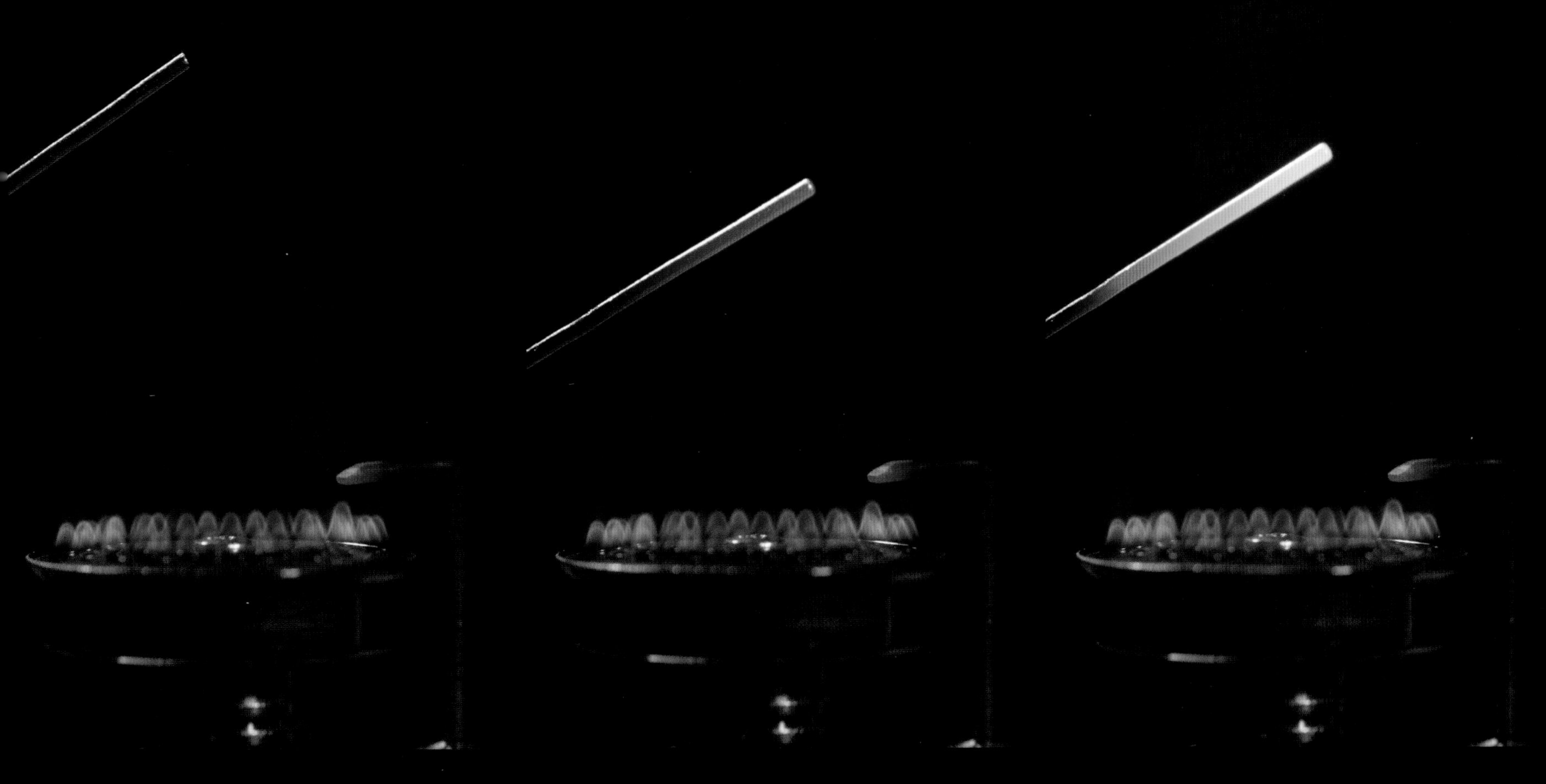

△▽ Before the invention of incandescent light bulbs, limelight was used in theaters to create focused beams of light. (Being "in the limelight" used to mean literally being in the spotlight on stage. Now it's just a metaphor for being the center of attention.) Limelights have a 5,100°F (2,800°C) oxy-hydrogen flame—a miniature hydrogen blowtorch—pointed at a cylinder of quicklime (calcium oxide). The main qualification of quicklime for the job is that it is impervious to such high temperatures, and glows with a particularly pleasing creamy soft light not equaled by modern technology until the invention of high-power white-light LEDs.

◁ The surface of the sun is very hot—about 10,000°F (5,500°C). That makes the sun's light almost pure white.

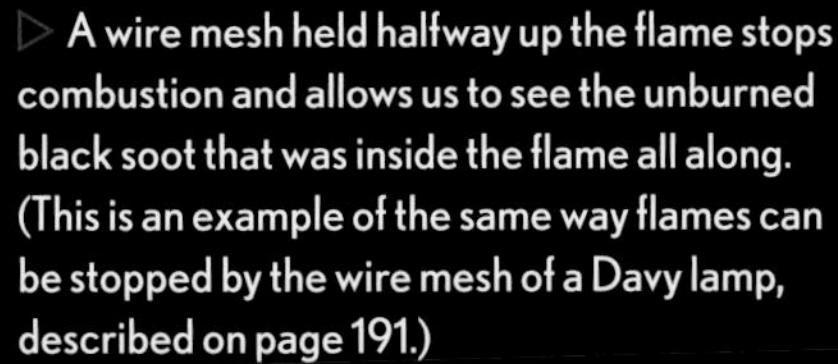

◁ The flame of a candle seems like it's made entirely of vapor, yet it glows like an incandescent solid. That's because a candle flame is actually filled with tiny particles of black soot (carbon residue from the burned wax). The soot doesn't look black inside the flame because it's so hot that it is glowing just like a hot light bulb filament. We don't see it coming out the top of the flame because (in a well-made candle) is it completely consumed by the flame before it reaches the top.

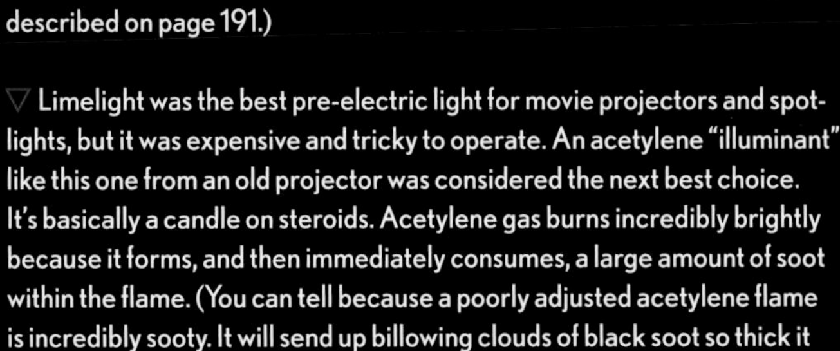

▷ A wire mesh held halfway up the flame stops combustion and allows us to see the unburned black soot that was inside the flame all along. (This is an example of the same way flames can be stopped by the wire mesh of a Davy lamp, described on page 191.)

These are all examples of incandescence, the glowing of very hot solid objects. Any time you see a nice yellow-white light coming from a hot object, chances are it's a result of this kind of incandescence. This is true even when it doesn't seem like the object is a solid.

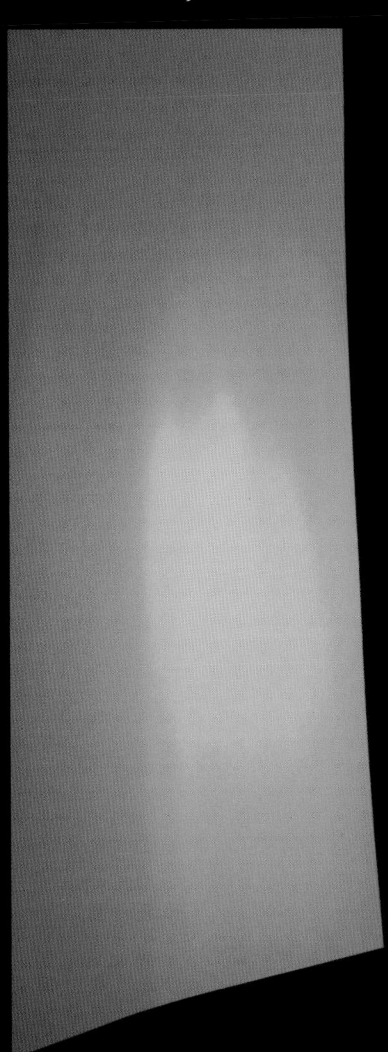

▽ Limelight was the best pre-electric light for movie projectors and spotlights, but it was expensive and tricky to operate. An acetylene "illuminant" like this one from an old projector was considered the next best choice. It's basically a candle on steroids. Acetylene gas burns incredibly brightly because it forms, and then immediately consumes, a large amount of soot within the flame. (You can tell because a poorly adjusted acetylene flame is incredibly sooty. It will send up billowing clouds of black soot so thick it forms strings and sheets.)

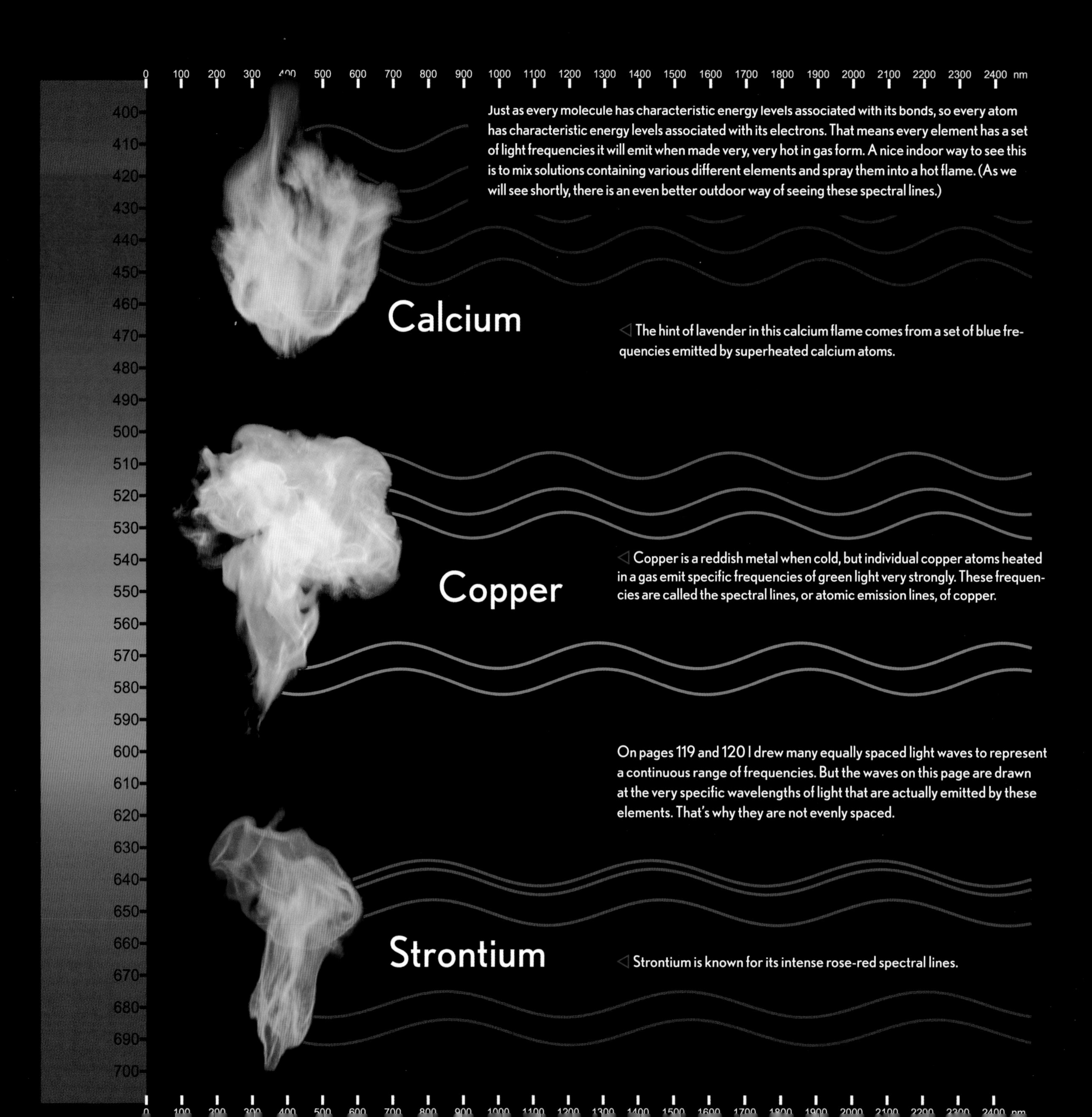

Just as every molecule has characteristic energy levels associated with its bonds, so every atom has characteristic energy levels associated with its electrons. That means every element has a set of light frequencies it will emit when made very, very hot in gas form. A nice indoor way to see this is to mix solutions containing various different elements and spray them into a hot flame. (As we will see shortly, there is an even better outdoor way of seeing these spectral lines.)

◁ The hint of lavender in this calcium flame comes from a set of blue frequencies emitted by superheated calcium atoms.

◁ Copper is a reddish metal when cold, but individual copper atoms heated in a gas emit specific frequencies of green light very strongly. These frequencies are called the spectral lines, or atomic emission lines, of copper.

On pages 119 and 120 I drew many equally spaced light waves to represent a continuous range of frequencies. But the waves on this page are drawn at the very specific wavelengths of light that are actually emitted by these elements. That's why they are not evenly spaced.

◁ Strontium is known for its intense rose-red spectral lines.

Our ears are very good at hearing the individual "spectral lines" in a piece of music: These are the notes that make up the composite sound. We can't do this with light, but fortunately there is a simple instrument called a spectroscope that breaks up a beam of light into its constituent "notes".

Isaac Newton famously split white light into a rainbow using a glass prism, but these days it's more common to use what's called a diffraction grating. (These things are cheaper and smaller than prisms.)

Seeing individual "notes" of light is very useful, because those notes correspond to specific energy levels within the structures of atoms and molecules. Just as a person with perfect pitch can tell you what notes make up a piece of music, so a spectrometer can show you the specific frequencies present in a beam of light, and thus tell you about the energy levels present in the atoms or molecules responsible for creating or filtering that light.

If you point a spectroscope at something very hot, say the filament of an old-fashioned incandescent light bulb, you'll see a smooth, uniform rainbow of color. This is classic white light: a mixture of all different frequencies. The spectrum of the sun looks very similar, which is of course why we prefer to spend time in this kind of light. It's familiar.

Some sources of light, such as this cheap fluorescent tube, create light that looks superficially like nice white light, but is in fact made up of several narrower ranges of frequencies. The light looks white only because our eyes are so bad at seeing color. Light like this tends to distort the colors of objects it is illuminating, which is why photos or videos taken under fluorescent light sometimes come out with a weird color cast.

LED light bulbs combine the efficiency of fluorescent (or better) with a much smoother spectrum. Not perfect, but really pretty darn good! These lights have taken over professional photo lighting for good reason: They are robust, cool, powerful, and render color very well. (The spectra of professional lights are even better than this home model.)

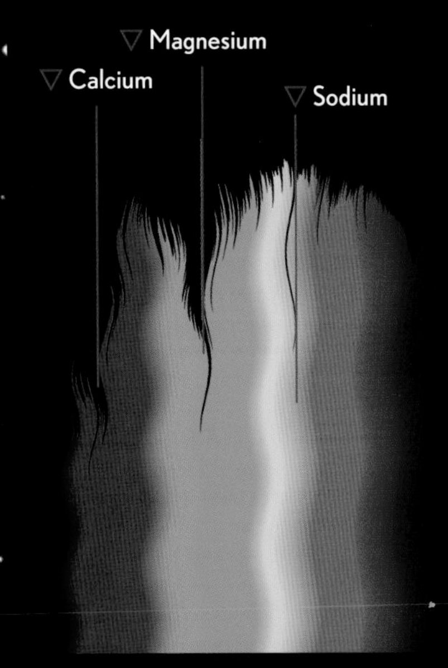

◁ Here's the mind-blowing part. Because a spectroscope can identify elements or molecules based only on the light coming from an object, you don't have to touch the object. You don't have to be in the same room with it. You don't even have to be on the same planet. By looking at the spectrum of a star, we can say without a shadow of a doubt what elements it does or doesn't have in its atmosphere. We can do this on Earth, with stars in our galaxy, and even with galaxies far more distant. Vast and ancient stars are singing to us in the music of light, and we understand them.

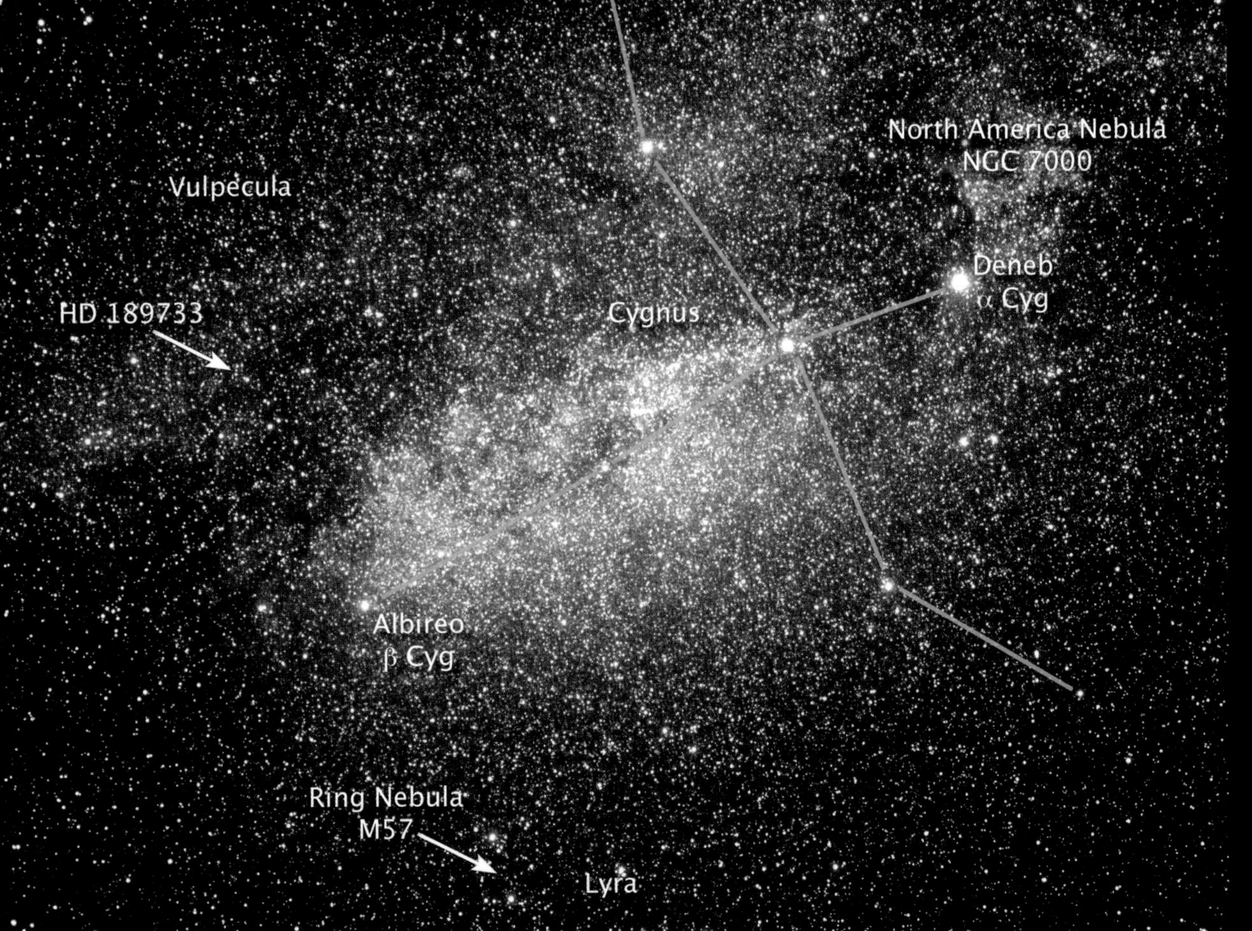

◁ If you have a powerful enough telescope and a sensitive enough spectroscope, you can even see spectral proof of elements and molecules in planets orbiting other stars. When you read in the news about yet another alien world discovered around yet another distant star, and you hear that it has oxygen in its atmosphere, or sulfuric acid rain, or crystals made of methane, that was all learned from the spectrum of light coming from that planet. For example, the planet HD 189733b, in orbit around the star we call HD 189733, is known to contain oxygen, water, methane, and carbon monoxide in its atmosphere. We know this because a spectroscope listened with its perfect pitch to the light of this distant world, and told us what it heard.

▽ This page shows the light emitted by the atoms of all the different chemical elements. (The few spectral lines we showed you a few pages ago for calcium, copper, and strontium were simplified to show only a few of the strongest lines. Most elements emit dozens or even hundreds of different spectral lines.)

A few elements are known for having particularly strong emission lines in just one region of the spectrum, meaning that these elements create intense hues of color when activated. These are the fireworks elements.

▷ Blue was long a difficult color to achieve in fireworks. Copper gives you a nice blue emission line, but also has a strong green line and is easily overpowered by brighter yellow light from the burning of the pyrotechnic fuel that creates the heat necessary to generate the emission lines in the first place. The development of PVC and chlorinated rubber fuels that burn at a relatively low temperature finally allowed fireworks to achieve rich, intense blue colors.

▷ Green sparkles typically come from barium salts.

▷ There is no more compelling demonstration of atomic emission lines than vividly colored fireworks. This red burst gets its color from strontium atoms, which were probably added to the mixture in the form of strontium carbonate. (The exact formulation of commercial fireworks like these are trade secrets, but the spectrum gives away the element used.)

Fireworks are not only a good illustration of atomic emission spectra, they are also a great example of the controlled fury that chemical reactions are capable of when one applies what I like to call the Ancient Chinese Art of Chemical Arranging.

The Ancient Chinese Art of Chemical Arranging

CHEMISTS TEND TO think of chemical reactions as things to be separated out and done one at a time (just as we, a few pages ago, separated out different elements to see what colors each one created in isolation). The fireworks designer sees those same reactions as a palette with which to paint a beautiful picture. The art is in the arrangement and sequencing of individual chemical reactions, one after the other, to produce a beautiful and thrilling experience.

The obvious place you see chemical arranging at work is in the distribution of colors in a fireworks display. But it starts well before then, with the fuse.

▽ Ah, the classic bottle rocket, illegal where I live in Illinois but shamelessly sold to the public thirty miles east in that lawless land we call Indiana. Rockets are popular in backyard fireworks displays, but virtually never used in professional shows.

△ The chemists' version of flower arranging: everything neatly lined up and separated out for detailed study.

▷ The fireworks maker's version of flower arranging: Artistic composition, balance, and variety are what matter, even if the chemistry is a mess.

Large fireworks displays use mortars, not rockets, because mortars are safer and more efficient. Small mortar shells like these, just 2 inches (5 centimeters) in diameter, are available, no questions asked, in the same places that sell bottle rockets. Professional shows use shells that are 3, 4, 6, 8, 10, or even 12 inches across (7.5 to 30 centimeters). Those are harder to get.

▽ A fuse made with gunpowder starts the action. (We'll learn much more about gunpowder in Chapter 6.)

Why mortars? A rocket, which is under power during its whole flight, can veer off course and head toward the audience. A mortar is shot out of a tube pointed in a particular direction—it has no choice but to go where it's sent. And a mortar shell is nearly all explosive payload. There's no heavy rocket motor to add weight and fall down on someone's head.

Here we see the anatomy of what I like to call an "Indiana-grade" mortar shell. (Technically it's a class 1.4G consumer–grade firework. This kind is legal in various other states, but as far as I'm concerned what matters is that I can buy it in Indiana.) Turn to page 195 to see what it looks like inside the tube when one of these is lit.

▽ In the center of the shell is the "break charge" made of flash powder, a more powerful explosive than gunpowder. (More on the types of explosives in Chapter 6.)

▽ Dried rice hulls are used as a packing material to keep everything tight. The amount of explosive is limited by law—and by cost—so they fill up the extra space with rice hulls, cotton seeds, or something else that's dry and compressible.

▽ These "stars" are the payload. Getting them up in the air and flying in all directions is the purpose of the rest of the firework. Each star is a packed-together bead of some combination of fuel and elements designed to burn with a particular color. Sometimes they are made with alternating layers of chemicals, allowing them to twinkle or change color. The stars are what make the beautiful shower of sparks spreading out in all directions.

▽ The fuse leads to a "lift charge" made of gunpowder. This material is responsible for lifting the shell up into the air.

▽ Cutting the mortar shell in half lets us see the time-delay fuse that connects the lift charge on the bottom to the break charge in the center of the ball. It takes just the right length of time to burn so the shell bursts at the highest point of its brief but glorious flight.

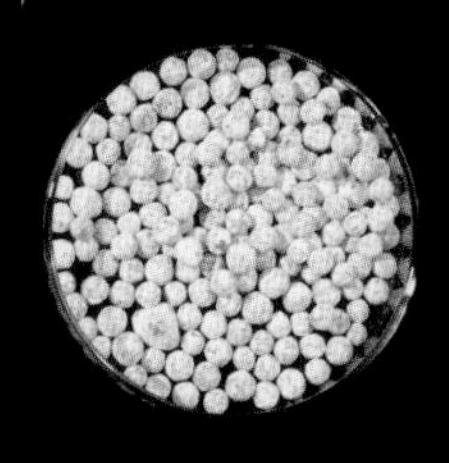

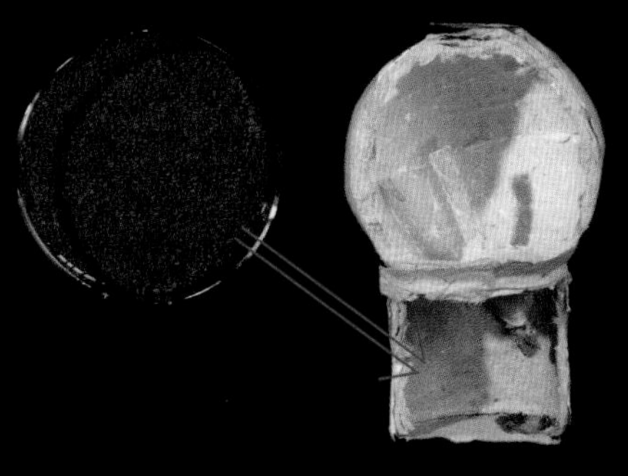

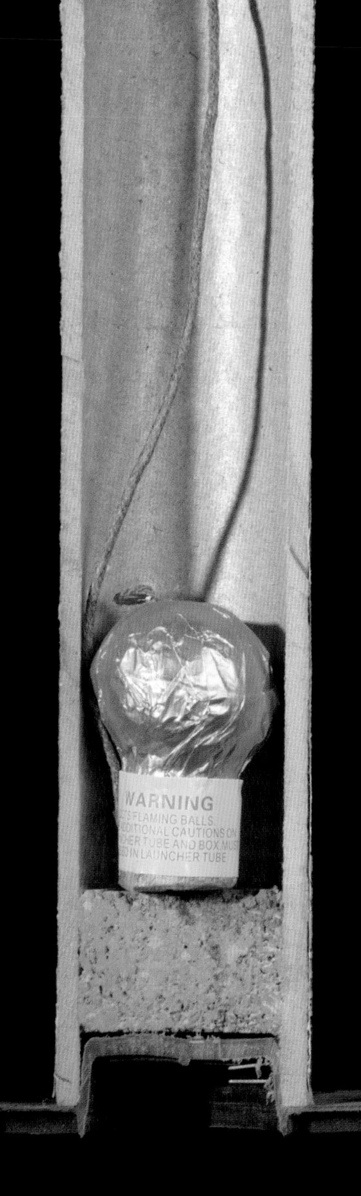

◁ Cotton balls soaked in alcohol solutions of metal salts are a good way to preview what color you might get if you include that salt in a firework.

▷ Coloring the stem of a candle is easy using any common sort of dye or pigment mixed into the wax. It's harder to color the candle's flame to match the color of the stem. The pigments in the wax cannot color the flame, because they simply burn up. These pretty birthday candles do it using the atomic emission lines from a mixture of the same sorts of elements used in fireworks. These elements don't change the color of the wax, but they survive the heat of the flame, and by matching pigments to elements, the candles' makers were able to color-coordinate wax and flame.

▽ Half a dozen elements, pure and in combination, create most of the color in these large displays. People sometimes describe fireworks as a symphony of light and sound. I would never do that because it's corny, but technically, it's pretty accurate. Each of these bursts is like an instrument contributing its frequencies of light, its notes, to the mix that is the whole. The result is raw elemental power, literally.

5

The Boring Chapter

THIS CHAPTER IS going to be about as interesting as watching paint dry. Well, maybe not everything will be quite that exciting. For example, I am aware of only *one* way of dying while watching grass grow, and watching water boil is a bit of a one-trick pony in comparison to the rich subjects of paint drying and grass growing. Although it is also potentially fatal.

Let's start by watching paint dry.

Paint drying has little to do with drying. Yes, it does sometimes involve water (or some other solvent) evaporating, but the important part is what happens later when the solvent is nearly gone and the paint has to do its magic of turning from a goopy liquid into a tough solid that will stay put for many years doing battle with rain, snow, hail, small children, and harsh sunlight.

If paint did nothing more than literally dry out, without undergoing some kind of chemical transformation, then it could just as easily be redissolved. There would be no way for water-based paint to be waterproof, or for oil-based paint to resist solvents.

Such paints do exist, but the best paints are far more clever in how they "dry."

Watching Paint Dry

WASHABLE KIDS' PAINT, the kind designed to come out of clothes easily even after it's dry, is the one example of paint "drying" that actually is mostly just water evaporating. Because the paint has to release right away when you make it wet again, it can't undergo any irreversible chemical reactions while drying. Instead it uses short-chain, water-soluble polymers to bind pigment particles together. This kind of paint isn't really serious paint, it's more of a toy, and not actually useful for anything other than being washed out of clothes, hair, carpets, tablecloths, smocks, blouses, socks, and small dogs. (By the way, "washable" is a bit of a relative term. The dog came out nice and clean, but those clothes are never going to be truly white again.)

Lacquers

YOU CAN MAKE SERIOUS, reasonably good paint that only dries by using a solvent other than water which the paint is not likely to ever encounter in the future. Such solvents include acetone, toluene, benzene, or any of a variety of other smelly, toxic, flammable, carcinogenic, or just polluting organic solvents. Since these solvents must, by the very nature of the situation, be encouraged to evaporate into the air, it's difficult to avoid their negative effects. These kinds of paints (primarily lacquers and shellacs) have been banned for large-scale use in many parts of the world.

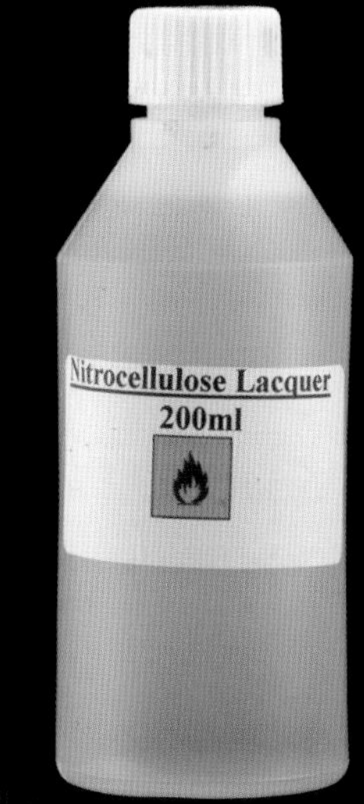

▷ Lacquers are examples of real paints that dry by evaporation only. But the solvent evaporating isn't water, it's something like acetone or toluene. If you put that same kind of solvent on the coating even years later, it will redissolve, ruining the finish. This can sometimes be a good thing. Damaged lacquer finishes can be repaired by redissolving a patch and reapplying a new layer, which can't be done with chemically hardening paints.

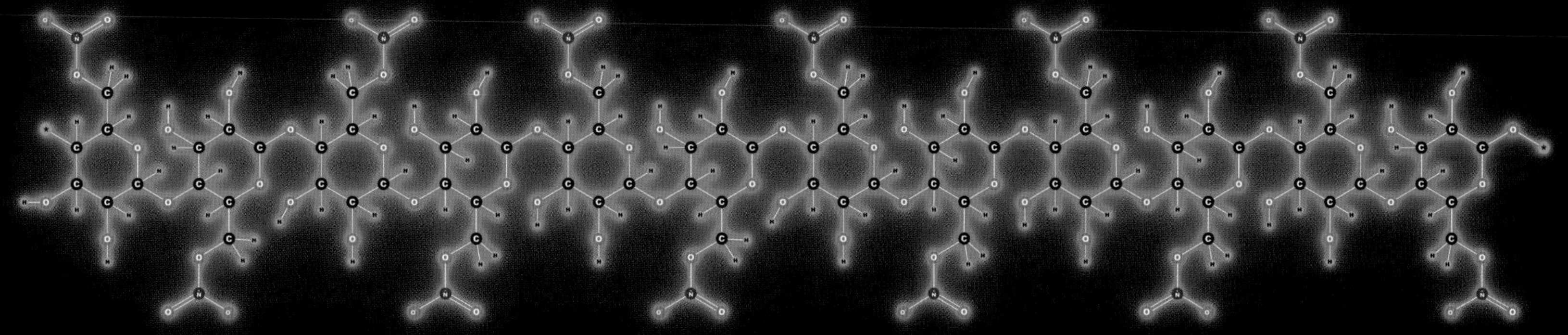

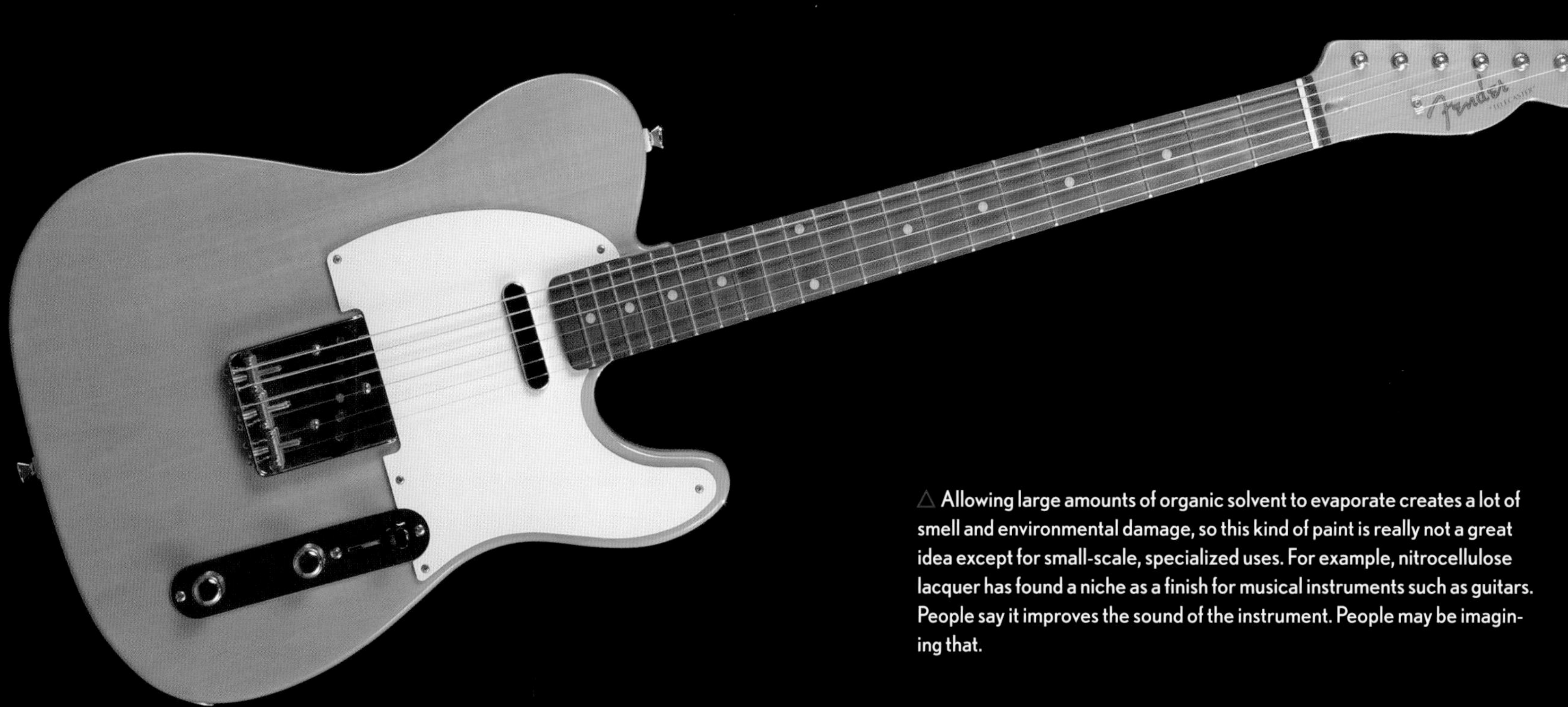

△ Allowing large amounts of organic solvent to evaporate creates a lot of smell and environmental damage, so this kind of paint is really not a great idea except for small-scale, specialized uses. For example, nitrocellulose lacquer has found a niche as a finish for musical instruments such as guitars. People say it improves the sound of the instrument. People may be imagining that.

▷ Nitrocellulose is quite flammable. It won't burn when it's coated on a solid surface, but the situation is different when a thick film of it stands on its own. Old "nitrate" movie film was notoriously flammable, which isn't surprising when you consider that it was just a thick layer of dried nitrocellulose lacquer made flexible with camphor oil (also flammable). Combine this film with an extremely hot arc light in the projector and you have a sometimes-fatal combination of explosively flammable film and nearby source of ignition. Many people died in historic theater fires because of nitrate film.

Why is nitrocellulose lacquer so flammable? It has to do with the "nitro" part of the name.

▷ Rag paper and cotton fabric are nearly 100 percent ordinary cellulose. They burn because cellulose molecules can react with oxygen molecules, turning into carbon dioxide (CO_2) and water (H_2O) with the release of enough energy to create a hot flame. It burns slowly (as we expect paper to burn) because the rate of the reaction is limited by the need for oxygen to come in from the surrounding air. (This is why blowing on a flame will often make it burn brighter: You're supplying additional oxygen.)

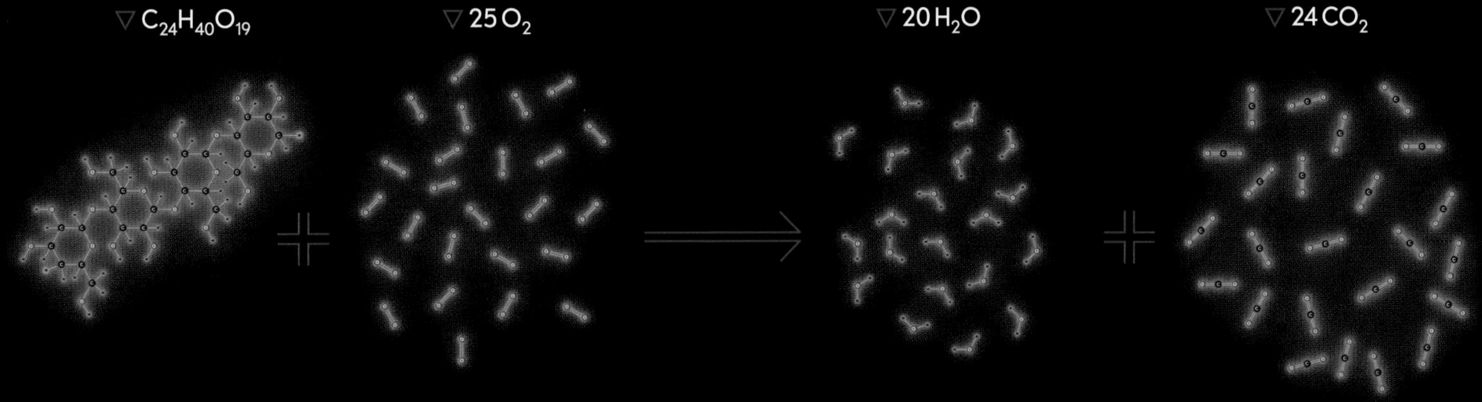

▽ $C_{24}H_{40}O_{19}$ + ▽ $25\,O_2$ ⟶ ▽ $20\,H_2O$ + ▽ $24\,CO_2$

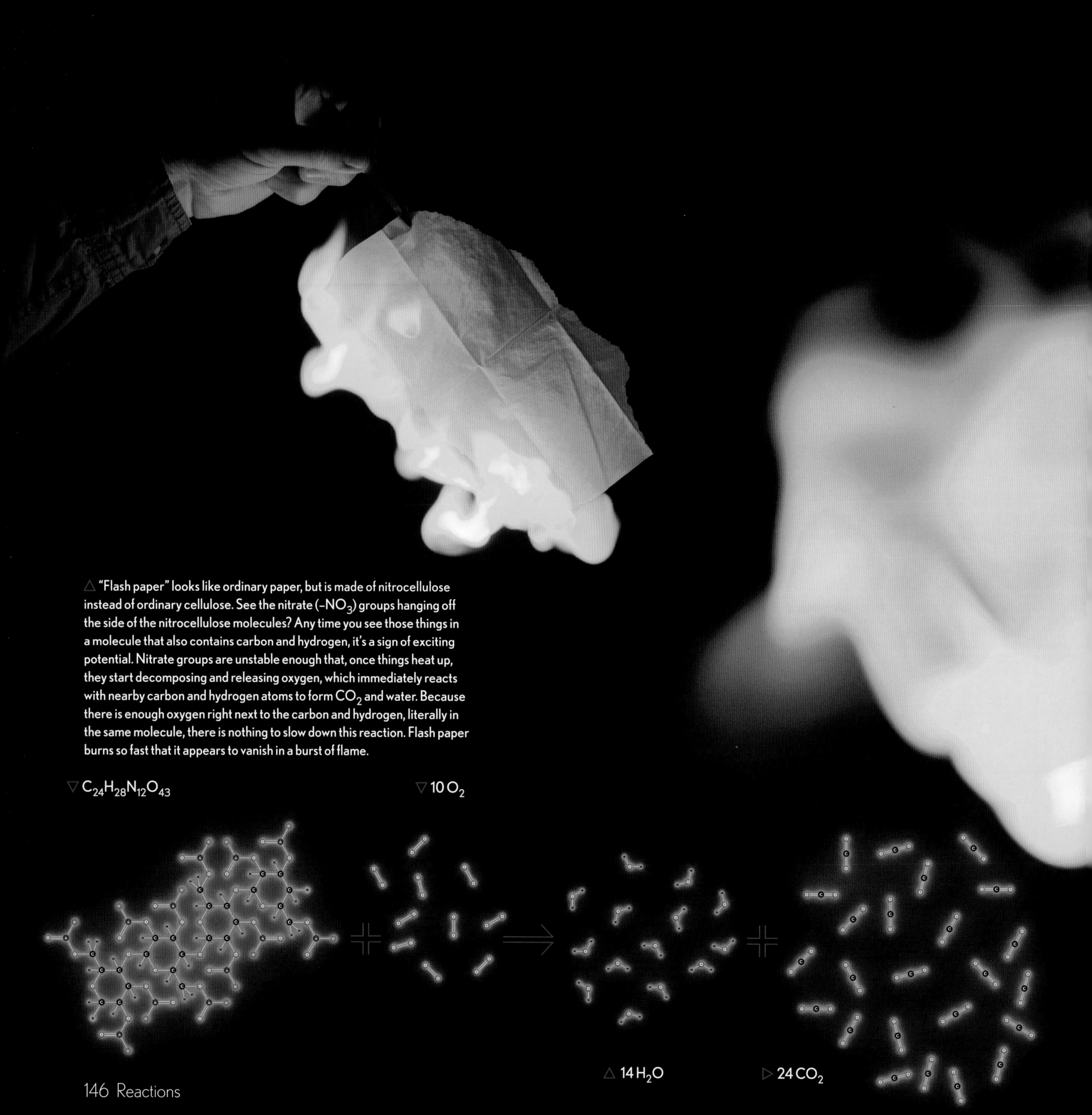

△ "Flash paper" looks like ordinary paper, but is made of nitrocellulose instead of ordinary cellulose. See the nitrate (–NO_3) groups hanging off the side of the nitrocellulose molecules? Any time you see those things in a molecule that also contains carbon and hydrogen, it's a sign of exciting potential. Nitrate groups are unstable enough that, once things heat up, they start decomposing and releasing oxygen, which immediately reacts with nearby carbon and hydrogen atoms to form CO_2 and water. Because there is enough oxygen right next to the carbon and hydrogen, literally in the same molecule, there is nothing to slow down this reaction. Flash paper burns so fast that it appears to vanish in a burst of flame.

△ $6 N_2$

△ Magicians use this stuff for magic tricks where, for example, a $100 bill (fake, printed on flash paper) vanishes into thin air in a burst of flame. Unlike most magic tricks this one is not fake in any way. When flash paper burns, the only reaction products are CO_2, water, and nitrogen gas. All three of these are natural components of the air. So the bill really does vanish into thin air—it literally *becomes* air.

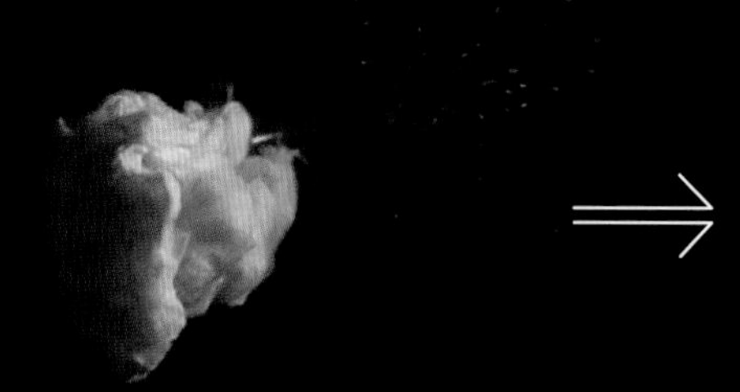

△ Another name for nitrocellulose when it's in the form of cotton wool is "guncotton." Confined in the barrel of a gun, it explodes with enough force to work as a superior alternative to gunpowder. Not only do the reactions between oxygen, carbon, and hydrogen supply energy and gas (CO_2), but the nitrogen atoms from the nitrate groups snap together to form nitrogen gas (N_2), which releases even more energy and contributes more gas to build up pressure behind the bullet in the barrel.

If nitrocellulose lacquer and guncotton are both made of the same thing, does that mean you could shoot a gun with dried paint? Well, I tried, but sadly it does not appear to be possible to do this. When ordinary cellulose is "nitrated" to turn it into nitrocellulose, you can control how many nitrate groups are added to the molecule. Guncotton is made by adding a *lot* of nitrate groups, while nitrocellulose lacquer has many fewer per unit of cellulose. This makes it flammable, but not explosive in the same way as guncotton. There's nothing I would have liked better than to be able to shoot a gun with dried paint! Do I get partial credit for getting a bullet jammed halfway down the barrel? Because that was the best I could do.

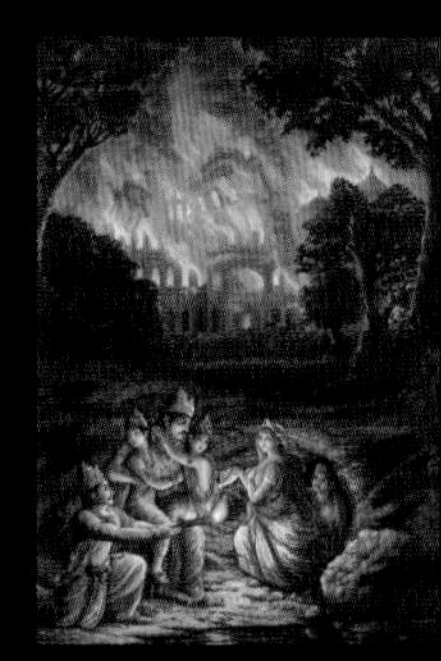

▷ In really big guns it's common to load the shell (the thing that gets thrown out the end) separately from the propellant charge (the thing that does the throwing). This is one of several charges intended for loading into a 155-millimeter (6-inch diameter) barrel Howitzer. It's relevant to a chapter on paint because what looks like a cardboard tube holding granules of something explosive is actually made almost entirely of nitrocellulose. Even the cardboard tube is made of nitrocellulose paper (basically flash paper). Why? So everything, including the tube, can contribute to the explosion, and so that nothing is left in the barrel after it's fired. (OK, truth be told, this is a training dummy made of entirely inert materials, but it's meant to look just like the real thing.)

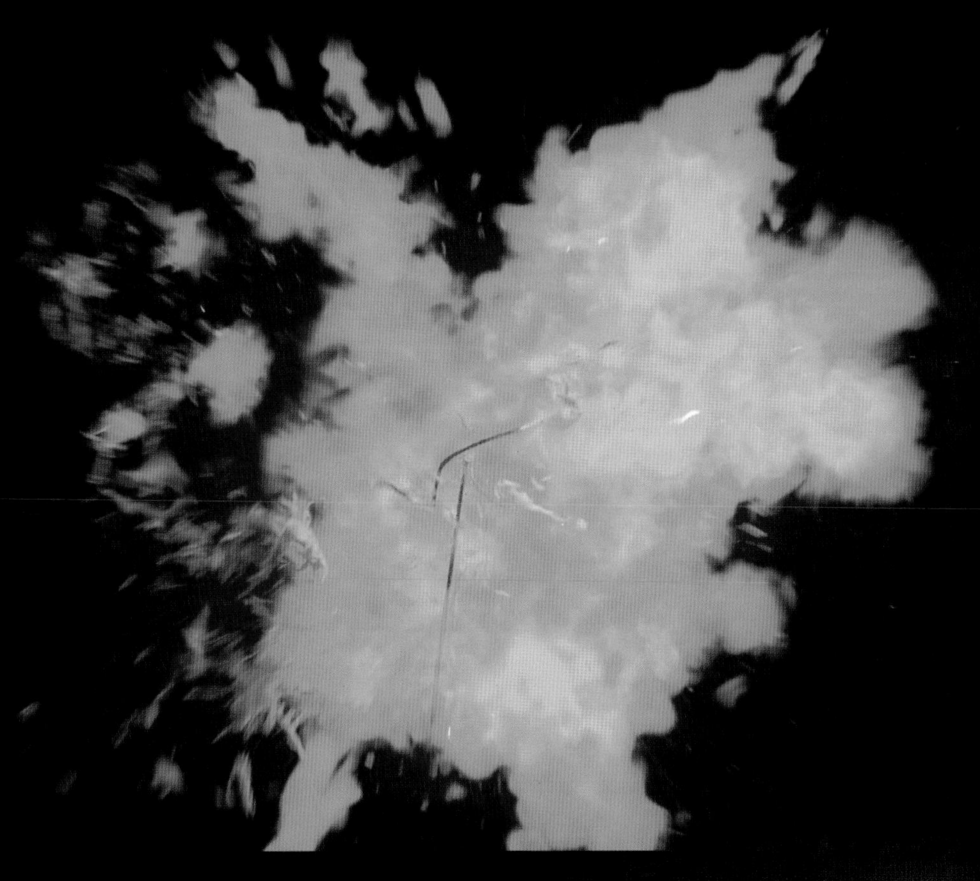

◁ Lacquer has been famous for being flammable for a very long time! The *Lakshagraha*, an ancient Sanskrit epic poem and companion to the better-known *Ramayana* epic poems, includes the story of a house built by Duryodhana for the sons of Pandu. It was built entirely of lacquer, because he intended to burn it down with them in it. It is not recorded exactly what kind of lacquer it was, but since nitrocellulose lacquer was only invented in the 1920s, it was presumably the old-fashioned kind. Some versions of the epic include saltpeter in the construction, which would certainly be a very effective way of making ordinary shellac far more flammable. Saltpeter is potassium nitrate, the key ingredient in gunpowder—see page 193.

▷ Shellac, made from a resin secreted by the lac bug, is another evaporation-only kind of paint, but we're not going to talk about it very much because you can't make it into an explosive. The chance of a shellac-related fatality is limited to the flammability of whatever solvent it's in. These flakes are how it's typically sold: It's up to you to decide what to dissolve them in. "Button lac" is shellac in the form of individually stamped buttons of crude shellac exactly as it comes from the factories in India.

Latex Paint

WATER-BASED LATEX paint is by *far* the most common kind of paint used by homeowners, construction companies, hobbyists, and really anyone who wants to easily and cheaply cover large areas with large amounts of paint. It's easy to use, doesn't smell too bad, and is considered pretty harmless for the environment (more or less so depending on what, other than water, is in the particular variety).

Latex paint is also an example of drying-only paint, but with a clever twist.

◁ The great thing about writing a book is that you get to clarify things that have been bothering you for years, when you're forced to research them so as not to mislead your readers. For example, did you know that there's no latex in latex paint? I didn't.

▷ To my ears the word "latex" means one thing: latex rubber, the kind of stuff you make medical gloves and Halloween masks out of. But latex rubber is just one way in which the word is used. "Latex" can be used to mean fine particles of any type of polymer (long-chain molecule) dispersed in water. Latex rubber is just the special case when the dispersed polymer is a particular kind of plant rubber. Latex paint is instead made with acrylic, vinyl, polyvinyl acetate, or several other types of polymer. But never latex rubber.

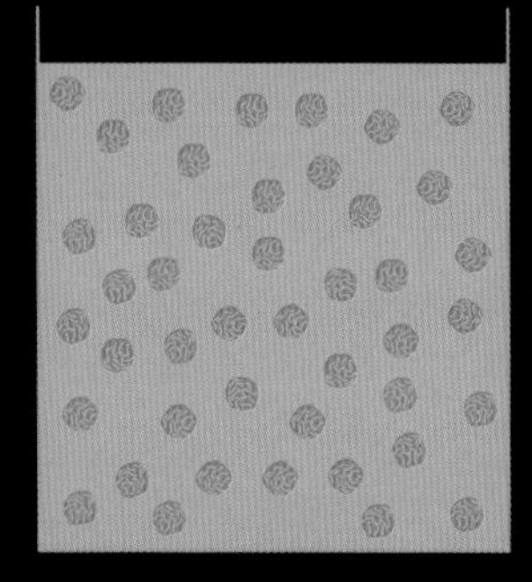

△ Latex paint in the can is a bit like Italian salad dressing that's been shaken really hard. Just as olive oil droplets are dispersed in water-based vinegar, latex paint contains microscopic droplets of oil dispersed in water. Because oil and water don't mix, the oil droplets do not dissolve into the water. And unlike salad dressing, the paint contains special ingredients to make sure the droplets repel each other slightly, ensuring that they never coalesce together and settle out into a separate layer, as tends to happen with salad dressing.

The "oil" in each tiny droplet is the solvent Texanol, a compound selected for its very particular rate of evaporation. Dissolved in the Texanol are many medium-length acrylic polymer chains. These polymers are soluble in Texanol, but not in water, so they remain trapped inside the droplets as long as those droplets are surrounded by water.

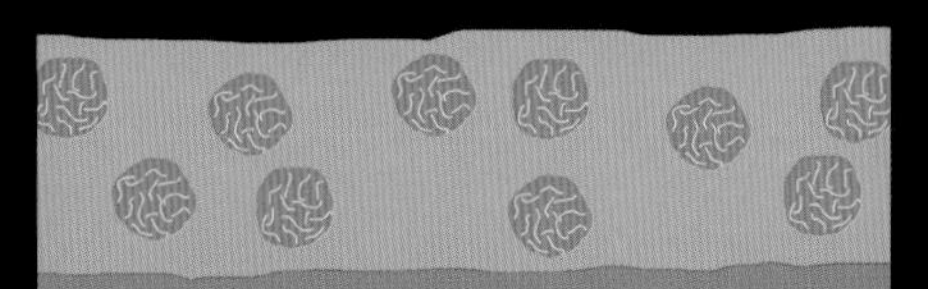

△ Right afer the paint is put on, it's still mostly water, with polymer-containing droplets of Texanol floating in the water.

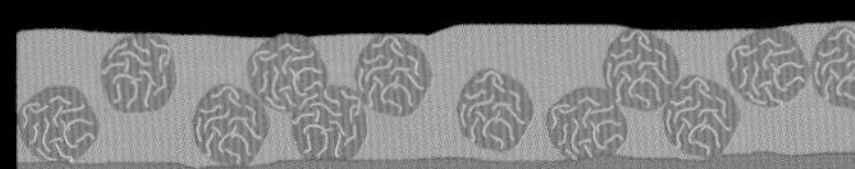

△ As the water evaporates, the Texanol drops get closer and closer. They repel each other, so they don't merge yet.

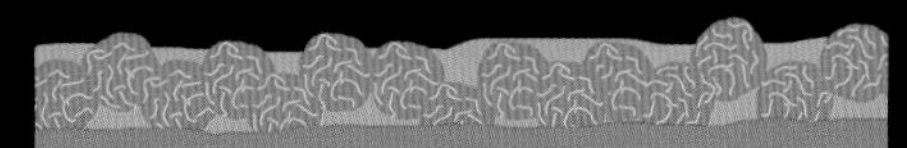

△ Eventually so much water has evaporated that the droplets are forced to touch and merge.

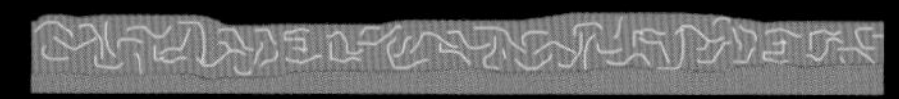

△ When all the water has evaported, the result is a continuous layer of Texanol, which evaporates much more slowly. The polymers mingle with each other during this time.

△ As the Texanol slowly evaporates, the polymers compact and interlock with each other.

△ After all the Texanol has evaporated, what's left is nothing but tangled-up polymer strands (and pigment particles trapped by them).

◁ Texanol is used because it evaporates significantly more slowly than water. After the water is all gone, you are left with a much thinner layer of still-liquid paint. But now instead of being a water-based liquid, it is Texanol-based liquid. The acrylic polymers that had been trapped in the individual droplets are now free to spread out and tangle up with each other throughout the whole Texanol layer.

As the Texanol slowly evaporates, the layer gets thinner and thinner until only the acrylic is left, hardened into a tough, solid film. At no point are any new strong chemical bonds formed, but the weaker bonds between the tightly entangled strands of acrylic polymer lock those strands together as tightly as possible in all but the toughest plastics. They can, however, be dissolved again by various organic solvents, making latex paint less chemical-resistant than cross-linking paints (see next section).

If latex paint dried only by evaporation of water, it could be redissolved by water just like washable kids' paint. This would make it pretty useless, particularly outdoors. But instead it is vulnerable only to organic solvents, which it's not likely to encounter in common household situations.

To put it another way, latex paint is a bit like older nitrocellulose or shellac finishes in that it involves evaporation of an organic solvent. But it performs the clever trick of using water as a second-level carrier to hugely reduce the *amount* of organic solvent needed, so much so that in jurisdictions where volatile organic solvent use is severely restricted, latex paint slips under the radar.

▷ Artist's impression of a Texanol droplet in water. Not to scale! In the droplet you see a mixture of long polymer molecules (which are actually much, much longer), and smaller Texanol molecules. A real droplet would have many millions of both.

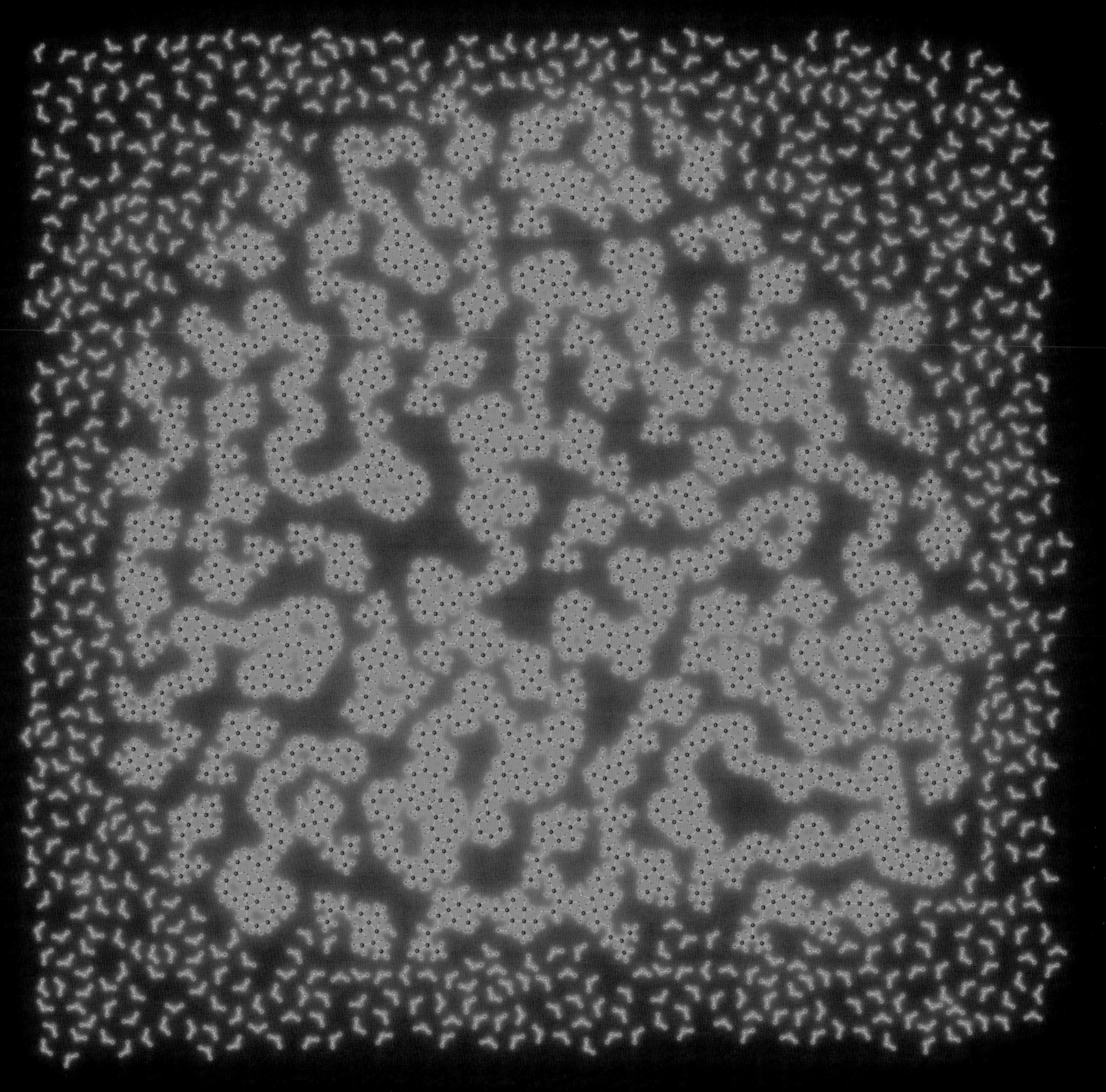

Oil-Based Paints

PAINT THAT ONLY literally dries, without further chemical hardening, has its place. But for a really tough end result, you want a paint that undergoes an irreversible chemical reaction to become permanently hard. This kind of paint is said to "cure" rather than just dry. One of the central mysteries of paint is how it manages to undergo this hardening reaction only *after* it's been painted onto something, not while it's still in the can. There are two fundamentally different ways this is achieved, only one of which typically results in deadly house fires.

Oil-based paints (natural and synthetic) harden when compounds in them react with oxygen from the air. They don't harden in the can, not because the can prevents anything from evaporating, but rather because the can prevents oxygen from reaching the oil and reacting with it. The skin that often forms on the surface of paint in a closed can is due to it reacting with oxygen in the air gap at the top.

Because this kind of paint requires oxygen to harden, it always hardens from the outside in. When the outermost layer hardens, this greatly slows down the rate at which oxygen can penetrate, which means there's a limit on how thickly oil paint can be applied before it takes nearly forever to harden. That's why several thin coats dry much faster than one thick coat.

△ Linseed oil, which comes raw or "boiled," has been used in paints for about 1,500 years. (The plant it comes from has been known for at least 30,000 years and the oil for at least 9,000.) Amusingly, the "organic flaxseed oil" that people eat for its healthy omega-3 fatty acid content is one and the same thing. "Flax" and "linseed" are two names for the same plant, so when you're eating flaxseed oil, you're literally eating paint. (But please don't drink linseed oil sold for painting purposes! It almost always contains metal salt additives that speed up the hardening process, and those things are not healthy.)

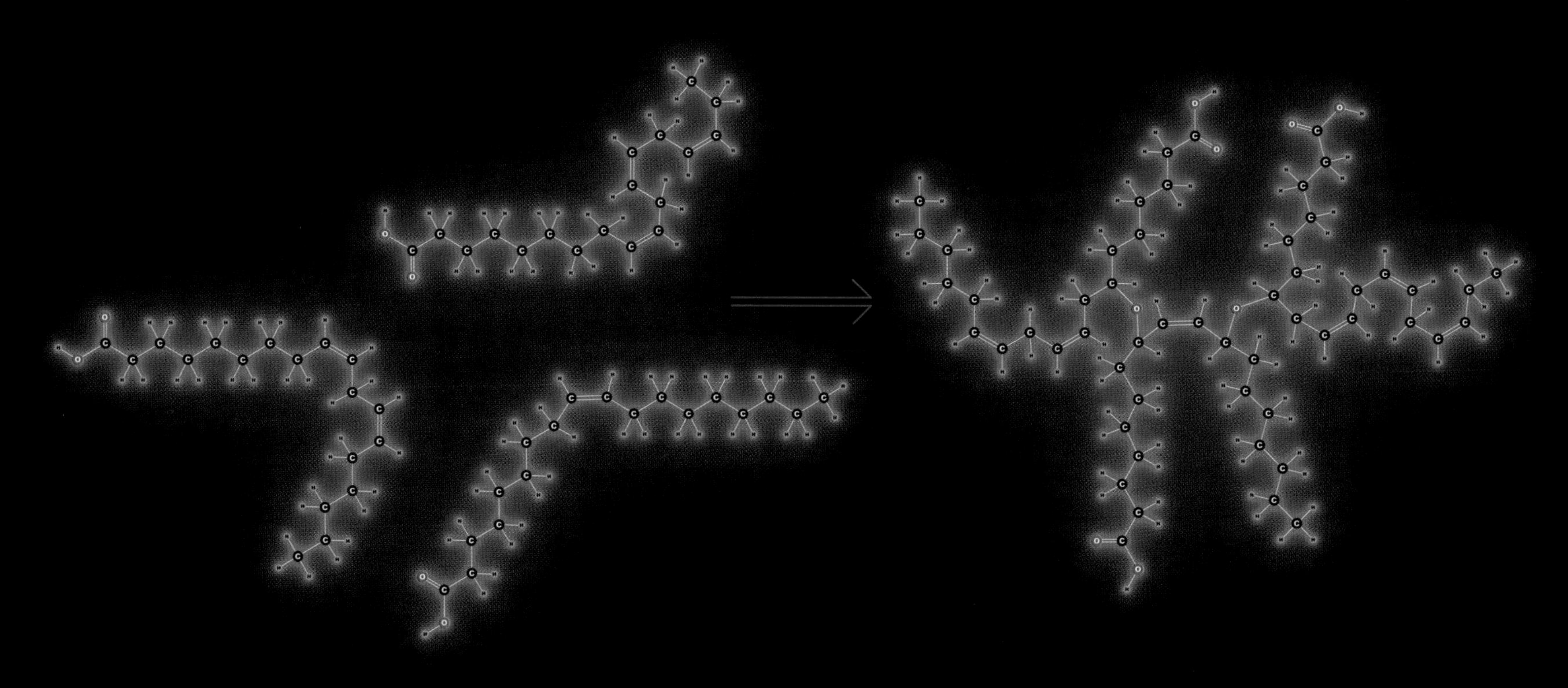

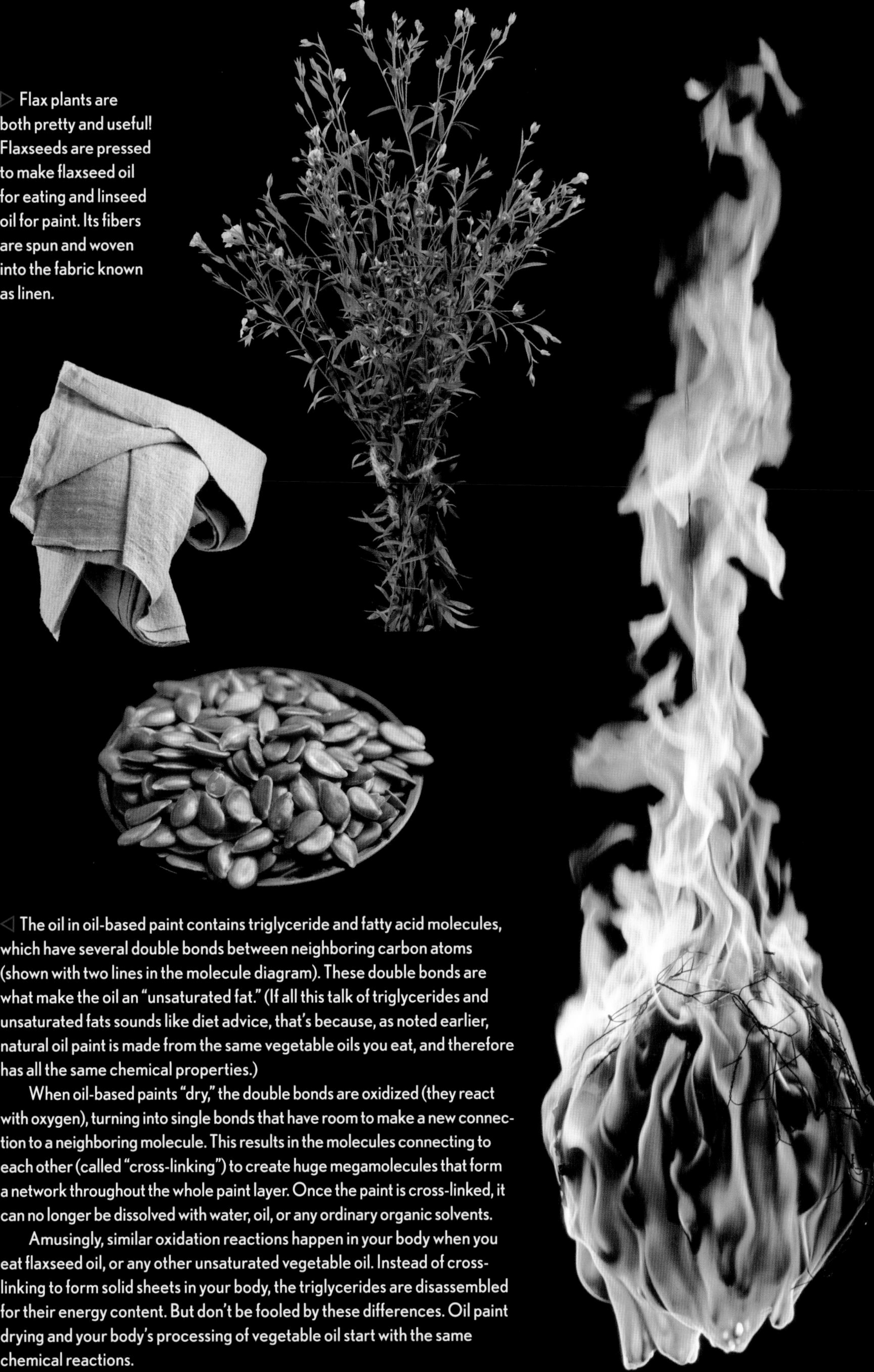

▷ Flax plants are both pretty and useful! Flaxseeds are pressed to make flaxseed oil for eating and linseed oil for paint. Its fibers are spun and woven into the fabric known as linen.

◁ The oil in oil-based paint contains triglyceride and fatty acid molecules, which have several double bonds between neighboring carbon atoms (shown with two lines in the molecule diagram). These double bonds are what make the oil an "unsaturated fat." (If all this talk of triglycerides and unsaturated fats sounds like diet advice, that's because, as noted earlier, natural oil paint is made from the same vegetable oils you eat, and therefore has all the same chemical properties.)

When oil-based paints "dry," the double bonds are oxidized (they react with oxygen), turning into single bonds that have room to make a new connection to a neighboring molecule. This results in the molecules connecting to each other (called "cross-linking") to create huge megamolecules that form a network throughout the whole paint layer. Once the paint is cross-linked, it can no longer be dissolved with water, oil, or any ordinary organic solvents.

Amusingly, similar oxidation reactions happen in your body when you eat flaxseed oil, or any other unsaturated vegetable oil. Instead of cross-linking to form solid sheets in your body, the triglycerides are disassembled for their energy content. But don't be fooled by these differences. Oil paint drying and your body's processing of vegetable oil start with the same chemical reactions.

◁ Sometimes watching paint dry is a matter of life and death. For example, between 2005 and 2009, seven people died from not properly watching their paint dry. OK, no, you don't actually have to watch all your paint dry. It's not going to catch fire on a wall you just painted. But that statistic is absolutely real. If you leave certain kinds of oil-based paint, including linseed oil, soaked into a pile of cotton rags, there is a good possibility that, about eight hours later, it will spontaneously catch fire.

What's happening is that the oxidation reaction, the one that causes it to harden, releases energy that heats up the paint. With paint spread thinly over a large area, this heating is completely insignificant. You could never feel it. But when paint is concentrated in a pile of soaked rags, which also provide insulation that traps the heat, energy can start to build up. And, as usual, when any chemical reaction gets warmer, it tends to start happening faster. So the hardening/oxidation reaction starts going faster and faster, releasing more and more energy, until eventually the whole thing smolders and then bursts into flame. This is why smart people have special fireproof metal containers for disposing of oil-soaked rags!

△ This is a highly engineered safety paint drying chamber (oily rag trash can). It has a ventilated ring on the bottom to thermally isolate the can from the floor (in case the contents catch fire), and a spring-loaded, fireproof lid to make sure the contents stay safely enclosed. Paint drying is serious business, OK? Not boring.

△ It's often said that the Golden Gate Bridge is constantly being painted from one end to the other, with the crew starting over from the beginning as soon as they're done at the far end. One imagines a small ceremony with cupcakes at the end of a multiyear job, the foreman saying, "Well guys, you know the drill, see you tomorrow back at the start!"

It's true that the bridge is constantly being painted, but not in order. The current crew of twenty-eight painters, five assistant painters, one head painter, and three dogs go around to wherever touch-up is needed most. It hasn't been repainted end-to-end since 1968 through 1995, because it took twenty-seven years. (I made up the three dogs, but the other unbelievable facts are real.)

△ The Golden Gate Bridge, finished in 1937, was originally coated with a linseed oil paint containing 68 percent red lead paste (made from lead tetroxide powder). They were really enthusiastic about lead in paint back then! No paint like that is available today. The Golden Gate Bridge is currently painted with a zinc-silicate primer and an acrylic top coat. The zinc provides corrosion resistance, the top coat provides the color and protects the primer.

△ The *whole enterprise* of endlessly painting and repainting the Golden Gate Bridge, and every other bridge, is necessary only because of the really unfortunate fact that iron rusts. If only iron—the cheapest, strongest, easiest-to-fabricate metal—didn't also rust, there's no way people would spend the huge sums of money needed to keep the bridge painted. It would be plain metal for sure. So if you like the Golden Gate's pretty International Orange color, just remember, it's there only on the back of the roughly $1 trillion a year that iron corrosion costs the world economy.

△ The most common kinds of serious modern paints and varnishes are made with synthetic variations of what is, basically, linseed oil. In fact, the manufacturing process for many of them starts with linseed oil or one of several similar vegetable oils, which are separated into their fatty acids and then combined with various other ingredients to make a resin that is more predictable and specific in its composition than what you get by just heating straight vegetable oil.

Epoxy Paint

YOU'RE PROBABLY MORE familiar with epoxy glue than epoxy paint, but both of these are quite common and work the same way. In both cases you have a "Part A" and a "Part B" that need to be mixed immediately before they are used. The hardening of epoxy is a chemical reaction between these two parts, so epoxy paint solves the "don't harden in the can" problem by keeping the two parts separate so they can't react.

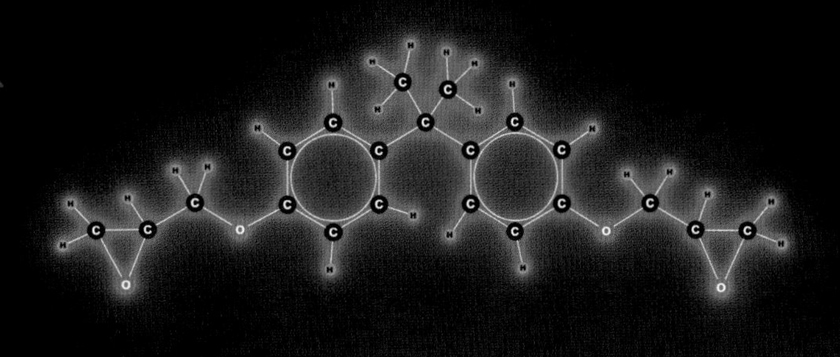

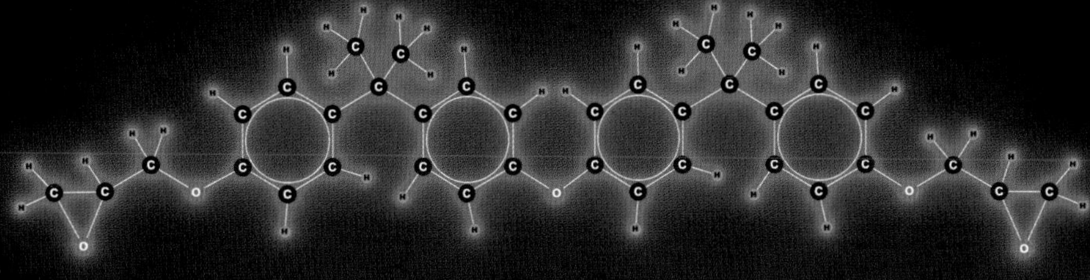

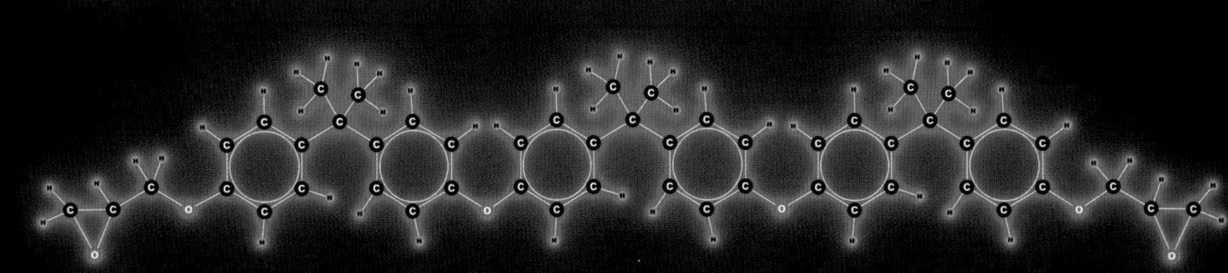

▽ A typical epoxy Part A contains a mixture of several different lengths of this compound, the beautifully named bisphenol A diglycidyl ether. Notice the highly strained epoxide groups at both ends. That's a sign that this molecule is primed and ready to react with something to create longer chains.

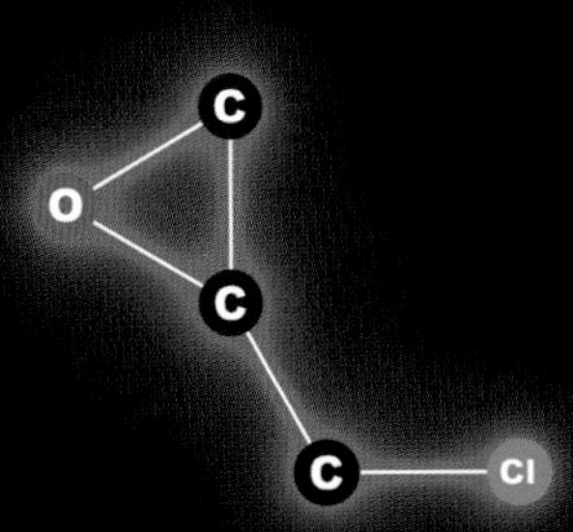

△ The term "epoxy" comes from the presence of epoxide groups in the chemicals that make up epoxy glues and paints. An epoxide is a very intense thing! Rings are common in chemistry, with six-member carbon rings being the most common by far. Larger rings just get floppier, but smaller rings become more and more strained, because the angles between the bonds are forced further and further away from their preferred ranges. The smallest ring you can possibly make is, of course, a three-member ring, and the very high strain of such rings tends to make them highly reactive. Anything with a three-member ring in it will always be eager to react with *something* to break open the ring. That's what an epoxide is: a three-member ring containing one oxygen and two carbon atoms. The specific example shown here is epichlorohydrin, a common precursor of many epoxy glues and paints. (Please note that, despite the similarity in the name, epichlorohydrin is not related to midichlorians and does not grant telepathic powers.)

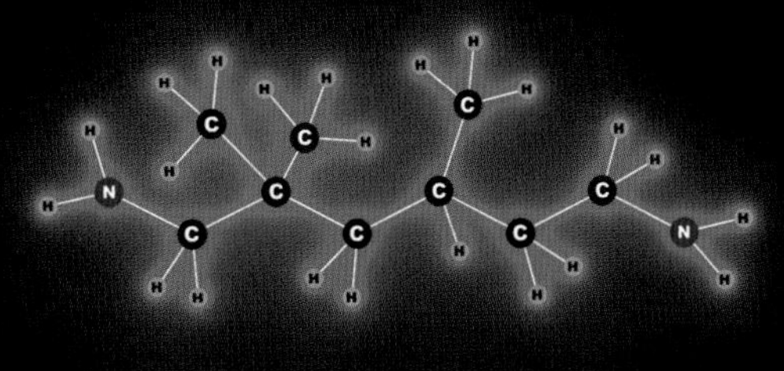

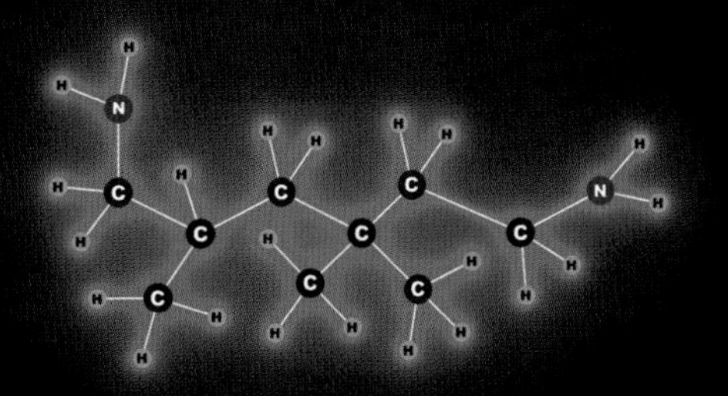

◁ A typical epoxy Part B contains some mixture of chemicals like these two, trimethylhexamethylenediamine and isophoronediamine. What matters about these chemicals isn't their names or their particular structures, it's that each of them has two $-NH_2$ (amine) groups stuck on it. These groups are able to link up with epoxide groups on the molecules in Part A.

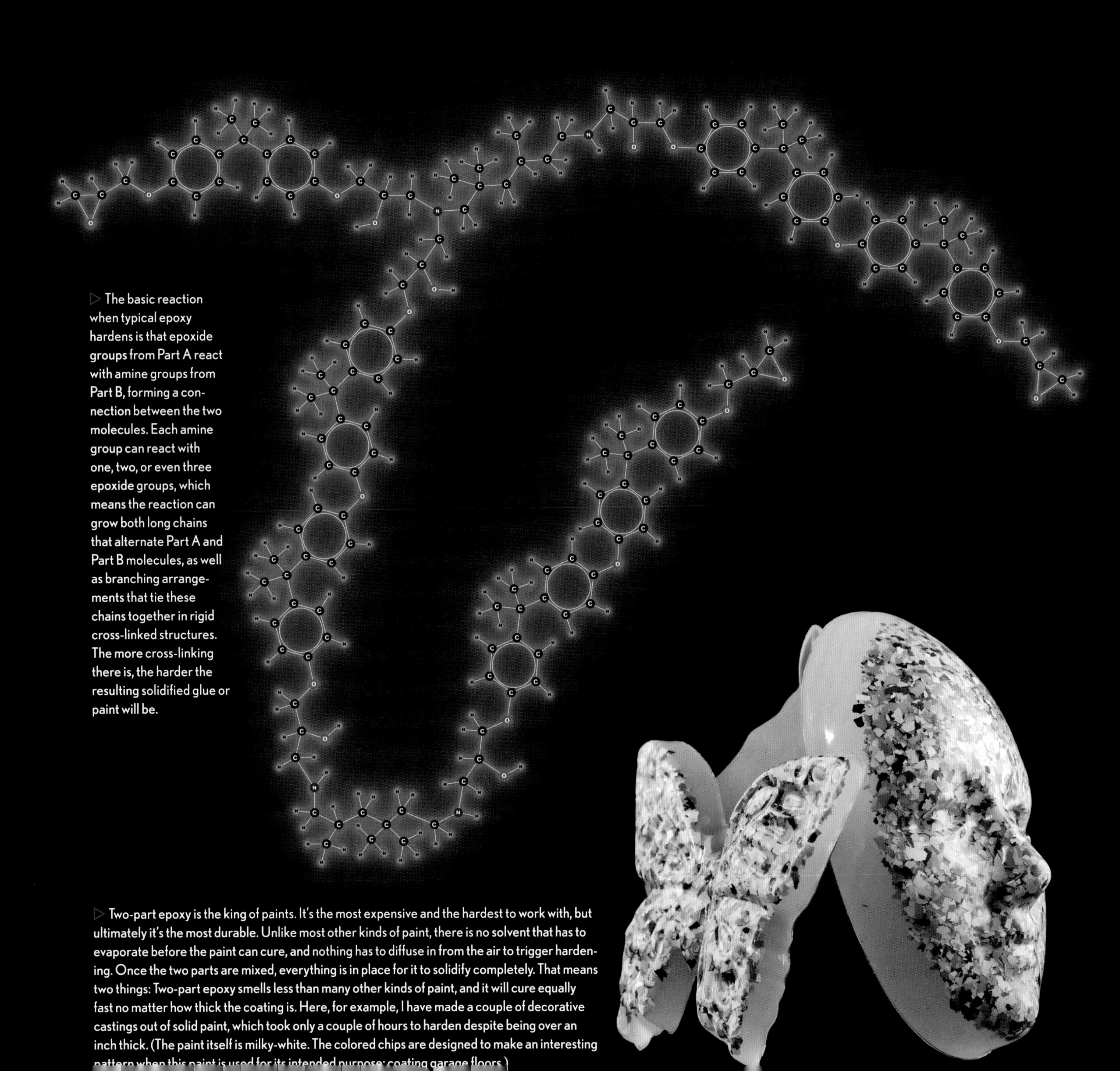

▷ The basic reaction when typical epoxy hardens is that epoxide groups from Part A react with amine groups from Part B, forming a connection between the two molecules. Each amine group can react with one, two, or even three epoxide groups, which means the reaction can grow both long chains that alternate Part A and Part B molecules, as well as branching arrangements that tie these chains together in rigid cross-linked structures. The more cross-linking there is, the harder the resulting solidified glue or paint will be.

▷ Two-part epoxy is the king of paints. It's the most expensive and the hardest to work with, but ultimately it's the most durable. Unlike most other kinds of paint, there is no solvent that has to evaporate before the paint can cure, and nothing has to diffuse in from the air to trigger hardening. Once the two parts are mixed, everything is in place for it to solidify completely. That means two things: Two-part epoxy smells less than many other kinds of paint, and it will cure equally fast no matter how thick the coating is. Here, for example, I have made a couple of decorative castings out of solid paint, which took only a couple of hours to harden despite being over an inch thick. (The paint itself is milky-white. The colored chips are designed to make an interesting pattern when this paint is used for its intended purpose: coating garage floors.)

▷ Because the curing of epoxy releases heat, and heat speeds up reactions, thicker layers can actually harden faster, and a too-thick layer can become hot enough to cause trouble. Epoxy paint left in the bucket, for example, can get hot enough to burn your fingers, and result, as we see here, in a solid casting of the bucket. You *definitely* can't save this kind of paint for later, no matter how tightly you try to seal it up. Once you mix it, you need to use it immediately.

△ This product straddles the line between epoxy glue and epoxy paint. It's pure epoxy resin without any pigments, and dries perfectly clear. In some ways it's like a varnish, but unlike a varnish it can be applied to any thickness, since it is able to harden all the way through solid areas.

◁ I made this table out of an old, cracked slice of a tree trunk. With gaps up to two inches wide that needed to be filled, clear epoxy was the only possible kind of "varnish" that could do the job. The layer on top is about $^1/_8$ inch (3 millimeters) thick, and, yes, it costs a fortune to pour about a gallon of pure epoxy into the cracks in a tree trunk. Worth it, don't you think? It's a lovely effect and impossible to achieve with any other product.

▽ When people pay too much for a bicycle, it often has a "carbon-fiber" frame. The fine carbon fibers are held together by epoxy resin, so basically you're riding on thread and paint.

Watching Grass Grow

WHERE I COME FROM, people spend a lot of time watching grass grow. Billions of dollars and much of the world's population depend on our grass. During the summer there are daily updates on the radio reporting on its progress, and discussions on the ways in which the people watching it are unsatisfied with the weather on that particular day.

We even have a saying for how fast it's supposed to grow each year: "Knee high by the Fourth of July." (This is actually a very out-of-date saying. Thanks to generations of careful selective breeding, it's more like, "Above your eye by the Fourth of July.")

A quite significant fraction of the average American citizen consists of carbon, oxygen, and hydrogen fixed from the air and water by the vast stands of this grass that spreads out over the immense flatness of the Midwest.

At 10 to 15 feet (3 to 5 meters) tall, corn, also called maize, is a midsized species of grass. The tallest kind of grass, bamboo, can be over 150 feet (50 meters) tall and can grow more than a foot a day, among the fastest of any living organism.

But it's not just the tremendous scale, speed, and economic importance of grass that make its growth worth watching. Inside even the humblest of lawn grasses is machinery of almost unimaginable complexity at work building huge molecules, atom by atom.

Much of the paint we watched dry in the last section was made of plant oil. These oils are dead things, undergoing quite simple chemical reactions in paint. Growing grass is actually *building* those sorts of chemicals, literally out of thin air. And while it's building the chemicals, it's also building the machinery needed to build the chemicals, and the machinery needed to make new copies of the machinery, and to defend the machinery against breakdown, predators, and infections.

What's going on inside a blade of grass growing is more complicated than anything happening in a car factory. It's just too small to see without the aid of computer simulations.

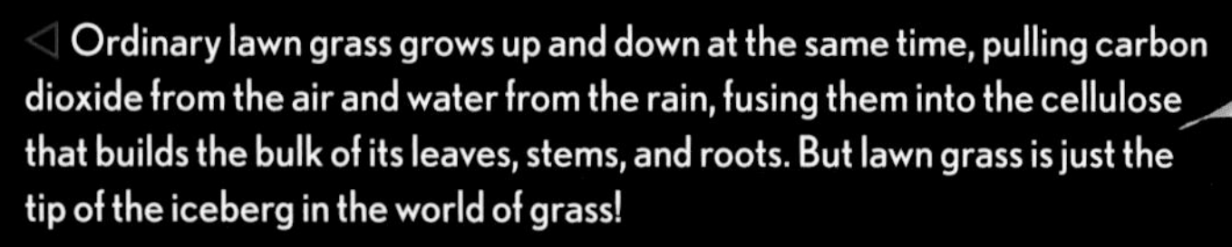

◁ Ordinary lawn grass grows up and down at the same time, pulling carbon dioxide from the air and water from the rain, fusing them into the cellulose that builds the bulk of its leaves, stems, and roots. But lawn grass is just the tip of the iceberg in the world of grass!

▷ This gorgeous corn plant, a representative of the finest Midwestern stock, was stolen from a neighbor's field by my photographer Nick on July 20th. It stands nearly twelve feet tall! Sadly it has only one ear, while many would have two or more. (Don't say that too loud, they get insecure about this kind of thing, and they are bigger than you.) See how all the leaves are pretty much in the same plane, alternating on opposite sides of the stalk? That means corn plants have an orientation. How well they can gather light from the sun depends on what direction they are facing. People worry about this. In fact, one research study found that by carefully arranging the orientation of each seed in the ground when it was planted, they could determine what direction the leaves in the full-grown plant would face. If this is done over a whole field, it results in a 10 to 20 percent increase in yield (the size of the harvest). If a machine could be devised to do that to whole fields at a time, the economic impact would be *huge*. I'm not kidding when I say people here take grass growing very, very seriously, and for good reason!

▽ Cellulose chain ▽ Glucose molecule ▽ Longer cellulose chain

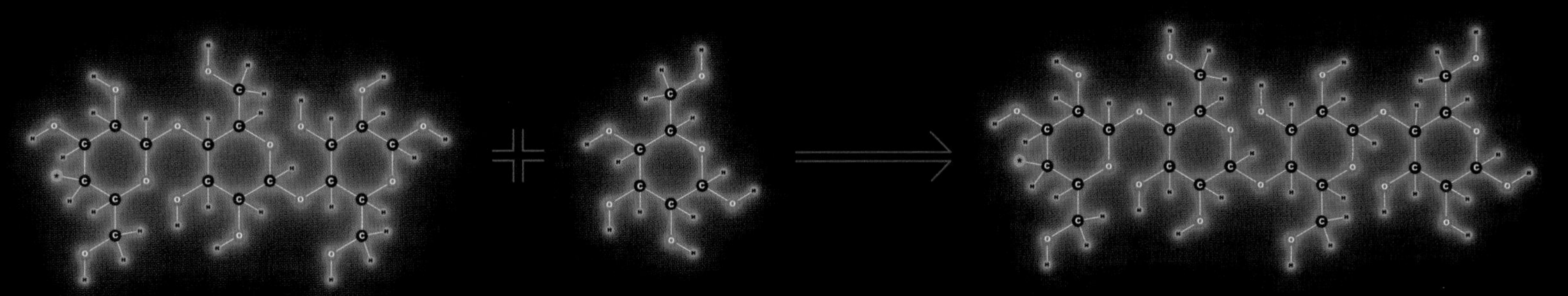

As we saw in Chapter 2 (see page 48), plants have the remarkable ability to capture energy from sunlight and turn it into chemical energy.

Plants use a lot of the energy they capture this way to grow bigger (in many cases because they are in a race with other plants to see who can get tallest fastest, and thus capture the most sunlight). When plants grow, they get bigger by creating new cells, new chlorophyll light-capturing structures, and new cellulose fibers to connect everything together. Wood, leaves, stems, and roots are all made mostly of cellulose.

Remarkably, cellulose itself is made entirely of sugar—glucose, to be specific. Glucose is a small molecule, but when tens of thousands of copies are linked together, the result is a gigantic cellulose molecule. Billions of cellulose molecules together form a thin fiber. Billions of fibers together form a whole plant.

This is done using a complex factory filled with protein machines.

◁ To see how the cellulose-building reaction is carried out by protein machines, we need to switch to a 3D view. So you'll know what you're looking at, here is what a cellulose chain (the same molecule having a glucose unit added to it below) looks like in 3D.

▽ In plants, glucose units are added to a growing cellulose chain by an enzyme which aligns each new glucose molecule with the growing chain, then triggers the reaction that joins them together. (See page 86 for an explanation of how enzymes help make chemistry work in living things.)

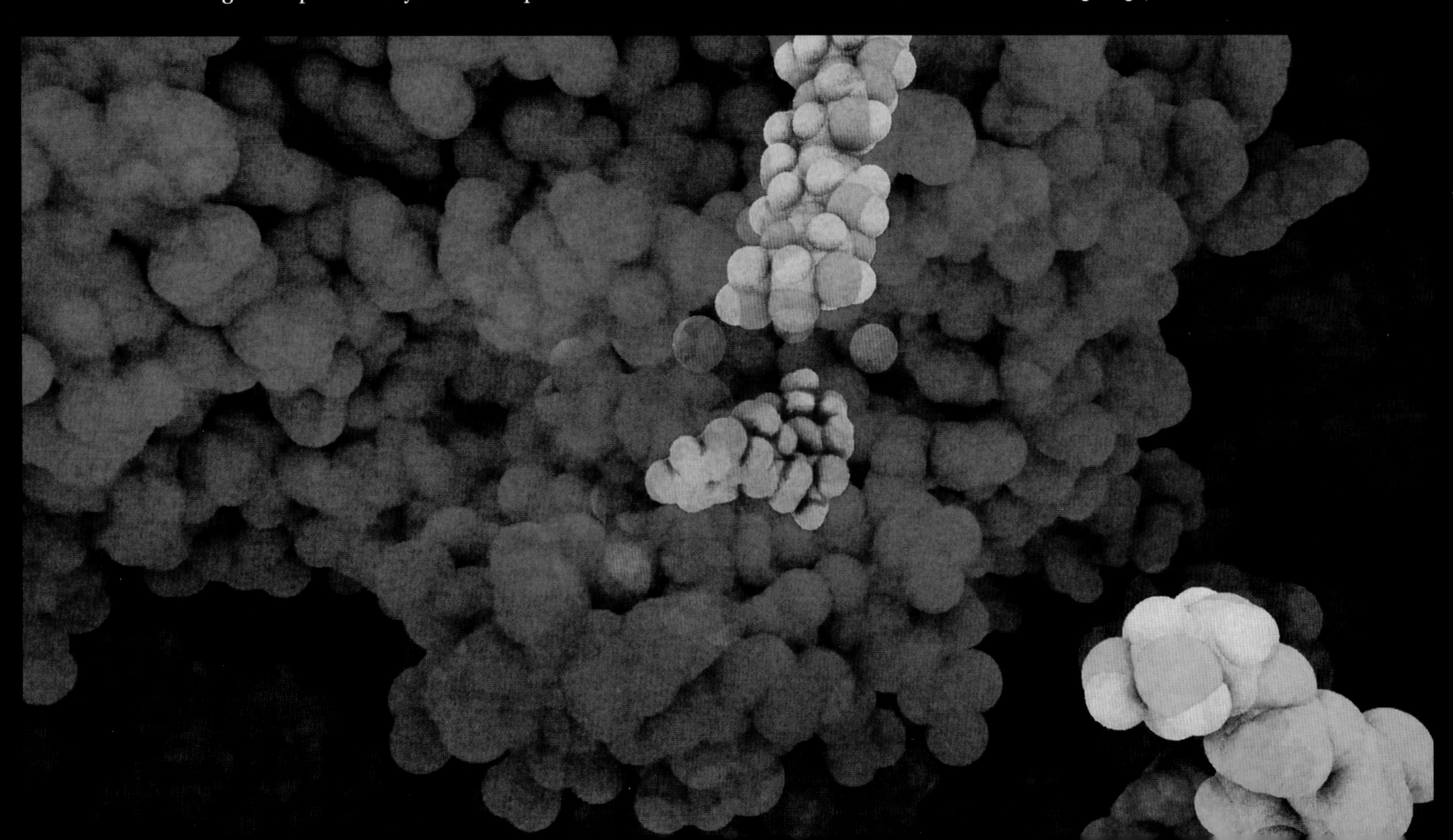

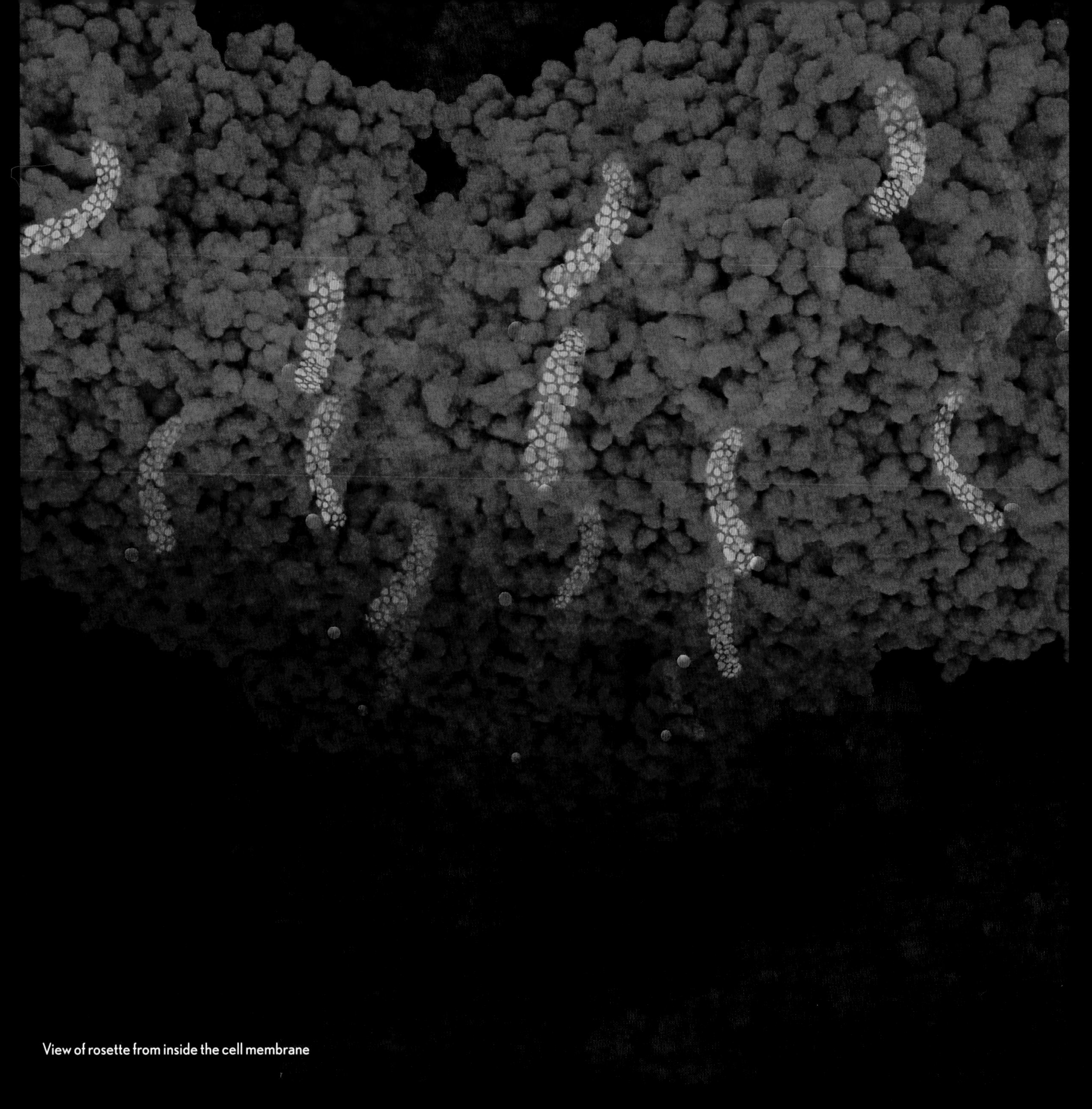

View of rosette from inside the cell membrane

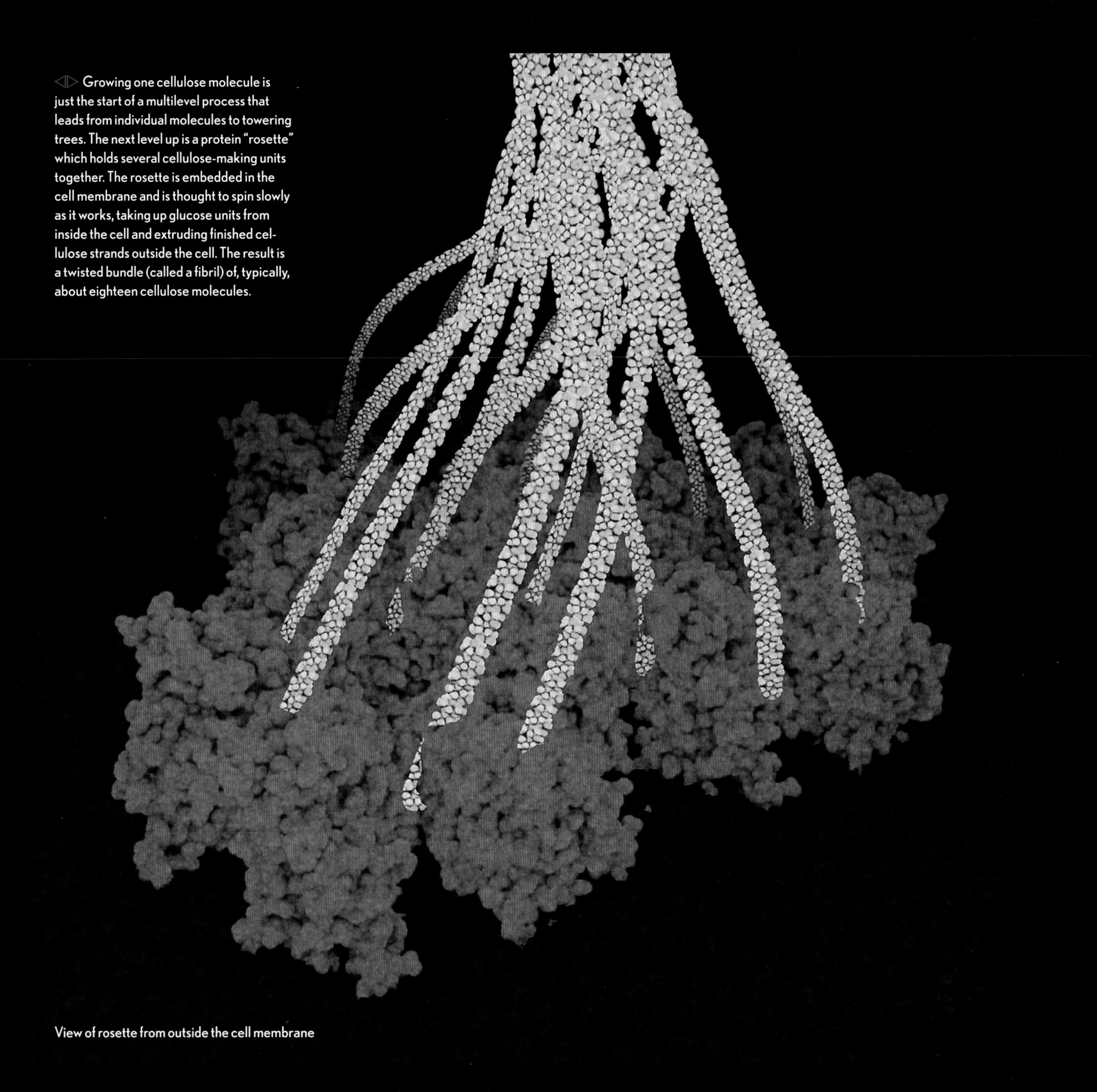

◁▷ Growing one cellulose molecule is just the start of a multilevel process that leads from individual molecules to towering trees. The next level up is a protein "rosette" which holds several cellulose-making units together. The rosette is embedded in the cell membrane and is thought to spin slowly as it works, taking up glucose units from inside the cell and extruding finished cellulose strands outside the cell. The result is a twisted bundle (called a fibril) of, typically, about eighteen cellulose molecules.

View of rosette from outside the cell membrane

△ Multiple rosettes working at the same time create thicker strands of hundreds or thousands of cellulose molecules (still far too thin to see with the naked eye). In what is one of the most remarkable examples of nanomachinery yet discovered, the rosettes in some plant cells are actually running on tiny protein "train tracks" that carry them around and past each other in complicated patterns as they work. The shape of these tracks determines the "weave" of the resulting fiber. This is why plant fibers are able to display such amazingly varied properties. Nothing we can do in making artificial fibers yet comes anywhere close to this level of sophistication.

◁ Only once we've completely lost sight of the individual molecules do we reach the level where you can actually see, with a powerful microscope, a tiny fiber beginning to form.

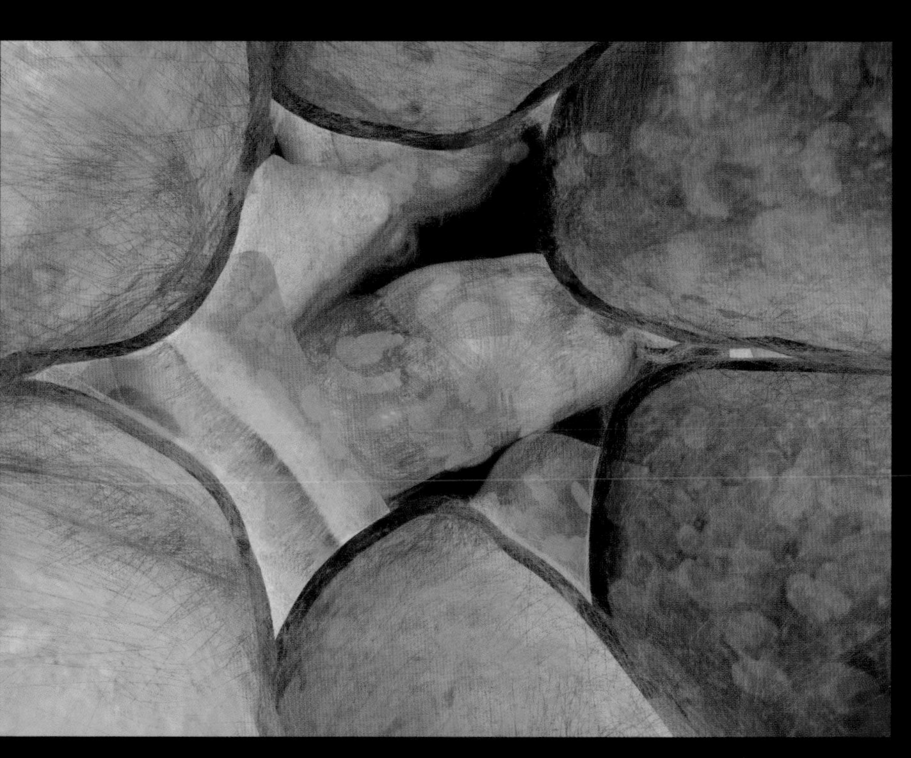

◁ At this level we're looking at a few individual plant cells, each with a tough cellulose wall that it has grown around it. Looking very closely with the naked eye, you can just about make out these cells.

▷ It takes millions of individual cells to make a single blade of grass. On each cell are thousands of rosettes, each making a dozen or more strands of cellulose. Multiply this together and you see that a single blade of grass has on the order of tens of billions of individual cellulose molecules under construction at any given time. Each one of them is having about ten new glucose units added to it per second, so this blade of grass is processing on the order of hundreds of billions of glucose molecules per second. One blade of grass. Just one.

◁ There is an absolutely mind-bogglingly vast amount of activity going on here. Trillions of cells, molecular machines beyond counting, all working away to build empires of cellulose, and people call this *boring*?

Not impressed yet? Here's a thought to chew on. The machinery at work here is stupendously *efficient*. Using just about ten hours of sunlight, the most efficient plants (technically certain species of photosynthetic bacteria) are able to capture enough energy to build a complete new copy of their light-gathering machinery and all necessary support services. Imagine if you had a solar panel so efficient that in just a few days it was able to generate enough energy to power the entire production process to create a new solar panel of the same size. Impossible!

Oh, wait, you already have one of those. It's called your lawn.

◁ It is said that if you tie someone down over a young bamboo shoot, the plant will grow right up through them over the course of a few days. Persistent rumors attribute this method of torture to the Viet Cong during the Vietnam War, but no hard evidence exists that it was ever actually used. (Slightly better evidence exists of entirely different groups of people using a kind of palm tree to do the same thing, but palms are not grasses, and this section is about grass, so I'm going to ignore this fact.)

Tests have determined that it *could* have worked. Bamboo will grow through simulated human flesh. Given that it's a bit hard to find other ways of specifically killing someone though the growth of grass, and that I promised death by each topic in this chapter, that's going to have to be good enough. One thing you can be quite sure of is that a person watching this kind of grass growing toward their belly is not going to be bored. Oh wait, yes, they are going to be bored. That's the point. But they won't be bored while being bored.

Bamboo shoots are not only hypothetically useful for torture, they are also delicious in stir fry. You can buy them cut up in cans, or as whole frozen shoots, which is what you see here.

◁ Almost all soda pop in the United States is sweetened with high fructose corn syrup, so if you're living off the calories in soda, you're living off corn, which means you're living off grass. And if you're drinking one of those fancy cane sugar sodas, you're still eating grass—sugarcane is just another kind of grass. The only sure way to avoid grass is to drink diet soda, which is better for you anyway.

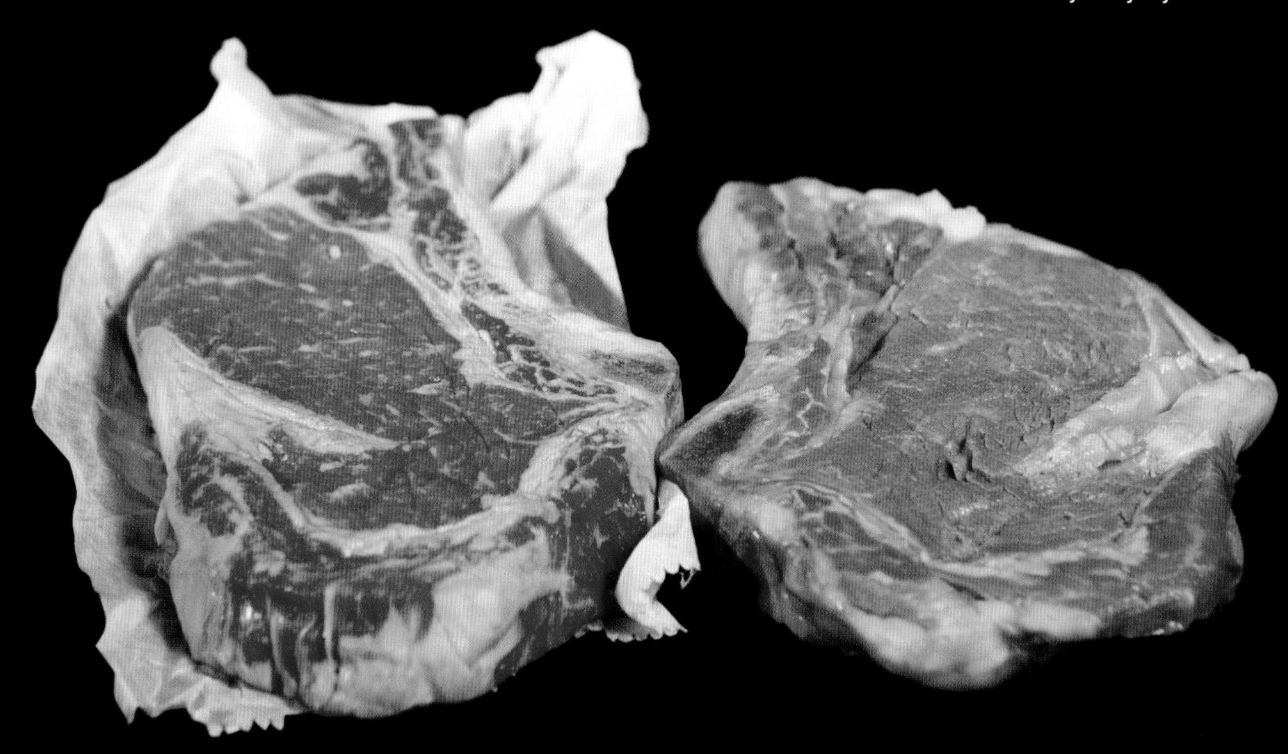

△ Beef cattle are fed grass or corn, which is also grass, so if you're eating beef, you're eating grass, indirectly. We are able to eat corn directly because we eat only the seeds (kernels), which contain starch, sugar, and protein as well as cellulose. We cannot eat lawn grass or the rest of the corn plant, the way cows can, because our stomachs are not able to carry out the chemical reactions necessary to convert the cellulose that makes up the leaves and stems into usable food energy. Cows can't do it either, but they have symbiotic bacteria in their rumens that do it for them.

Watching Water Boil

THERE ARE AGES named after metals, alloys, and compounds (Iron Age, Bronze Age, and Stone Age), but water is the only substance that has *two* ages named after it (Ice Age and Steam Age, both of which are phases of water).

OK, maybe the relatively brief Steam Age isn't quite the same *kind* of age, but it certainly was an important part of the history of mankind. It resulted in some of the most beautiful machines of all time, inspired steampunk hipsters two hundred years later, and did it all based on boiling water.

Liquid water turning into water vapor (water in gas form, called steam) isn't generally considered a chemical reaction, but it's definitely a chemical process. Hydrogen bonds in liquid water are broken as individual molecules are cast out of the group to seek their fortune in the air above (called evaporation). Surprisingly, exactly how this happens at the molecular level is not well understood. The best computer simulations give conflicting answers to basic questions about how precisely it happens.

But we can say that, at any given temperature, there will always be a certain amount of water vapor above a pot of water. This "vapor pressure" is low when the water is cool, and rises as you heat it. What's special about the boiling point is simply that it is the temperature at which the vapor pressure exactly reaches the pressure of the air surrounding the water.

When that happens, pockets of water vapor can start to form under the water's surface and have enough force to push the water away, forming a bubble of steam, which then rises to the surface. That's called boiling.

▷ The fluffy white stuff we think of as "steam" isn't really steam in the technical sense. It's tiny droplets of water condensed back into liquid form from the true steam. (In other words, if you can see it, it's not actually steam anymore.) Each of the droplets, though too small to see with the naked eye, contains about twice as many water molecules as there are people on Earth.

◁ Gaseous water (steam) at 212°F (100°C). True steam is an invisible gas. It forms first as bubbles under the surface of the water, then rises up and is released at the surface.

◁ Liquid water at very slightly below 212°F (100°C).

◁ Liquid water at 212°F (100°C). Bubbles rise from the bottom because that's where the heat is applied, making the water at the bottom slightly hotter than at the top. Because hot water rises, the temperature difference can never be more than a fraction of a degree.

This is a portable immersion heater used by lonely people to heat one cup of tea at a time. It's really quite sad. But it does help us see how individual bubbles of steam form. Bubbles like this can only form if the vapor pressure of the water at its current temperature is at least as high as the pressure of the air pressing down on the water. If the vapor pressure is lower, any such bubbles that tried to form would immediately get squeezed down to nothing. Once the temperature gets high enough, the pressure crosses that threshold and bubbles start forming. Once they reach a certain size, they leave the heater and race for the surface of the water, announcing by their popping sound that it's time to start drinking alone again.

Just because water is hot enough to form bubbles doesn't mean it will actually happen. Bubbles won't start forming unless they have some kind of seed to form around. All it takes is a tiny bit of dust, or a rough spot in the pot. But when clean water is boiled in a clean, smooth container, it can sometimes get quite a bit *above* the boiling point before any bubbles form. This kind of superheated water can be quite dangerous, because eventually something will trigger it and a *big* bubble will form very suddenly. This is called "bumping" and you don't want it to happen if your hand is anywhere near the pot. Because chemists are often boiling high-purity water in extra-clean glassware, bumping is a real problem. The solution is to put a couple of boiling stones or granules (often made of silica or Teflon) in the pot. Their only job is to have a rough surface that encourages bubbles to form as soon as the temperature reaches the boiling point.

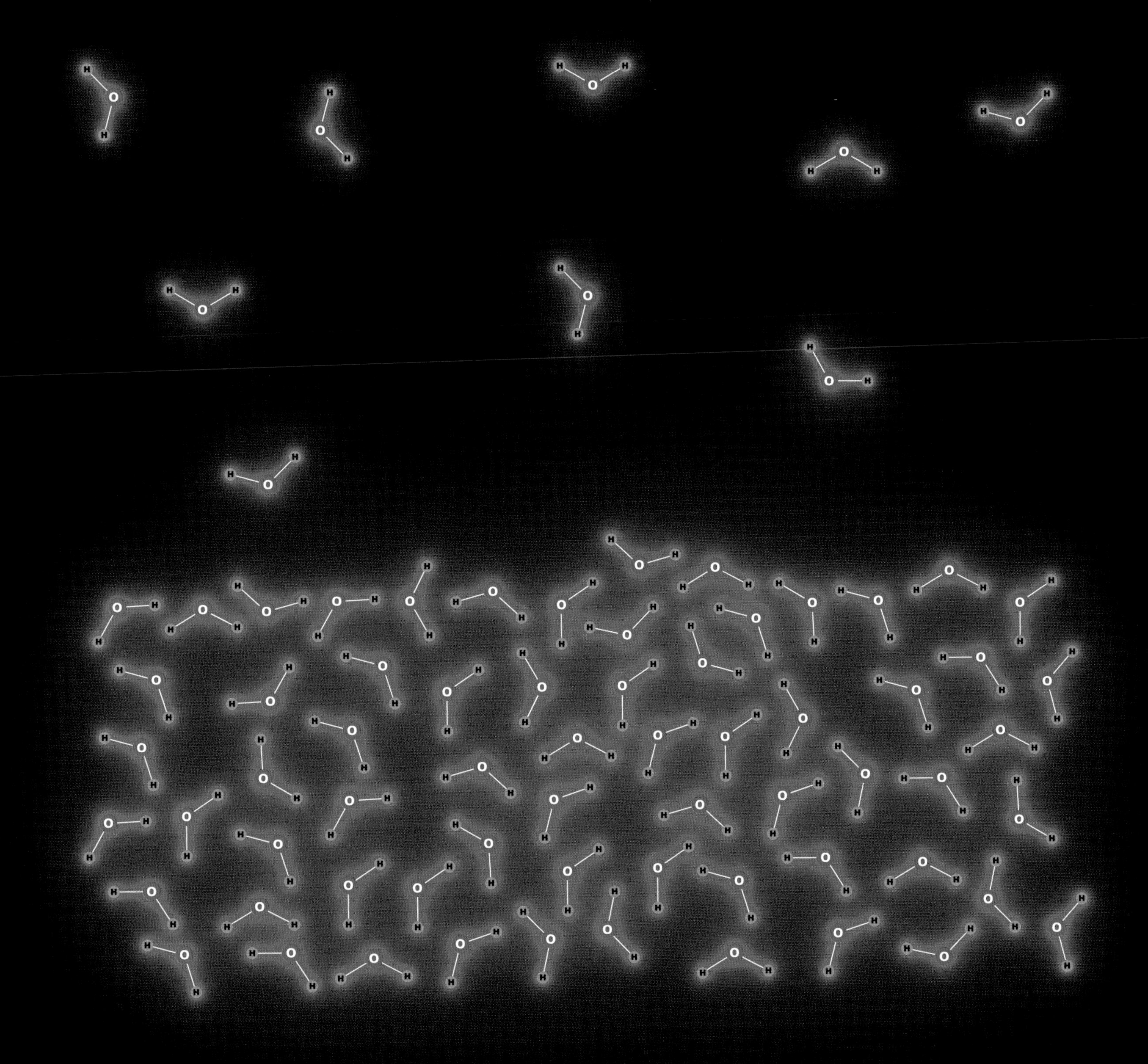

△ Water is a simple molecule, but a very complicated liquid. The hydrogen bonds that form between neighboring water molecules (see page 203) give liquid water fleeting connected structures on the nanoscale, with various groupings of molecules surviving for a few nanoseconds at a time. Near the surface of a body of water the molecules line up in an even more organized way, creating the "surface tension" that allows specialized bugs to walk on water. But none of this can prevent the inevitable random collisions that occasionally propel a water molecule away from its friends and out into the cruel world of nitrogen and oxygen gas we call the air. That's called evaporation. At higher temperatures where the molecules are moving faster and faster, this happens more and more often. When it gets out of hand, it's called boiling.

▷ One problem with theoretical simulations of water boiling is that all calculations indicate that it shouldn't happen until the temperature is much higher than what is observed. The bonds holding water molecules together are strong enough that they should be able to hold it in liquid form much longer. An intriguing theory, unproven at this point, is that molecular-scale waves on the surface concentrate thermal energy and conspire to throw water molecules free of the surface at the crests of waves. If you ever think the world is pretty well understood, remember that we actually have no idea why you don't have to wait a whole lot longer for a pot of water to boil.

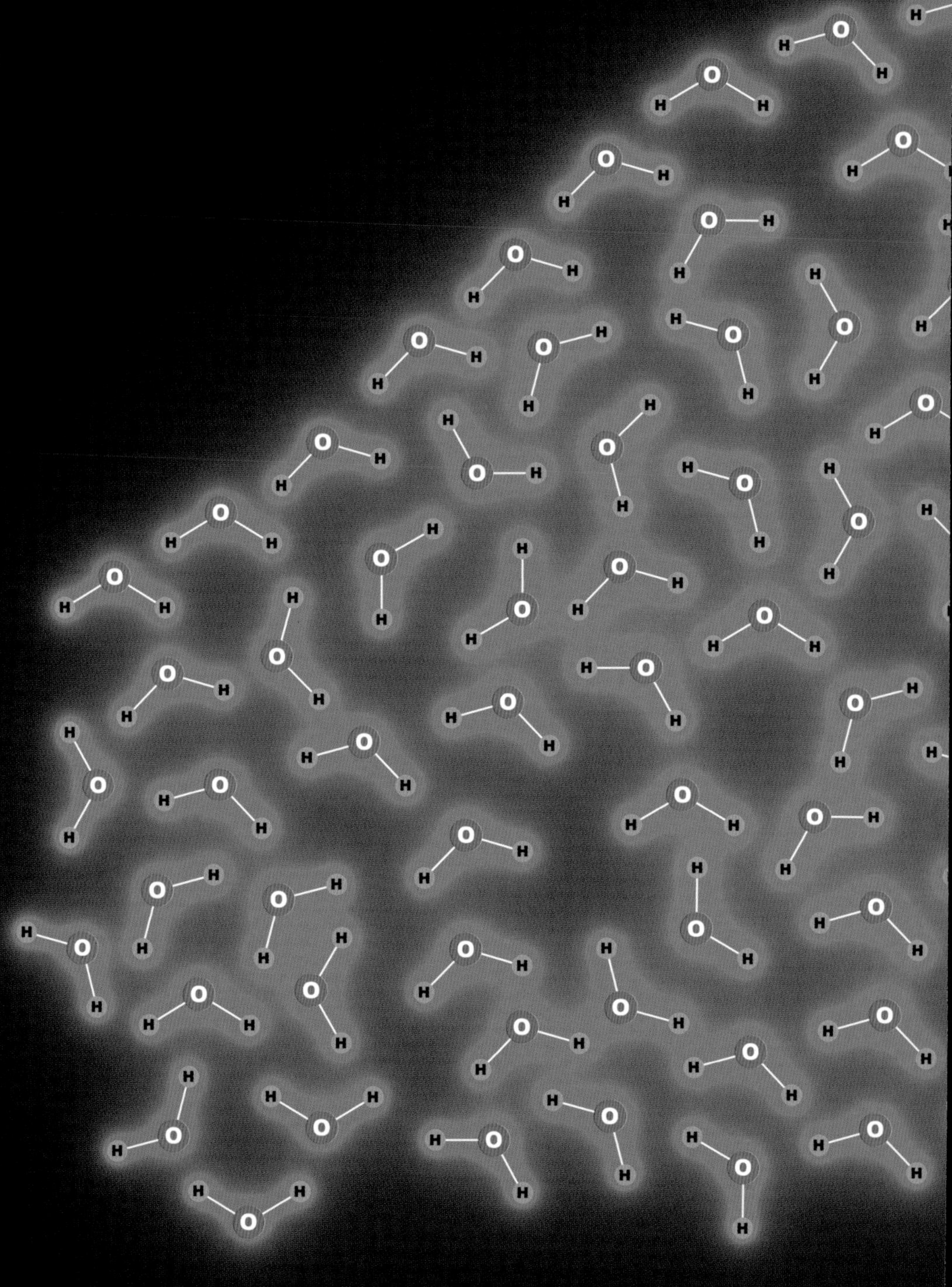

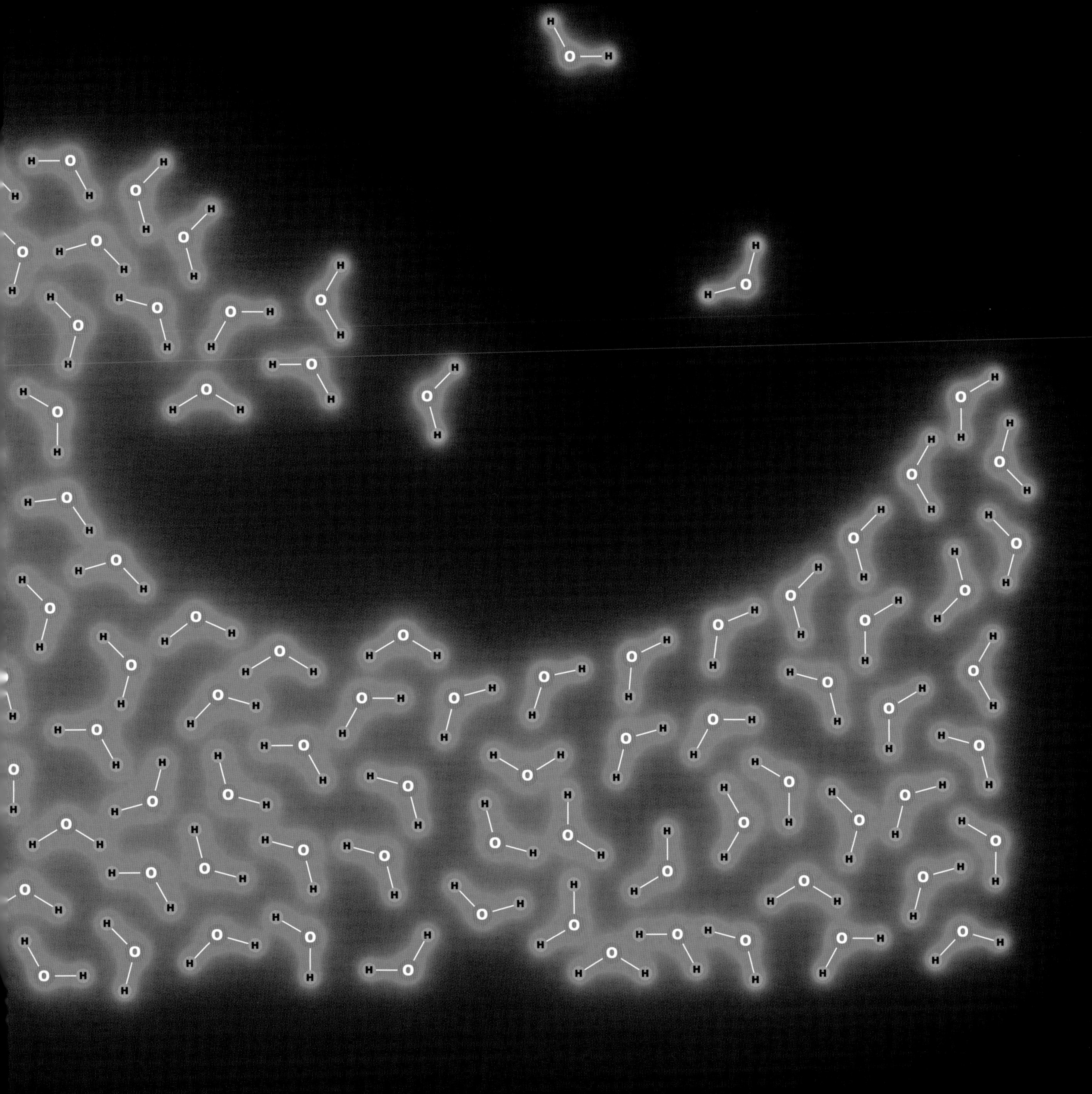

◁ You'll often see cooking instructions that say to boil things longer at higher altitudes. That's because at high altitudes the pressure of the atmosphere is lower, so the vapor pressure of water equals the air pressure at a lower temperature. (Which is another way of saying that the boiling point of water is lower at higher altitudes.) Since things cook more slowly at lower temperatures, you need to boil them longer. The effect is significant. In Denver at an altitude of 1 mile (1.6 kilometers), water boils 5°C (9°F) cooler than in New York City, which is close to sea level.

You can take this to extremes in a vacuum jar, where water can be made to boil at room temperature just by lowering the pressure enough. This is why you don't want to be caught in outer space without a space suit. The water in your eyes and skin would quickly start to boil, which is known to hurt quite a lot.

△ Does a hotter pan boil water more slowly? Sometimes, yes. A droplet of water can skitter about on the surface of a very hot frying pan for quite a while, protected from the heat by a thin layer of steam that keeps it levitated. At a slightly lower temperature, there isn't enough steam to keep the drop separated, and it evaporates very quickly as soon as it touches the metal. The skittering around is called the Leidenfrost effect.

▷ My hand isn't in boiling water, it's immersed in incredibly cold boiling liquid nitrogen. My hand is protected from freezing by the same Leidenfrost effect that prevents a drop of water on a hot frying pan from touching the pan. In this case, the liquid nitrogen plays the part of water, and my hand is the hot frying pan. To a liquid that boils at −195.79 °C (−320 °F), my hand is like a red-hot poker.

This footage from the launch of Apollo 11 was shot at 500 frames per second by a 16mm film camera mounted *very* close to the action. It shows the water deluge system on launch pad 39A dumping, at its peak, about 7,000 pounds of water per second onto the launch pad to protect it from the heat of the roughly 10,000 pounds per second of fuel being burned by the five F-1 rocket engines, all of which are pointing straight at the pad. Most of this water is almost instantly boiling into steam, which you can't see because steam is an invisible gas. It's only after the rocket leaves that you can see how much water really is entering the area.

▷ The rocket has just started lifting off, and the engines are about to clear the top of the launch pad.

1. Tremendous heat! You're looking at the full force of the most powerful machine ever built, the Saturn V moon rocket.

2. There are thousands of pounds per second of water being dumped here. You can't see any of it because it's all being instantly turned into steam.

3. Now that the rocket is farther away, you can start to see some water, while the protective coating on the hold-down arm shields burns away.

4. The arms are charred black (as intended, to protect the metal underneath), and the water is starting to show.

5. Moments later the rocket is gone, and the water is in still coming, cooling down the expensive structure after its trial by fire.

6. Toward the end of the 30-second sequence, you can really see just how much water you were not seeing before.

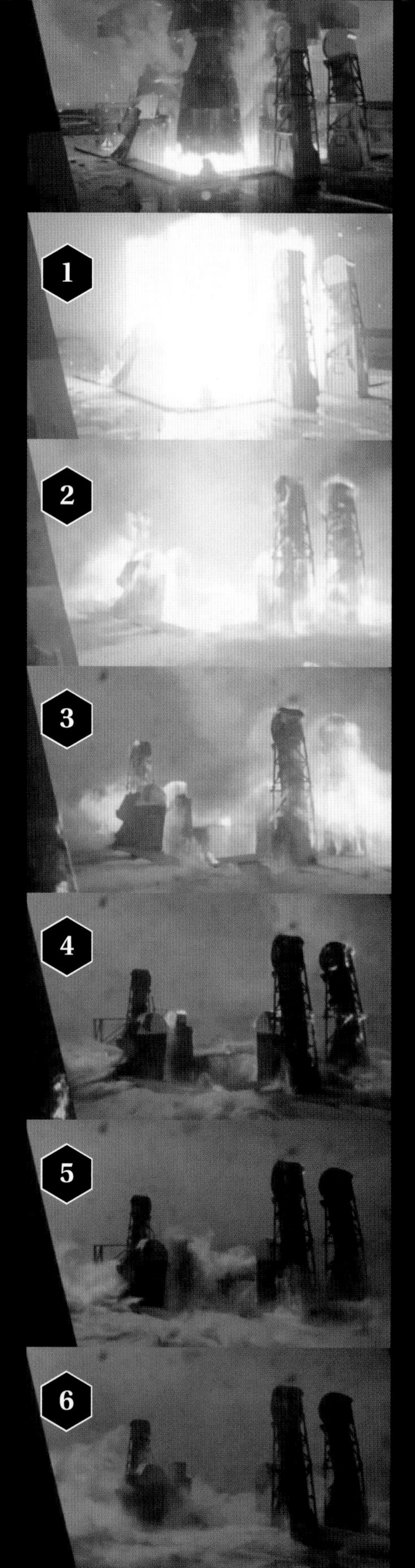

△ Great thundering beasts like this one are remnants of an age when power was raw and hot, when machines puffed and belched like all real machines should. The Steam Age is unmatched for pure mechanicalness, and unlike today's machines, you can see all the important parts and understand how they worked just as plain as day. The only reason we don't use these beautiful machines today is that they are really not very good engines: This 100-ton, bus-sized monster is less powerful than the average compact car of today.

△ Boilers blowing up is no laughing matter. Thanks to security cameras this has been recorded many times, and the damage done can be extreme, with multiple fatalities and entire buildings destroyed. The force of steam pressure must not be underestimated.

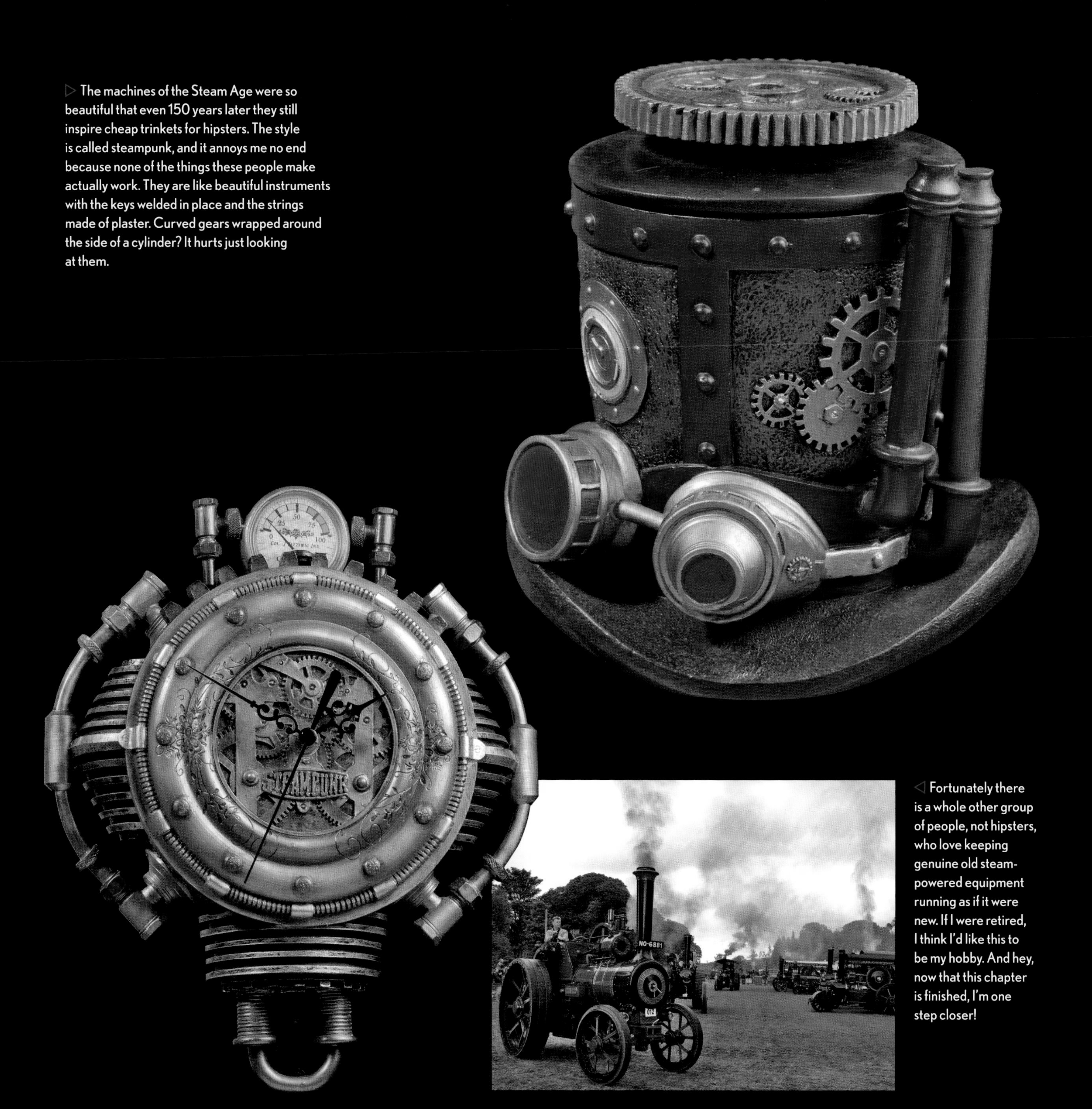

▷ The machines of the Steam Age were so beautiful that even 150 years later they still inspire cheap trinkets for hipsters. The style is called steampunk, and it annoys me no end because none of the things these people make actually work. They are like beautiful instruments with the keys welded in place and the strings made of plaster. Curved gears wrapped around the side of a cylinder? It hurts just looking at them.

◁ Fortunately there is a whole other group of people, not hipsters, who love keeping genuine old steam-powered equipment running as if it were new. If I were retired, I think I'd like this to be my hobby. And hey, now that this chapter is finished, I'm one step closer!

6

The Need for Speed

REACTIONS ARE chemical substances changing one into another. That necessarily happens *over time*. Reactions take place across a marvelously wide range of times, from the long sleep of geological time, to the blink of an eye, to times that make the blink of an eye feel as slow as mountains eroding.

The range of speeds is truly astonishing. The reactions in this chapter cover more than twenty-five orders of magnitude, by which is meant ten times faster, then ten times faster again, then another ten times faster, repeated twenty-five times. In total, the fastest reaction described in this chapter is 10,000,000,000,000,000,000,000,000 times faster than the slowest one.

Other than in the deep cold of space, or on planets distant from any star, molecules are always moving very fast. Even when they are participating in reactions that take centuries to progress, the individual molecules are still always moving very fast. So the first question to answer is, how is it even *possible* for a chemical reaction to be slow?

I'll start slow in explaining this. Literally. There are slower things going on out in space, and probably deep inside the earth as well, but I like things I can see and touch, so I'm starting this chapter with the slowest visible-on-Earth chemical reaction I can think of: weathering.

Weathering

WEATHERING AND EROSION are technical terms for two different things. Erosion is the mechanical battering down of rock into sand, or the washing away of sand or soil. It is caused by the force of rain, or ice, or wind, or water flowing. The Grand Canyon, for example, was eroded out of solid rock by rain, wind, and the Colorado River. Erosion is powerful over time, but not what this book is about.

Weathering, on the other hand, is a chemical process within the scope of our discussion. Much weathering occurs through the combined efforts of water and CO_2, so let's look at how these chemicals work with each other.

△ The slowest reactions we can see happening on earth naturally involve the oldest things we have on the planet: mountains. And the hills that used to be mountains, and the plains that used to be hills, and the canyons that used to be plains. I like this particular mountain because my children played on it before they, unlike the mountain, grew old and changed. In the time that I and my children have lived and aged, the mountain has changed little. While the glaciers are nearly gone now, the rock still stands. But mountains too grow old and change. It just takes longer.

△ The Swiss Alps my children played on, and I before them, are young ranges still sharp and steep. With the passage of time the mountains, like us, become more rounded, softer in their views and more settled in their ways. A billion years or more have passed since the ancient Blue Ridge Mountains, some of the oldest in the world, had the sharp tongues of their youth.

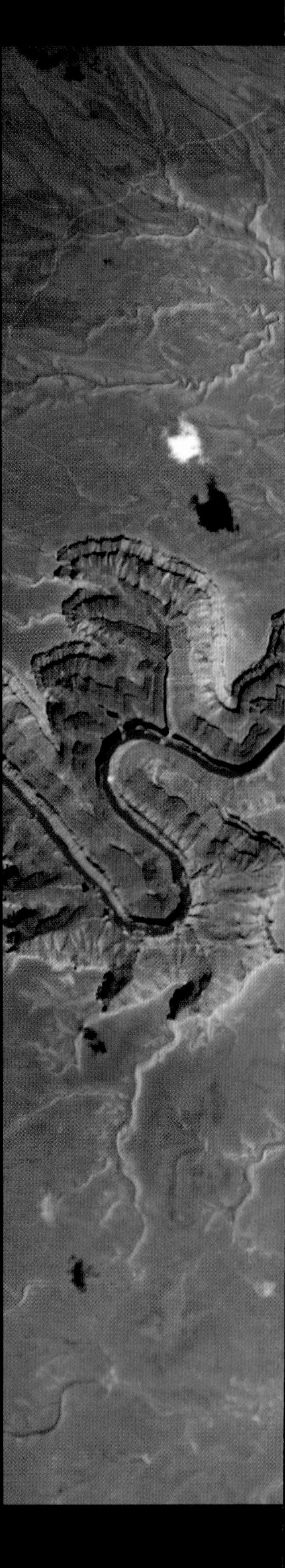

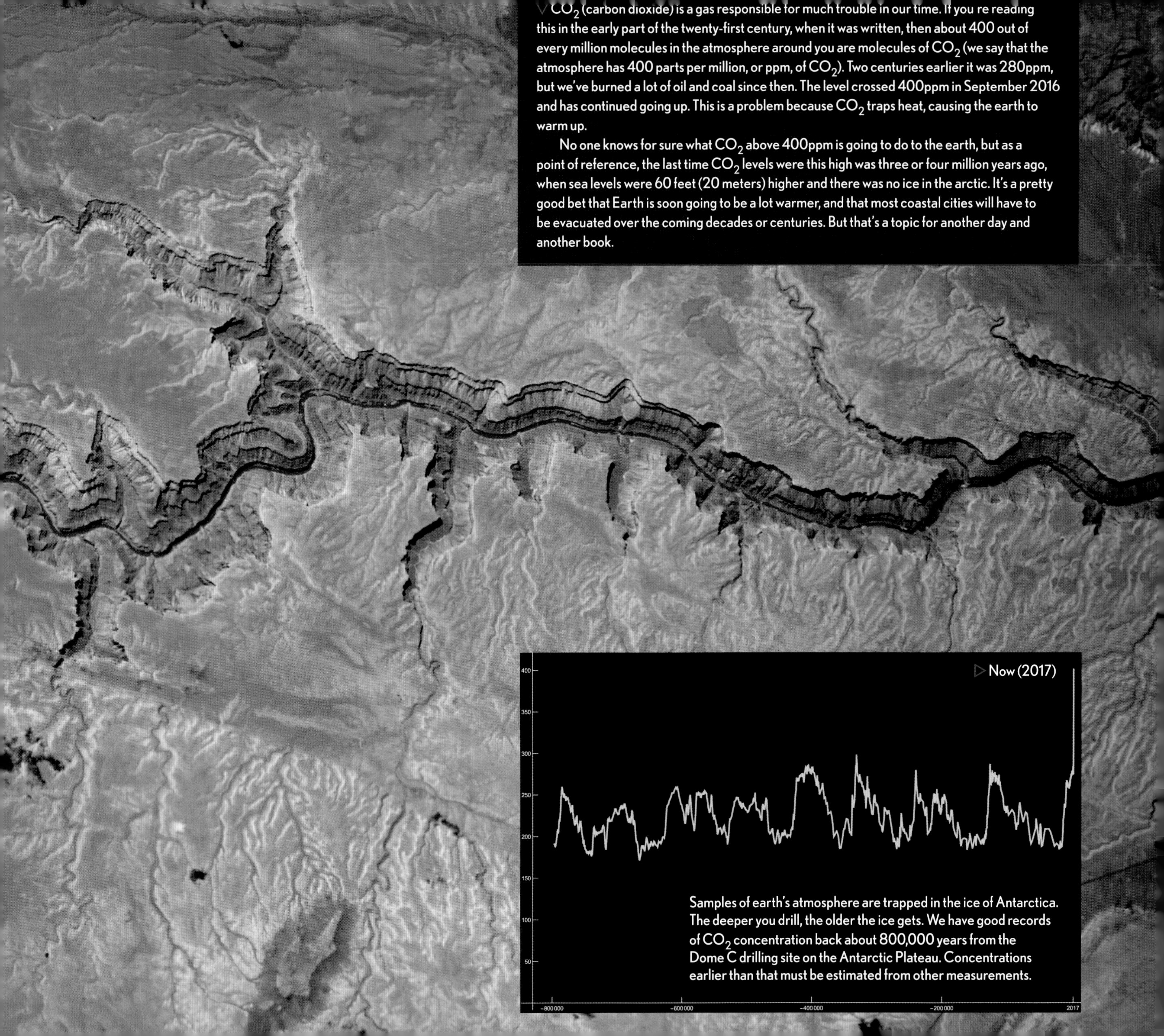

CO_2 (carbon dioxide) is a gas responsible for much trouble in our time. If you're reading this in the early part of the twenty-first century, when it was written, then about 400 out of every million molecules in the atmosphere around you are molecules of CO_2 (we say that the atmosphere has 400 parts per million, or ppm, of CO_2). Two centuries earlier it was 280ppm, but we've burned a lot of oil and coal since then. The level crossed 400ppm in September 2016 and has continued going up. This is a problem because CO_2 traps heat, causing the earth to warm up.

No one knows for sure what CO_2 above 400ppm is going to do to the earth, but as a point of reference, the last time CO_2 levels were this high was three or four million years ago, when sea levels were 60 feet (20 meters) higher and there was no ice in the arctic. It's a pretty good bet that Earth is soon going to be a lot warmer, and that most coastal cities will have to be evacuated over the coming decades or centuries. But that's a topic for another day and another book.

Samples of earth's atmosphere are trapped in the ice of Antarctica. The deeper you drill, the older the ice gets. We have good records of CO_2 concentration back about 800,000 years from the Dome C drilling site on the Antarctic Plateau. Concentrations earlier than that must be estimated from other measurements.

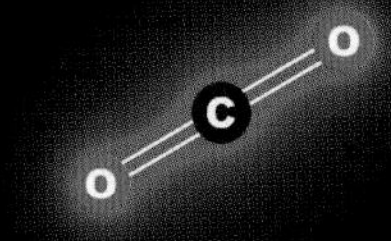

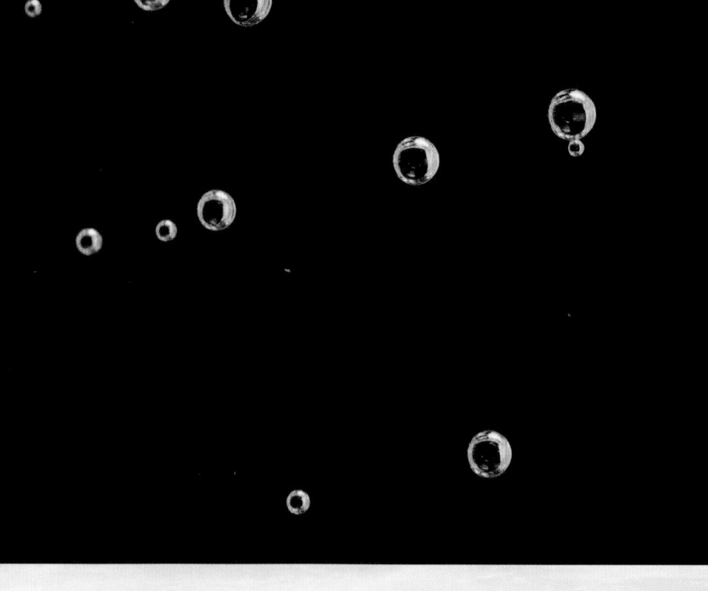

◁ The bubbles in this soda pop are made of escaping CO_2. When CO_2 is forced into water under pressure, some of the CO_2 molecules combine with a water molecule, then immediately throw off a hydrogen atom. The result is called carbonic acid: dissolved HCO_3^- ions (bicarbonate ions) and H^+ ions (hydrogen ions). The presence of H^+ ions is what defines this substance as an acid. It also contributes to the tangy taste of soda.

This reaction is easily reversible: H^+ and HCO_3^- can recombine, split off a water molecule, and release a molecule of CO_2. When enough CO_2 molecules have accumulated, they form a bubble and escape.

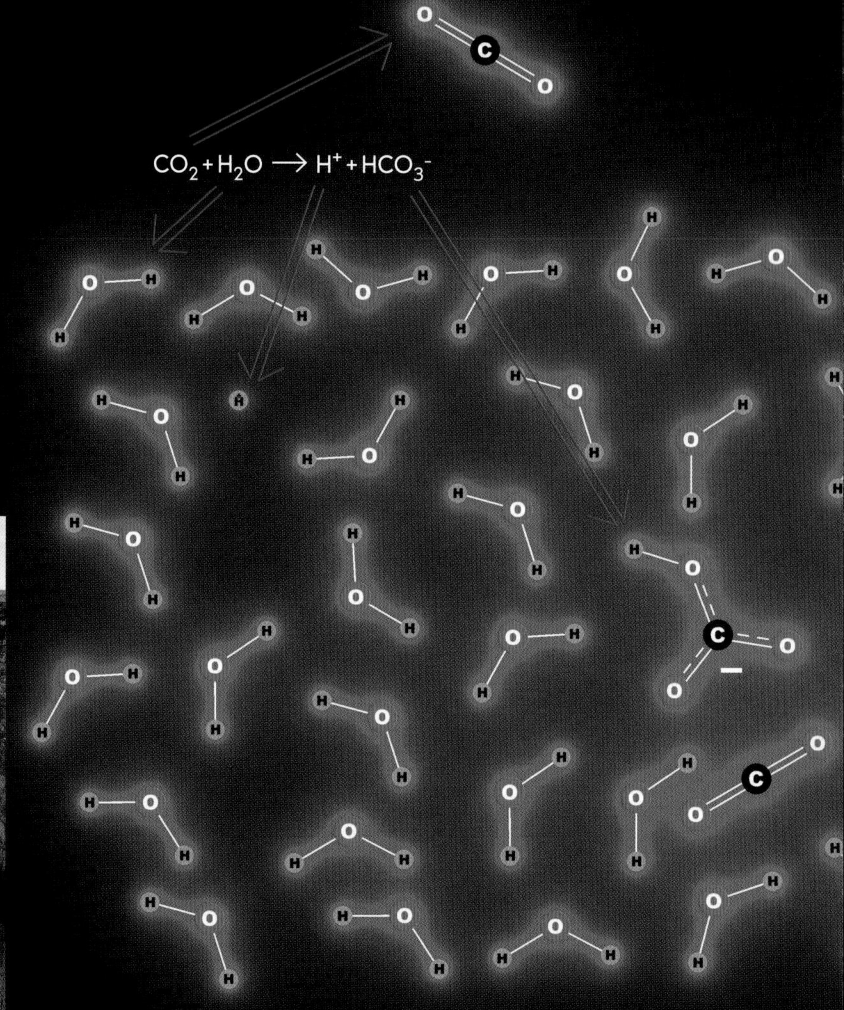

△ Limestone is made of calcium carbonate ($CaCO_3$), which exists as crystals of calcium ions (Ca^{2+}) alternating with carbonate ions CO_3^{2-}. Carbonic acid in rainwater, though weak, is strong enough to react, very slowly, with limestone, converting the carbonate ions (CO_3^{2-}) into bicarbonate ions (HCO_3^-). After this reaction, there are not enough free H^+ ions left in the water for all the bicarbonate ions to convert back into CO_2 gas and escape. The carbon and the calcium have been trapped in the water and will eventually flow down to the ocean. In this way whole mountain ranges are dissolved and washed away.

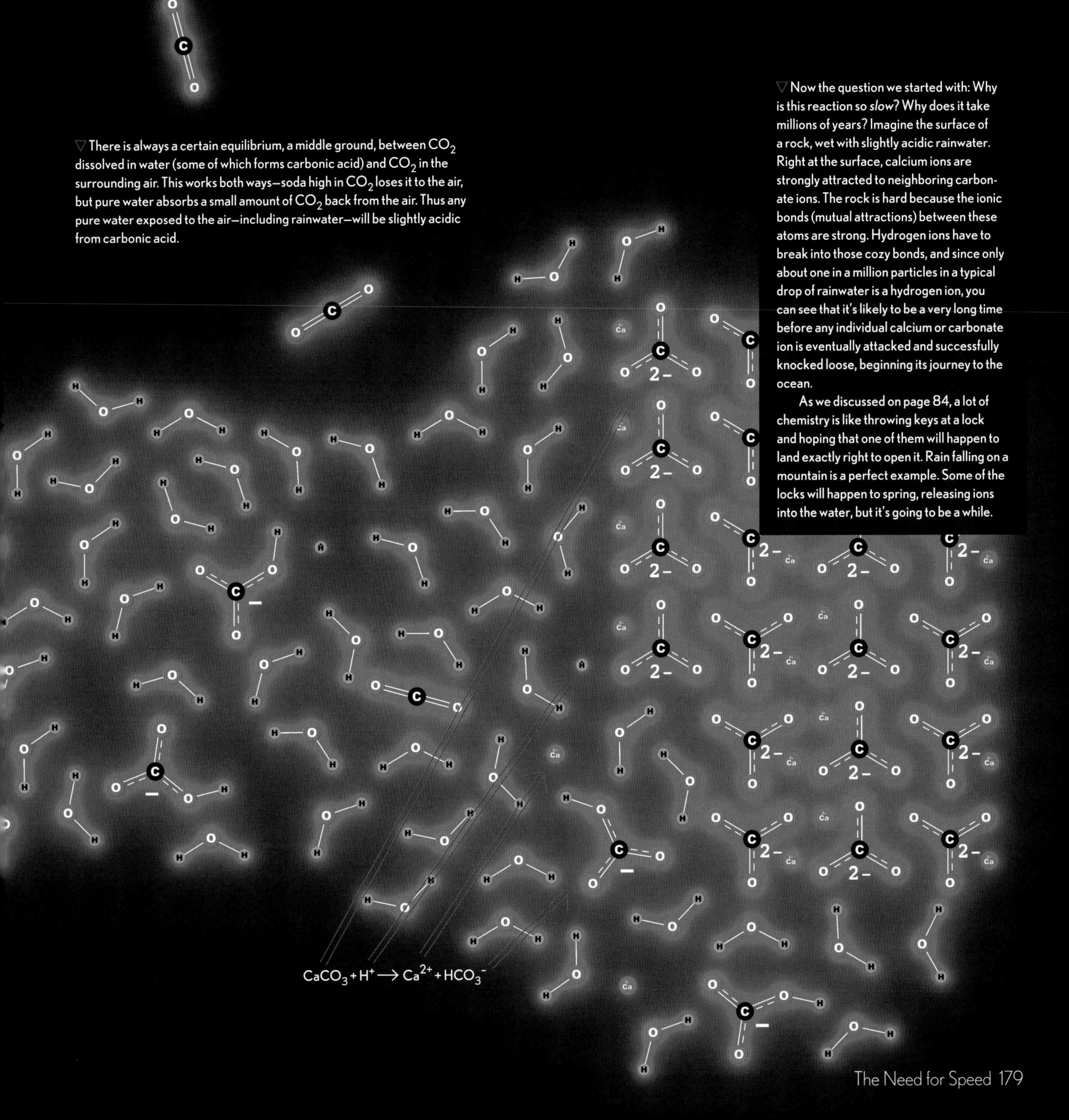

▽ There is always a certain equilibrium, a middle ground, between CO_2 dissolved in water (some of which forms carbonic acid) and CO_2 in the surrounding air. This works both ways—soda high in CO_2 loses it to the air, but pure water absorbs a small amount of CO_2 back from the air. Thus any pure water exposed to the air—including rainwater—will be slightly acidic from carbonic acid.

▽ Now the question we started with: Why is this reaction so *slow*? Why does it take millions of years? Imagine the surface of a rock, wet with slightly acidic rainwater. Right at the surface, calcium ions are strongly attracted to neighboring carbonate ions. The rock is hard because the ionic bonds (mutual attractions) between these atoms are strong. Hydrogen ions have to break into those cozy bonds, and since only about one in a million particles in a typical drop of rainwater is a hydrogen ion, you can see that it's likely to be a very long time before any individual calcium or carbonate ion is eventually attacked and successfully knocked loose, beginning its journey to the ocean.

As we discussed on page 84, a lot of chemistry is like throwing keys at a lock and hoping that one of them will happen to land exactly right to open it. Rain falling on a mountain is a perfect example. Some of the locks will happen to spring, releasing ions into the water, but it's going to be a while.

▽ All limestone exposed to rain is always eroding due to the carbonic acid formed when rainwater absorbs CO_2 from the air. Thus it has always been, and for this we are not to blame. But in our cities the rate of erosion is greatly accelerated by the much stronger sulfuric and nitric acids created when rainwater absorbs sulfur and nitrogen oxides from the air. This is mostly our fault. These oxides come in large part from human factories, power plants, and cars. The acid rain they create erodes sculptures, buildings, gravestones, and everything else made of limestone at a far higher rate than would nature alone.

Once washed down to the ocean, bicarbonate ions are taken up by sea creatures who use them to build shells made of . . . drum roll . . . calcium carbonate! Seashells are made of the same substance as the limestone from which the bicarbonate ions washed down.

And where did that limestone rock come from in the first place? It came from the shells of sea creatures laid down in the ocean floor over millions of years, and then thrust up into mountain ranges through great upheavals as the continents of Earth shifted and rearranged themselves.

Seashells are made of mountains, and mountains are made of seashells.

This cycle, known as the "slow carbon cycle" is the one of the grandest, and slowest, in all the world. It takes about two hundred million years to make one complete revolution from mountain to ocean and back to mountain. Millions of cubic miles of limestone, representing most of the world's carbon, have gone through this cycle over and over again.

An individual carbon atom may spend a few hundred years participating in the exciting, dynamic world of the air, and in the ocean, and in the colorful patterned shells of exotic sea creatures. Then it's back to sleep for a hundred million years or more, deep under the ocean, then deep under land, crushed, cracked, bent, and finally thrust high up above the water to start again another brief dance in the light above.

The chemical weathering of limestone over millions of years is one of the most important forms of weathering, not just because it transforms the landscape, but because it is a key part of the interconnected cycles that control the concentration of CO_2 in the atmosphere. In the long run, it is the rocks and the oceans that determine the fate of all carbon. But, as they say, in the long run we are all dead. In the short run, over times of less than tens of thousands of years, it's up to us to decide how much CO_2 there will be in our air.

▽ Taking things to the extreme, this small limestone sculpture eroded to this deplorable state in just a few minutes by dipping it in concentrated acid. Obviously that's not likely to happen outside a mad scientist's lair, but I did it as an illustration of what's known as "accelerated life testing." (OK, technically, I did it because I wanted to see what would happen. I made up the accelerated life testing bit as an excuse so I could put the statue in my book and make it tax-deductible.)

What is accelerated life testing? Suppose you are a manufacturer of chairs and you want to know whether your chairs are strong enough to last twenty years in customers' homes. But you don't want to wait twenty years to find out. So you build a machine that simulates someone sitting down and getting back out of the chair three times a day, every day, for twenty years. Except your machine does it three times a minute, and simulates twenty years of sitting in just fifteen days.

Similarly, if you want to predict the result of years or even centuries of acid exposure on limestone, but you want to know the answer before your great-great-grandchildren are born, you can do experiments with much stronger acids and extrapolate the results. Care, of course, must be taken to make sure you understand the effect of concentration on rate. If you make the acid half as strong, does the erosion go half as fast, or one-quarter as fast? Is there a lower limit below which the acid has no effect? All these questions have answers, and knowing them, you can make accurate long-term projections based on this kind of testing.

Moving up a little in the speed scale, ordinary rusting is also a form of chemical weathering. But it happens so fast that you rarely see it occurring in nature. If there were any iron mountains (made of actual iron) they would have rusted away long ago, both because rusting is fast, and because there is no natural process that converts iron oxide (rust) back into iron metal.

Rusting is an example of oxidation. Iron reacts with oxygen to form iron oxide. When oxidation happens rapidly, and releases a lot of heat, we have a different name for it—burning. When wood or other organic substances are burning, carbon is turning into carbon oxide (more specifically carbon dioxide, because its formula is CO_2, and "di" stands for 2). It's a fun fact that iron can actually rust fast enough to qualify as burning.

△ There may not be much *natural* rusting going on, but there is a huge amount of *unnatural* rusting happening. You see it afflicting all our fleeting creations—bridges, cars, railings, and the countless other things we vainly make of iron and optimistically place in harm's way on our roads and in our back yards. Out there, all these things inevitably rust away to nothing. "The graveyards of the rusted automobile," in the immortal words of Arlo Guthrie, are such a fixture of our landscape that it's easy to forget just how short-lived they are.

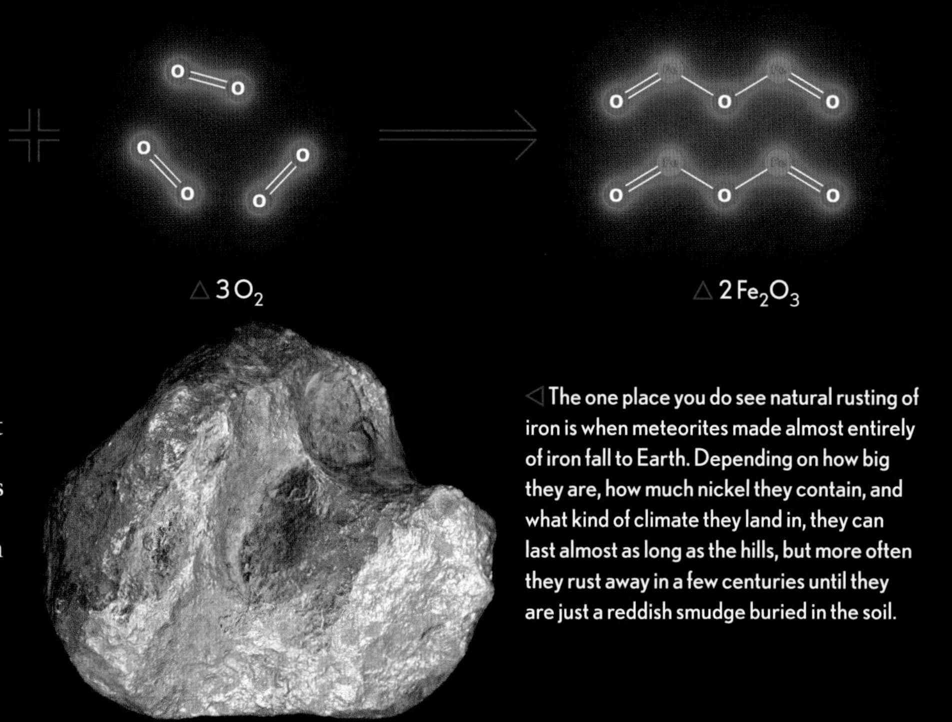

△ 4 Fe

△ 3 O_2

△ 2 Fe_2O_3

◁ The one place you do see natural rusting of iron is when meteorites made almost entirely of iron fall to Earth. Depending on how big they are, how much nickel they contain, and what kind of climate they land in, they can last almost as long as the hills, but more often they rust away in a few centuries until they are just a reddish smudge buried in the soil.

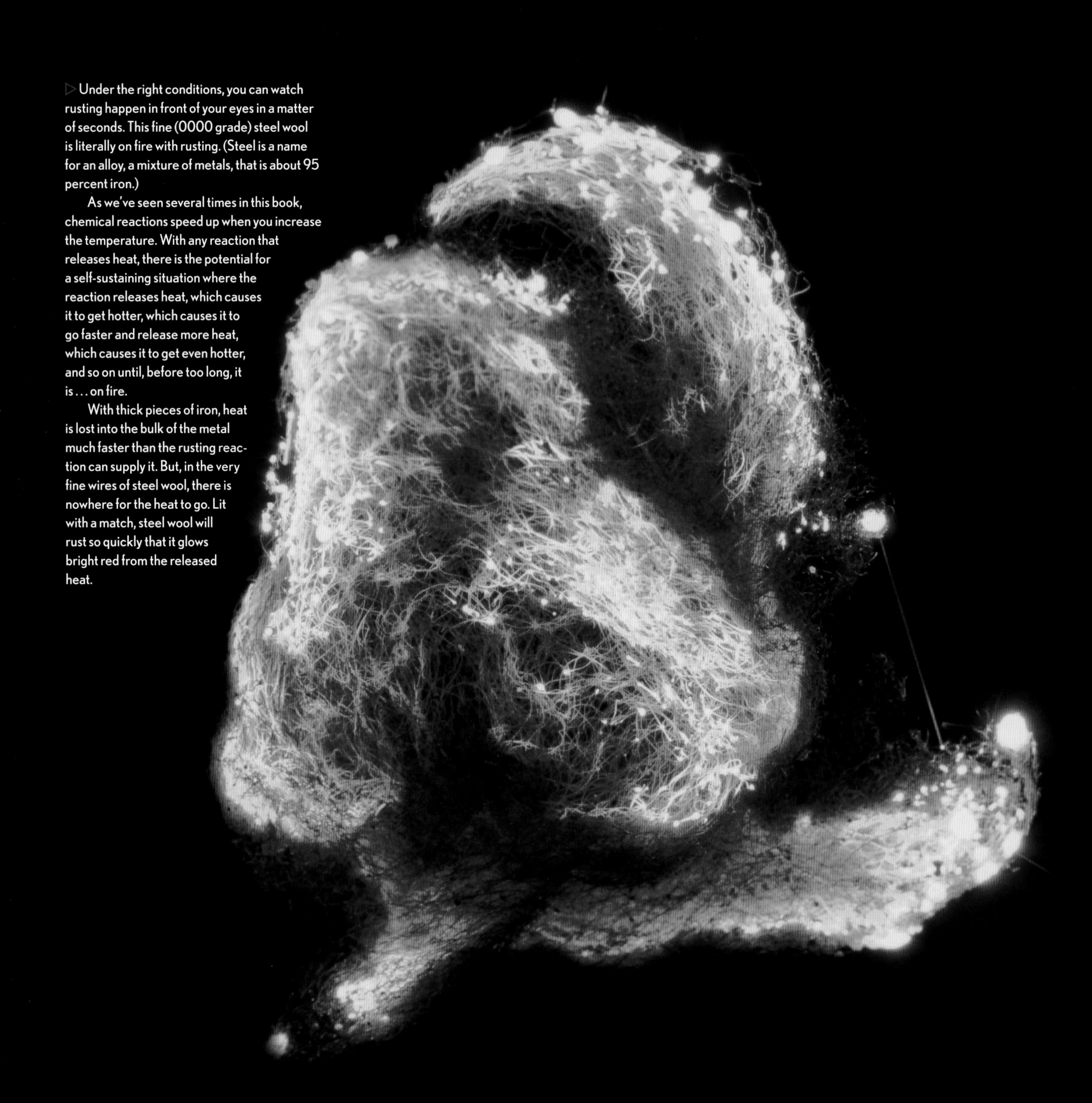

▷ Under the right conditions, you can watch rusting happen in front of your eyes in a matter of seconds. This fine (0000 grade) steel wool is literally on fire with rusting. (Steel is a name for an alloy, a mixture of metals, that is about 95 percent iron.)

As we've seen several times in this book, chemical reactions speed up when you increase the temperature. With any reaction that releases heat, there is the potential for a self-sustaining situation where the reaction releases heat, which causes it to get hotter, which causes it to go faster and release more heat, which causes it to get even hotter, and so on until, before too long, it is . . . on fire.

With thick pieces of iron, heat is lost into the bulk of the metal much faster than the rusting reaction can supply it. But, in the very fine wires of steel wool, there is nowhere for the heat to go. Lit with a match, steel wool will rust so quickly that it glows bright red from the released heat.

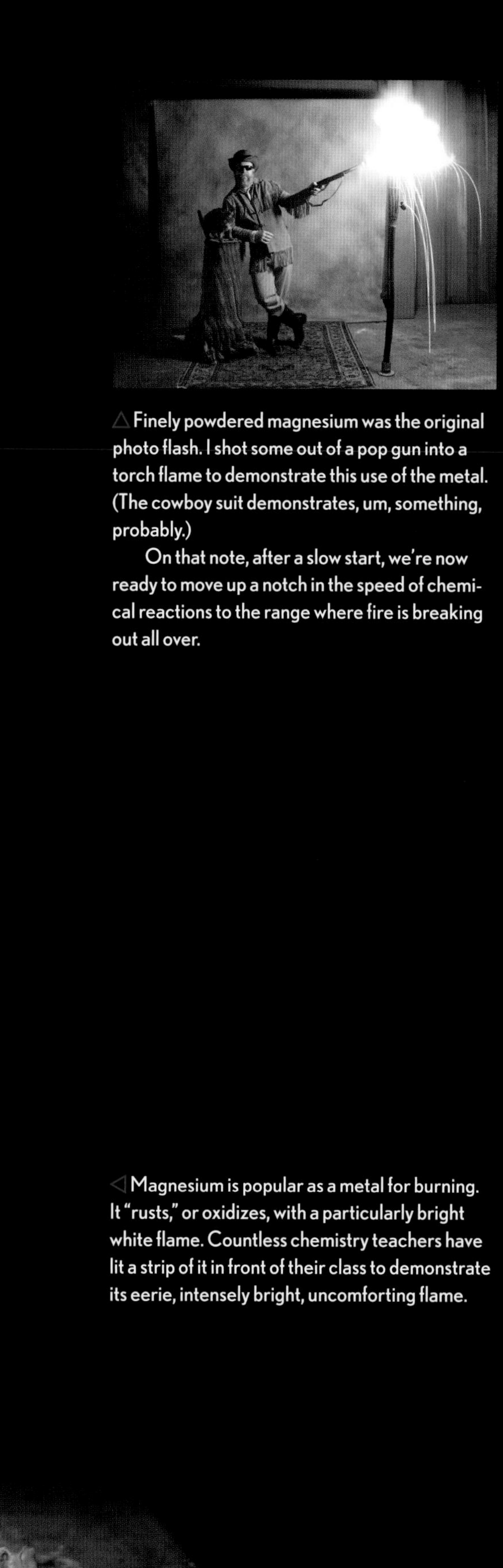

△ Finely powdered magnesium was the original photo flash. I shot some out of a pop gun into a torch flame to demonstrate this use of the metal. (The cowboy suit demonstrates, um, something, probably.)

On that note, after a slow start, we're now ready to move up a notch in the speed of chemical reactions to the range where fire is breaking out all over.

◁ Magnesium is popular as a metal for burning. It "rusts," or oxidizes, with a particularly bright white flame. Countless chemistry teachers have lit a strip of it in front of their class to demonstrate its eerie, intensely bright, uncomforting flame.

Fire

FIRE IS WHAT SPARKS (ha, ha) many future scientists' interest in chemistry. It's just so *exciting* to watch the world, or at least a small part of it, burn.

The common factor in all reactions normally defined as fire is that they involve a combination of some fuel—wood, gas, coal, even some metals—and one particular element, oxygen. Oxygen is the single most important factor in every fire, whether it's a forest of wood burning or a droplet of gasoline exploding inside an engine. We spend so much time talking about the fuel only because the fuel is the part that varies from one example to the next. We take the oxygen for granted because it's always assumed to be there.

Access to oxygen is what determines the rate and the fate of any fire. Without it, there is no fire. Little bit of oxygen, little bit of fire. More oxygen, more fire.

The whole edifice of life on dry land depends on exactly how much oxygen there is in the air. Right now air is just over 20 percent oxygen (O_2) by volume. Much less than this and we mammals would not be able to survive. Our high-energy metabolisms depend on oxygen just as much as fires do (see page 104 to learn how our bodies are as if on fire).

We need a good bit of oxygen in the air, but if there were much more oxygen we would be in trouble for a different reason.

Most of the air, the nearly 80 percent that isn't oxygen, is nitrogen gas (N_2), which is important because of what it doesn't do. Nitrogen does not react with anything except under a few special circumstances. In particular, it does not participate when things are burning with oxygen from the air. Nitrogen, being almost completely inert, just gets in the way and cools everything down. It slows and dampens every kind of fire that relies on air.

△ Dried grass burns powerfully. This is just a few acres of prairie grass in my back yard and it looks like the end of the world! When whole forests catch fire the result is truly awe inspiring. But the greatest fires on earth are *nothing* compared to what they would be like without the almost 80 percent nitrogen in the air to keep them in check. There would be no forests standing on Earth if the oxygen concentration was much higher than it is now. (Which is ironic, since it's green plants that release oxygen into the air and keep up its concentration.)

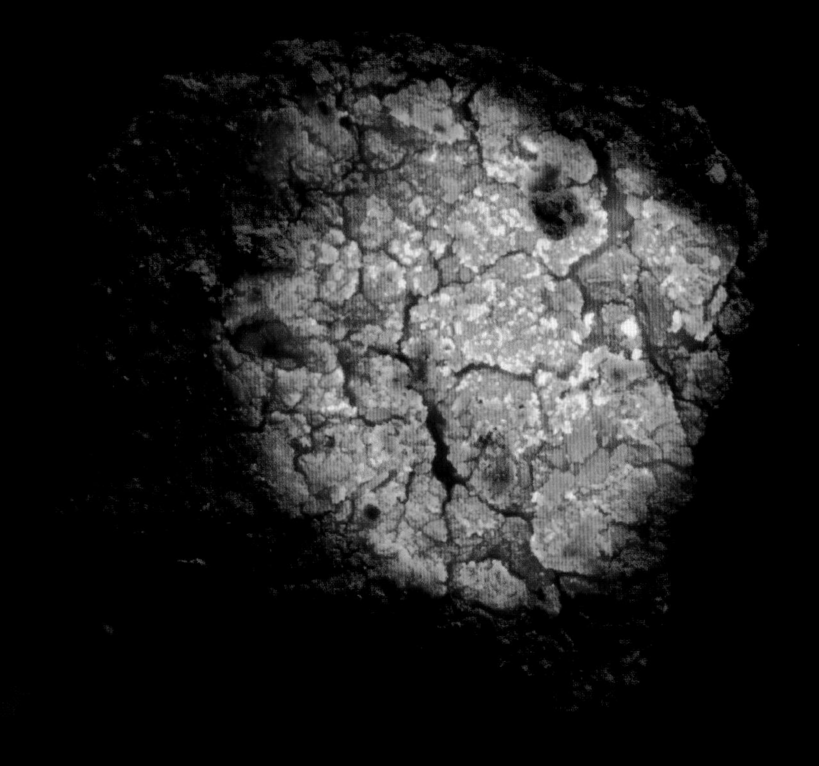

▷ Charcoal burns sedately, slowly, over the course of an hour. Blow pure oxygen gas at it and it leaps into action, crackling, sparking, and glowing furiously. A briquette will be gone in less than a minute under a steady stream of oxygen.

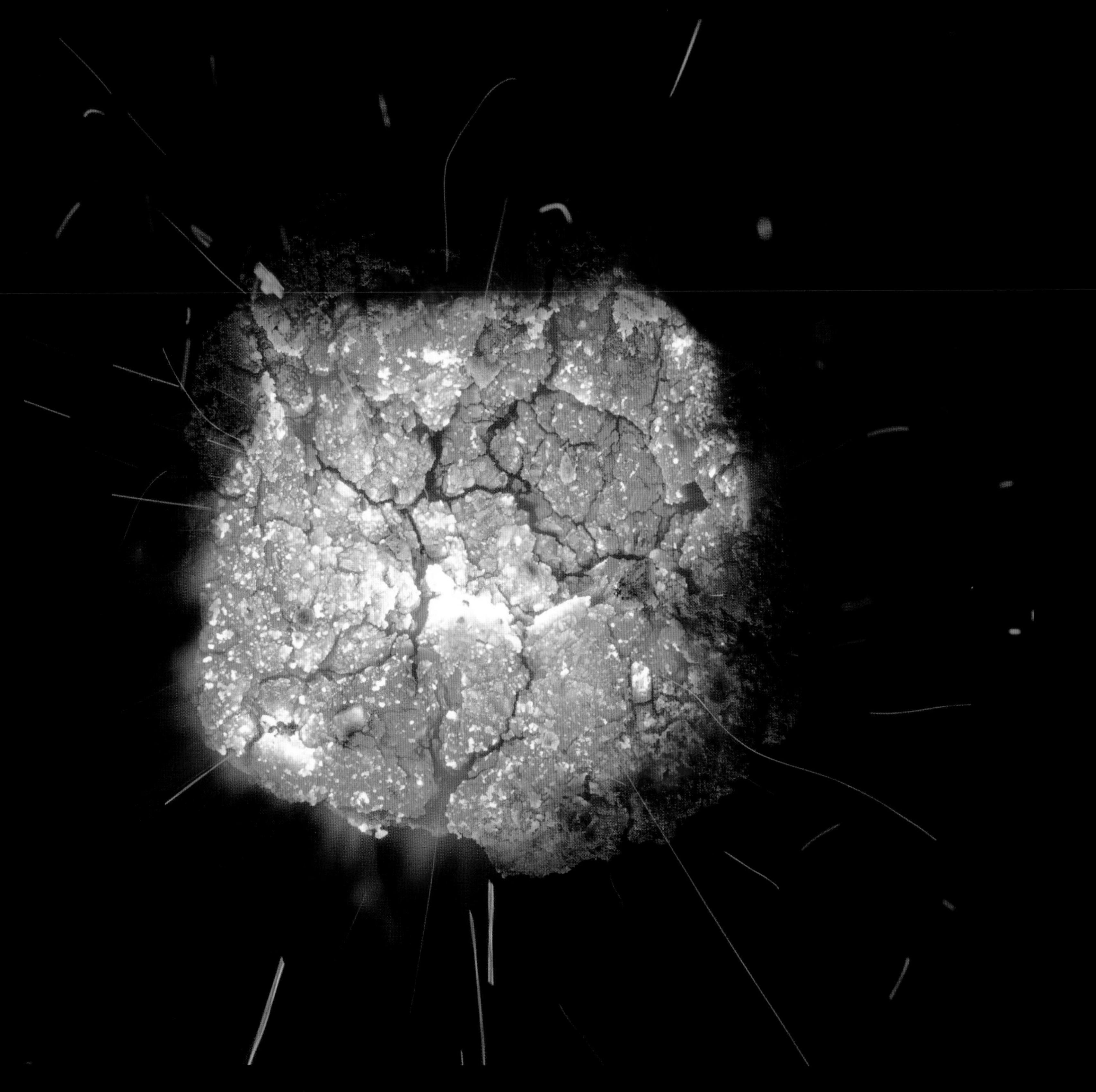

▷ Charcoal dust dropped into a container flushed with pure oxygen explodes into delightful starbursts.

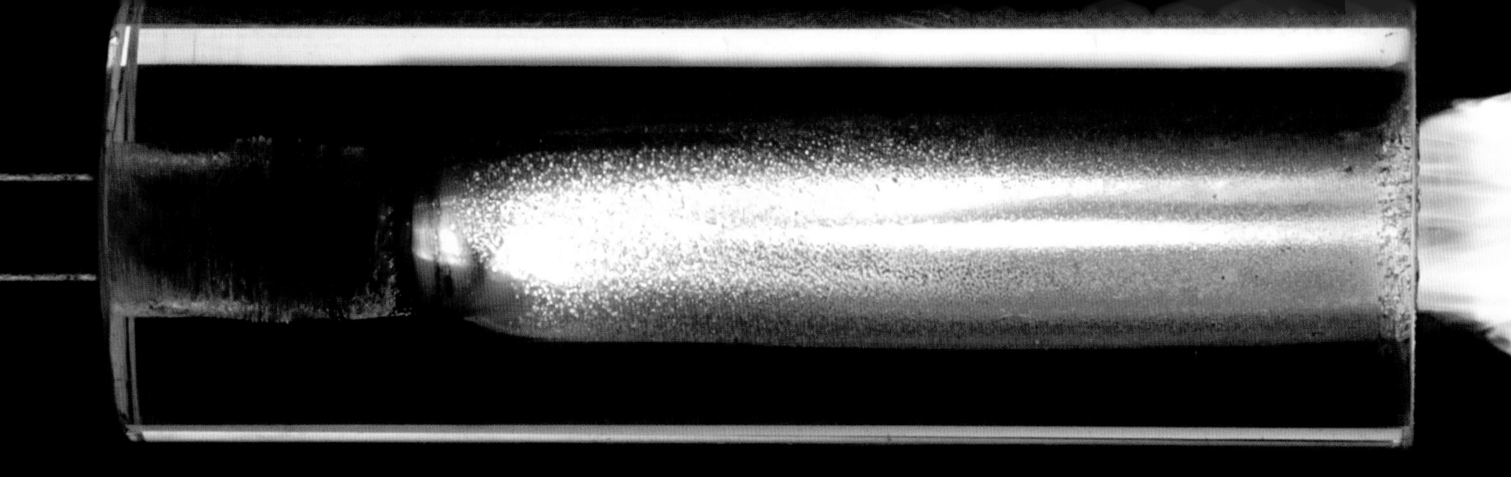

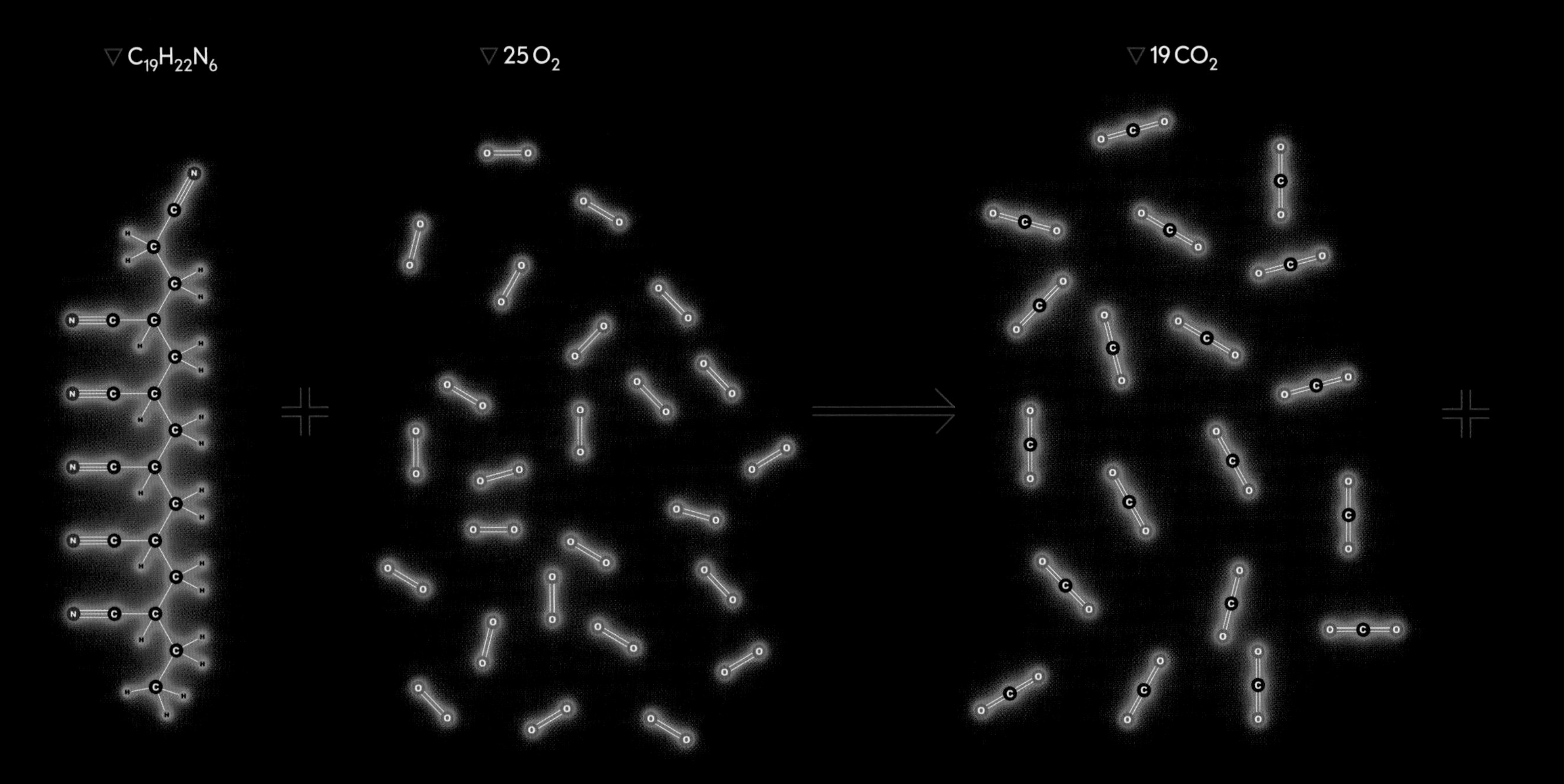

$C_{19}H_{22}N_6$
$25\,O_2$
$19\,CO_2$

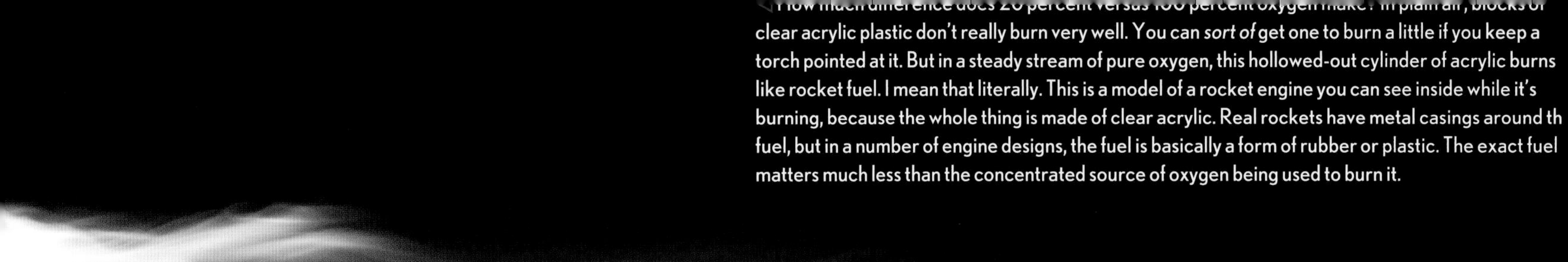

◁ How much difference does 20 percent versus 100 percent oxygen make? In plain air, blocks of clear acrylic plastic don't really burn very well. You can *sort of* get one to burn a little if you keep a torch pointed at it. But in a steady stream of pure oxygen, this hollowed-out cylinder of acrylic burns like rocket fuel. I mean that literally. This is a model of a rocket engine you can see inside while it's burning, because the whole thing is made of clear acrylic. Real rockets have metal casings around th fuel, but in a number of engine designs, the fuel is basically a form of rubber or plastic. The exact fuel matters much less than the concentrated source of oxygen being used to burn it.

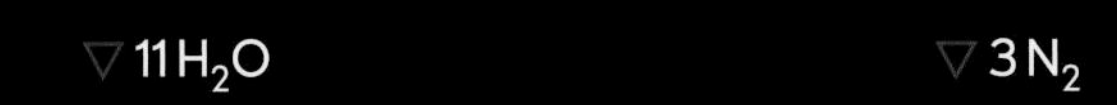

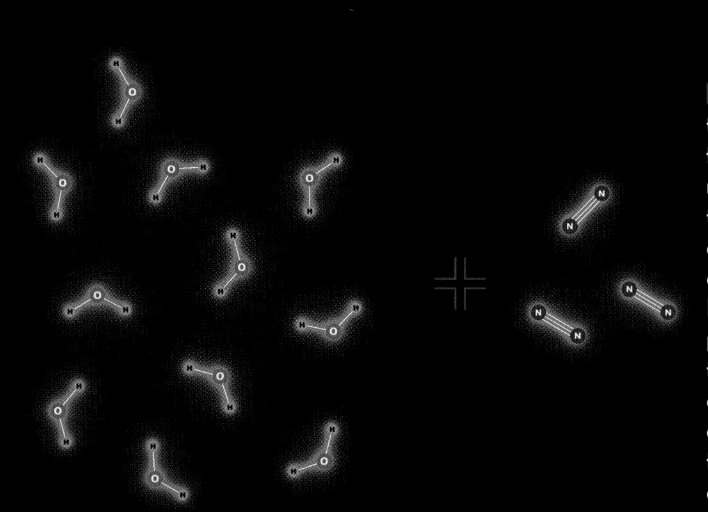

▷ The Saturn V rocket—the machine that took us to the moon—ran on diesel fuel (technically purified kerosene, which is more expensive but not much different). What gave it the power to reach the moon wasn't a special fuel, but the fact that ordinary kerosene was burned with pure liquid oxygen. Fun fact: Each of the five Rocketdyne F-1 engines in this rocket has a 55,000-horsepower gas turbine engine in it. That was just the *fuel pump* whose job was to deliver fuel and oxygen to the part doing the real work. The F-1 engine was the most powerful engine ever built for any purpose. Together the five in this rocket consumed about fifteen cubic yards (cubic meters) of fuel and liquid oxygen per second, and delivered 7.6 million pounds of thrust.

Fire is wild, and it is a dynamic process. It can only happen quickly, because it relies on a feedback loop where the heat generated by the fire is what keeps the fire going.

Ever since I saw this done in a James Bond movie I've wanted to try it for myself. Agent 007 used hair spray and a lighter to kill a deadly snake planted by some evil genius. Since no one is trying to kill me, I just did it for the photograph, which I wanted to illustrate the way in which fire is in a constant race to keep itself going.

Unlike many movie stunts, this one works exactly as advertised. Manufacturers have made some attempt to make hair spray less flammable and thus safer (for people who are in more danger from fire than snakes), but it's still able to give a good show. And canned spray ether, sold for starting cars, has to be extremely flammable because its whole reason for existing is to be so flammable that it can't help but catch fire in an engine's cylinders.

★Needless to say, *do not try this unless you are ready for the consequences.* This demonstration can easily throw fire to the other side of a large room, or flash back and burn you.

▽Still-cold propellent and ether shoot out of the nozzle.

▽The heat of the fire fights against the flow of the cold fuel.

▷In this region, highly flammable ether and propellant gas are not on fire. They are coming out of the can very fast, and they have not had a chance to mix with oxygen from the air yet. Nothing can burn without first coming in contact with oxygen.

△Here is where the battle rages. Heat from the fire to the right is trying to ignite the fuel, but the fuel is coming cold and fast. The battlefront shifts back and forth as random turbulence tips the balance one way or another. When a flame is "blown out," it is because fuel and/or air is coming too fast for the fire to be able to heat it up and sustain the reaction.

△Here everything is over but the burning. Fuel and air are mixing and burning vigorously, with no doubt about the outcome.

Most fires that rely on oxygen from the air are always very close to going out. The nearly 80 percent nitrogen ever-present in the air means that even highly flammable gases like methane or propane are just barely able to keep themselves aflame. A celebrated discovery by Humphry Davy in 1815 showed how this fact could be used to keep mine workers safe (or at least slightly safer) from underground explosions. He discovered that even fine wire mesh is enough to stop an explosive gas fire in its tracks.

▽ All that's left past the end of the fire is CO_2, water, and a few random molecules of uncombusted and partially combusted fuel. And a lot of heat. All the products of the reaction leave the area very hot (which is why you can use this reaction to chase away deadly snakes).

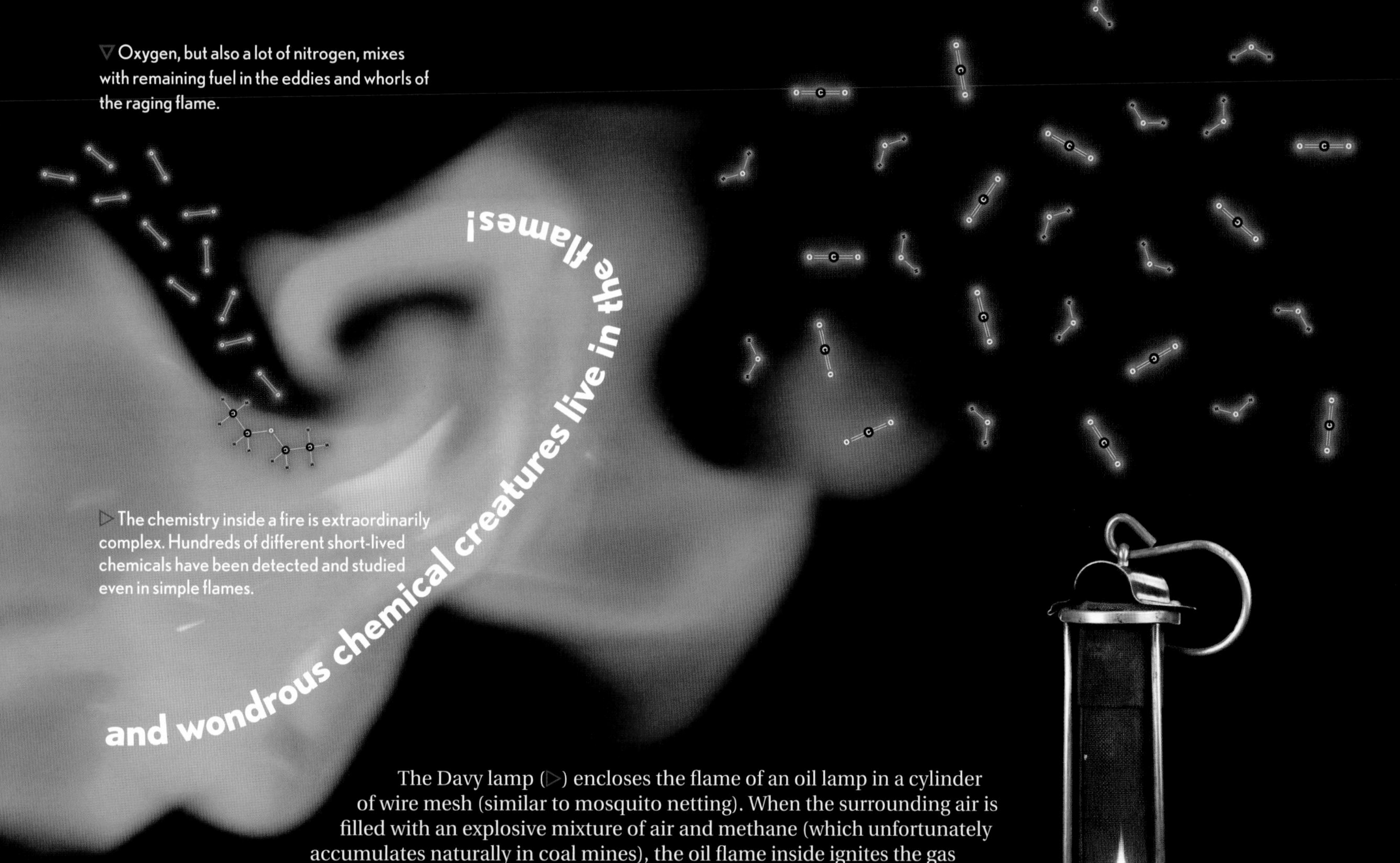

▽ Oxygen, but also a lot of nitrogen, mixes with remaining fuel in the eddies and whorls of the raging flame.

▷ The chemistry inside a fire is extraordinarily complex. Hundreds of different short-lived chemicals have been detected and studied even in simple flames.

The Davy lamp (▷) encloses the flame of an oil lamp in a cylinder of wire mesh (similar to mosquito netting). When the surrounding air is filled with an explosive mixture of air and methane (which unfortunately accumulates naturally in coal mines), the oil flame inside ignites the gas mixture inside the wire cylinder, resulting in a glowing halo around the flame. But the fire does not travel through the mesh, even though it's full of thousands of open holes. The slight cooling effect of the nearby wire is enough to extinguish any fire that tries to make it through the holes.

Fires in air may seem ferocious, but they are actually quite delicate, and they have one huge limitation—they can only ever burn as fast as the supply of oxygen allows. To ramp up the speed of a fire, we need to bring a better supply of oxygen closer to the fuel.

>Fast Fire

TO MAKE A FIRE burn faster, the first step is to increase its access to oxygen. We saw that with the plastic rocket engine. But that kind of fire is still limited by the rate at which fuel and oxygen come together and mix with each other. To make a fire *really* fast, you need to premix the fuel and the oxygen.

Speed is the only difference between desirable burning and violent explosion. Nowhere is this more often demonstrated than in the gas explosions that rock parts of the world where heating and cooking are done with propane gas.

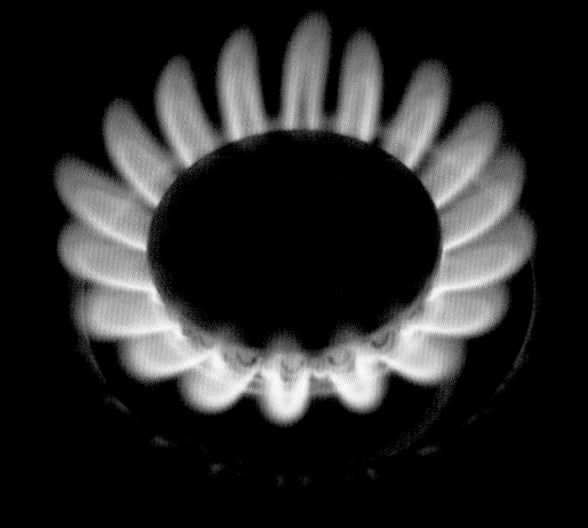

△ When propane is lit before it is allowed to mix with air, the result is a gentle cooking flame. The rate of the reaction is limited by the rate at which fuel and air are allowed to mix.

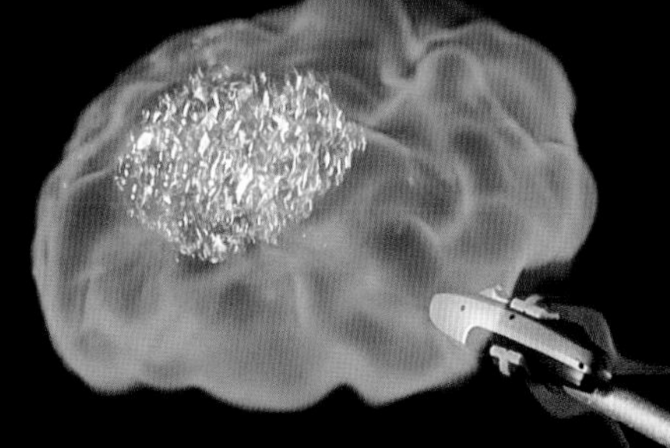

△ Earlier (see page 72) we saw what happens when you light bubbles filled with pure hydrogen, compared to bubbles filled with a mixture of hydrogen and oxygen. The premixed bubbles *explode* instead of just burning. There isn't any more energy released by the exploding bubbles than by the burning ones. The exploding ones just do it *faster*.

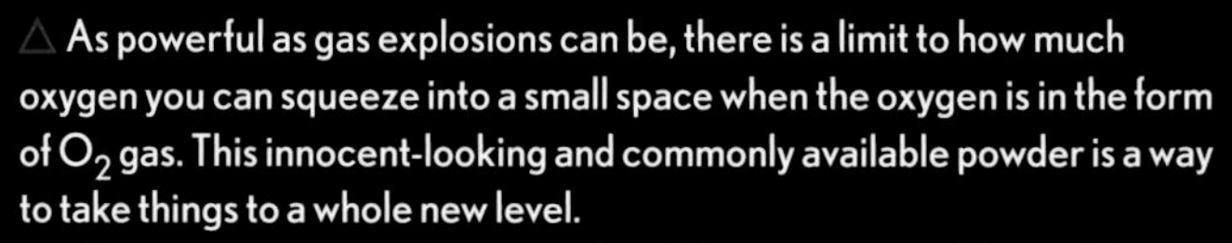

△ As powerful as gas explosions can be, there is a limit to how much oxygen you can squeeze into a small space when the oxygen is in the form of O_2 gas. This innocent-looking and commonly available powder is a way to take things to a whole new level.

Potassium nitrate, saltpeter in the old language, is in effect solid oxygen (not pure, but close enough). Per unit of volume it can supply over 700 times as much oxygen to a fire or explosion as pure oxygen gas at normal atmospheric pressure (and 3,600 times as much as the same volume of air).

As we've seen in earlier chapters, if you mix potassium nitrate with just about anything flammable, the result is something even more flammable. Mix it with sawdust and you get a road flare (page 96). Mix it with paper and you get trick paper that burns furiously (page 14). Mix it with charcoal and sulfur and you get gunpowder (pages 15 and 193).

▷ But when propane, which is heavy and slow-moving for a gas, accumulates in low areas of a house, mixes with air, and then accidentally catches fire, the result is a thundering explosion that tends to completely destroy the house that used to be heated and fed by this two-faced gas. The reaction is an explosion instead of a gentle flame because fuel molecules are right next to oxygen molecules, ready to go as soon as the necessary heat is supplied to initiate the reaction.

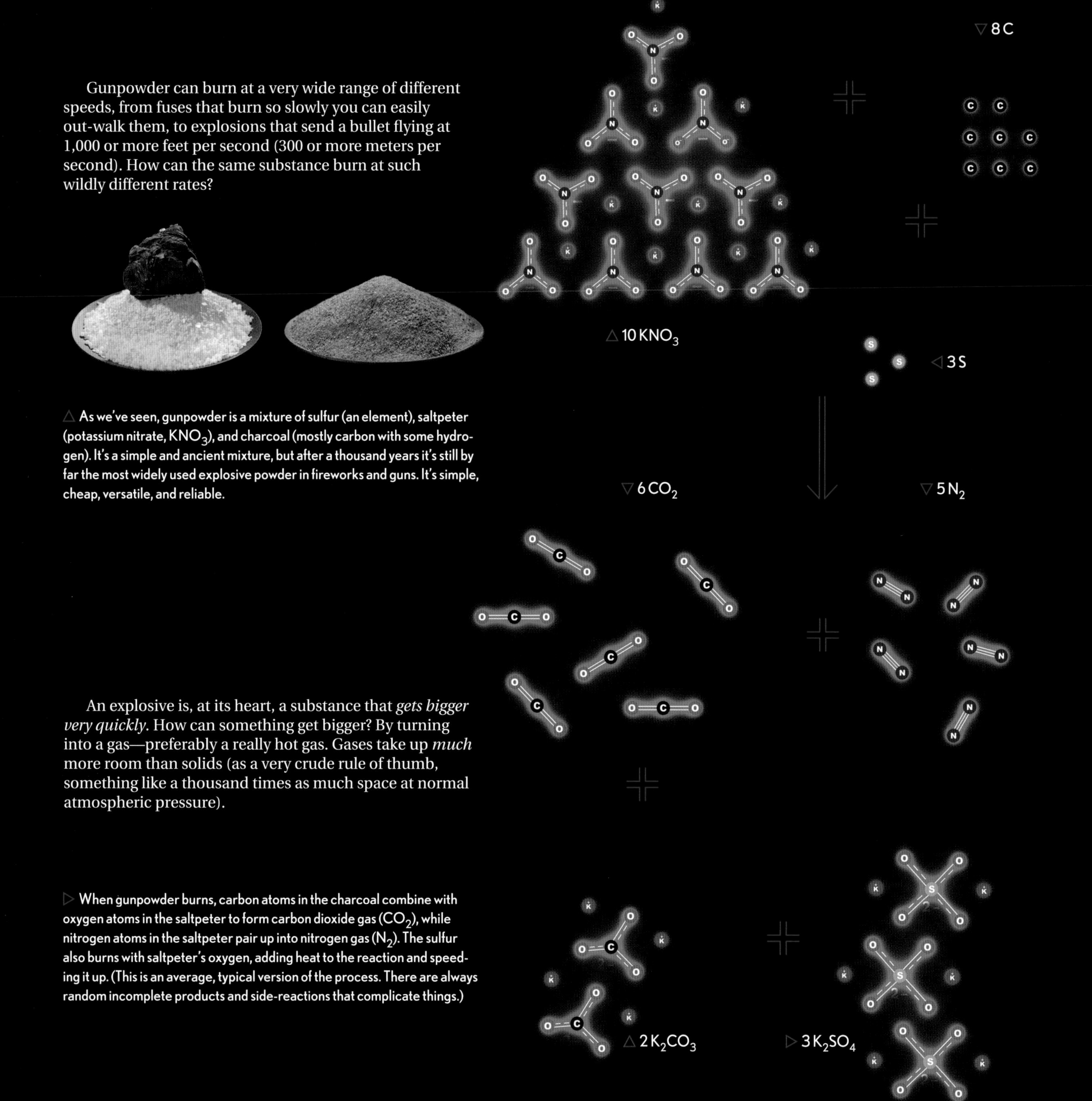

Gunpowder can burn at a very wide range of different speeds, from fuses that burn so slowly you can easily out-walk them, to explosions that send a bullet flying at 1,000 or more feet per second (300 or more meters per second). How can the same substance burn at such wildly different rates?

△ As we've seen, gunpowder is a mixture of sulfur (an element), saltpeter (potassium nitrate, KNO_3), and charcoal (mostly carbon with some hydrogen). It's a simple and ancient mixture, but after a thousand years it's still by far the most widely used explosive powder in fireworks and guns. It's simple, cheap, versatile, and reliable.

An explosive is, at its heart, a substance that *gets bigger very quickly*. How can something get bigger? By turning into a gas—preferably a really hot gas. Gases take up *much* more room than solids (as a very crude rule of thumb, something like a thousand times as much space at normal atmospheric pressure).

▷ When gunpowder burns, carbon atoms in the charcoal combine with oxygen atoms in the saltpeter to form carbon dioxide gas (CO_2), while nitrogen atoms in the saltpeter pair up into nitrogen gas (N_2). The sulfur also burns with saltpeter's oxygen, adding heat to the reaction and speeding it up. (This is an average, typical version of the process. There are always random incomplete products and side-reactions that complicate things.)

There are two key factors in how fast a given batch of gunpowder will burn: how close the fuel molecules are to the oxygen-bearing molecules, and how long it takes for the heat generated by the reaction in one spot to ignite the mixture next to it. Just as with hair spray or paper on fire, gunpowder on fire relies on a chain reaction where the heat of the reaction drives the reaction forward.

Let's start with the first problem: getting the fuel close to the oxygen.

▽ Gunpowder is a mixture of solid particles, each made of one of its three ingredients. Smaller particles, the result of the material being ground longer and finer, mean that, on average, fuel molecules are closer to oxygen. With smaller particles, it takes less time for a fuel molecule to find some oxygen to react with, and the powder burns faster.

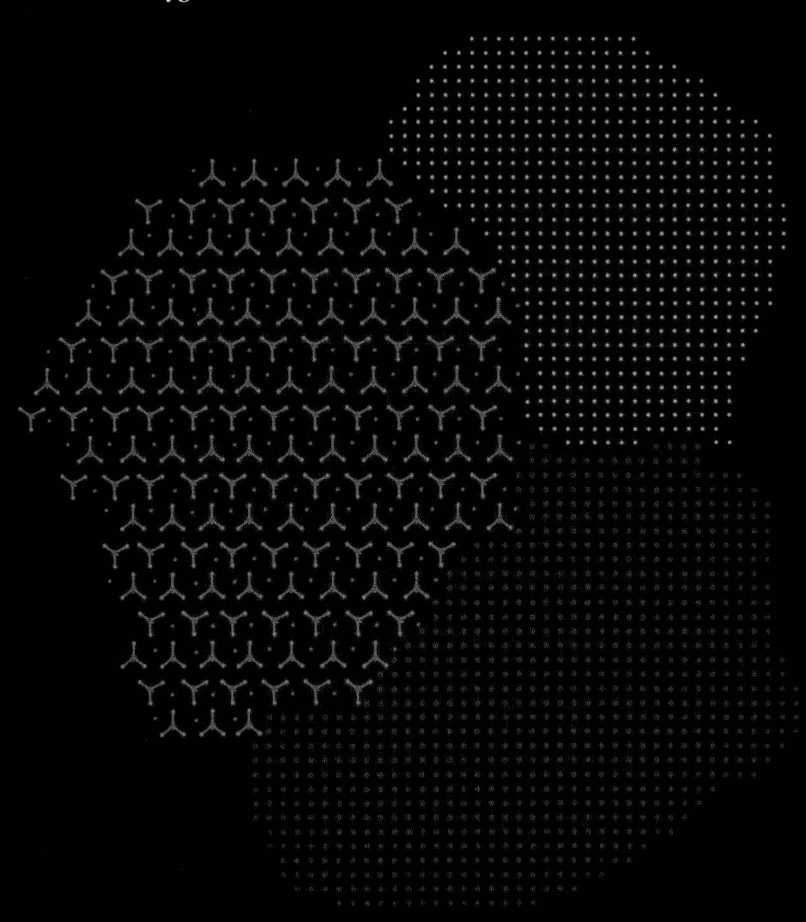

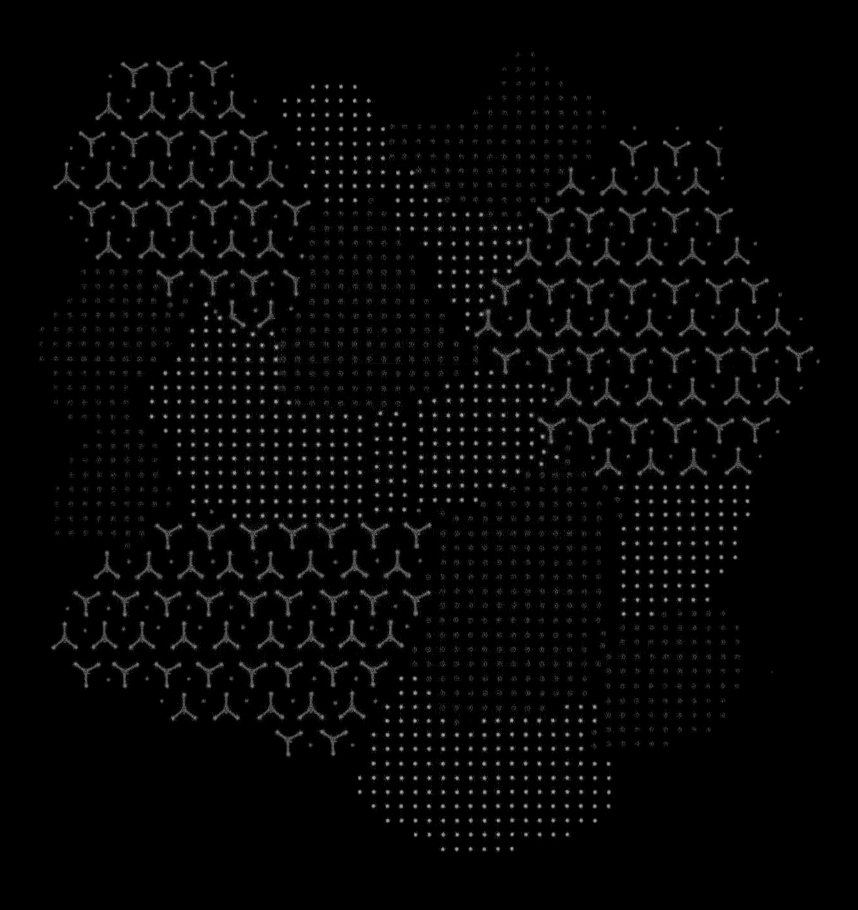

◁ Grinding gunpowder is serious business. Historic gunpowder mills along the Delaware River were built with solid stone walls on three sides, and a lightweight wooden wall on the side facing the river. When the gunpowder being ground accidentally exploded (which was pretty much taken for granted) the wooden wall would be blown out into the river while the three solid walls protected the rest of the factory. They would just build a new wooden wall and get back to business.

▷ Closer to home, hobbyists who grind their own gunpowder do it in small ball mills like this one. They place the mill a hundred feet or more away from anything, behind sandbags, and plug it in with a long extension cord. The clever part is that you can unplug the extension cord at the far end, stopping the mill, and the danger of explosion, from a safe distance.

Next problem: getting the heat transferred quickly so the whole batch can burn in as little time as possible. This is largely about containment—keeping the gunpowder in one place until it's all finished burning.

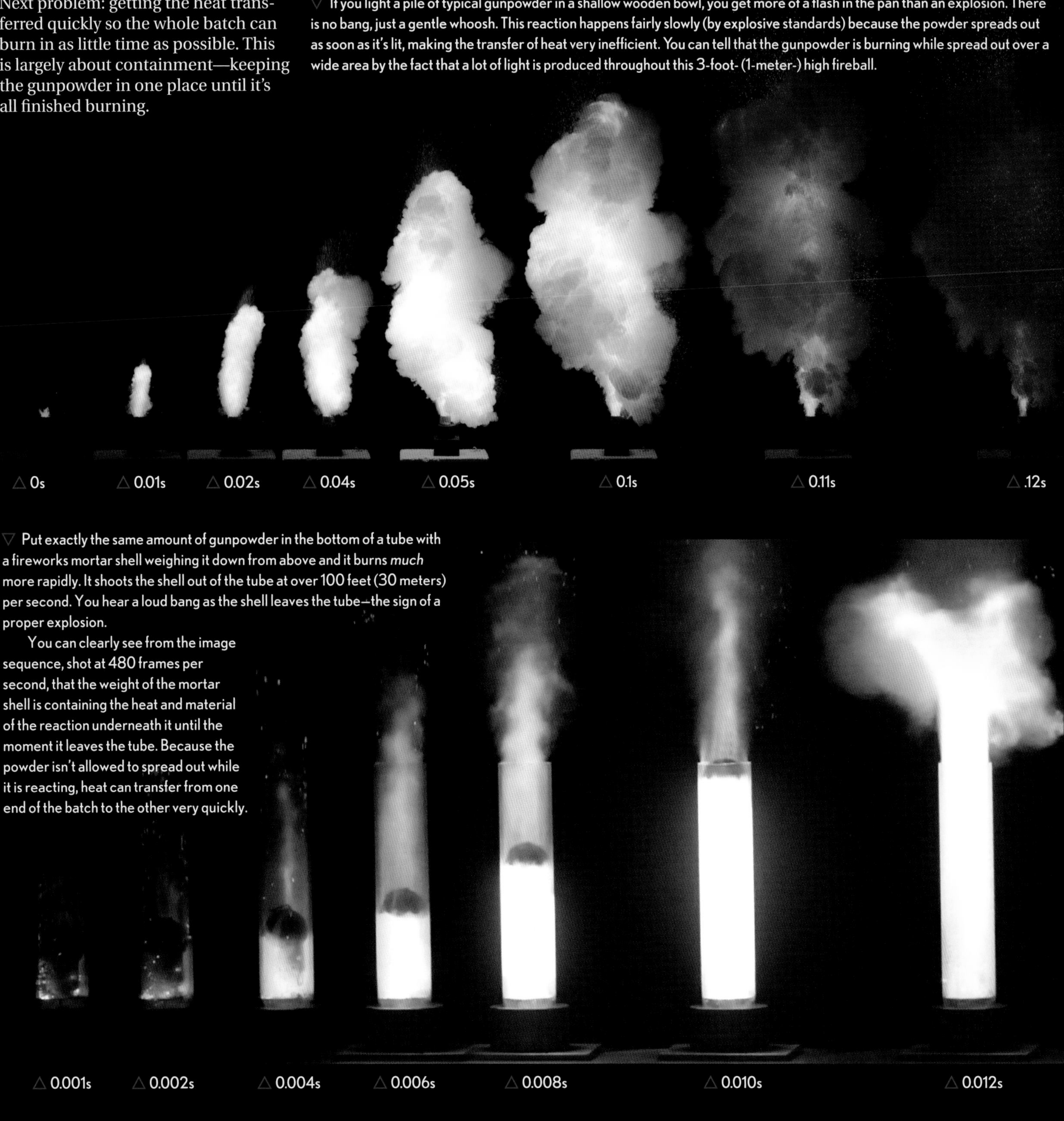

▽ If you light a pile of typical gunpowder in a shallow wooden bowl, you get more of a flash in the pan than an explosion. There is no bang, just a gentle whoosh. This reaction happens fairly slowly (by explosive standards) because the powder spreads out as soon as it's lit, making the transfer of heat very inefficient. You can tell that the gunpowder is burning while spread out over a wide area by the fact that a lot of light is produced throughout this 3-foot- (1-meter-) high fireball.

△ 0s △ 0.01s △ 0.02s △ 0.04s △ 0.05s △ 0.1s △ 0.11s △ .12s

▽ Put exactly the same amount of gunpowder in the bottom of a tube with a fireworks mortar shell weighing it down from above and it burns *much* more rapidly. It shoots the shell out of the tube at over 100 feet (30 meters) per second. You hear a loud bang as the shell leaves the tube—the sign of a proper explosion.

You can clearly see from the image sequence, shot at 480 frames per second, that the weight of the mortar shell is containing the heat and material of the reaction underneath it until the moment it leaves the tube. Because the powder isn't allowed to spread out while it is reacting, heat can transfer from one end of the batch to the other very quickly.

△ 0.001s △ 0.002s △ 0.004s △ 0.006s △ 0.008s △ 0.010s △ 0.012s

△ The goal with gunpowder, even in the barrel of a gun or at the bottom of a mortar tube, isn't to make it burn as fast as possible. The goal is to make it burn at the *correct rate* for the job. For example, the gunpowder shown here is the particular kind used in the "lift charge" of a fireworks mortar. It burns at just the right rate to push the mortar up out of the tube without tearing the shell or the tube apart.

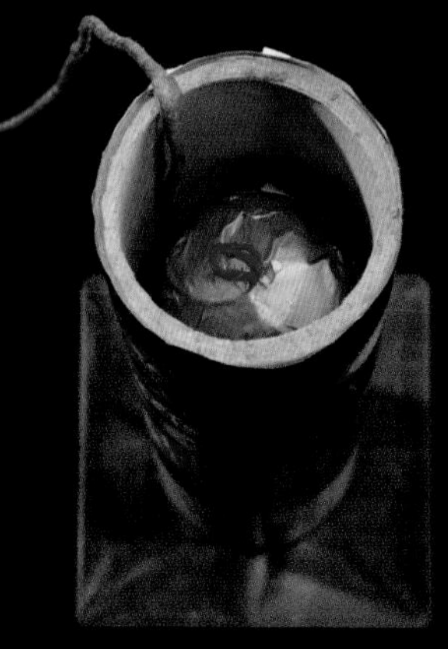

◁ Standard mortar launch tubes are not made of steel, they are made of . . . cardboard. Not even very thick cardboard. That's a sign that the explosion happening inside them must not build up very much pressure, which in turn means the explosion must not be very fast. Well, yes, it's fast on a human time scale, and fast compared to open-air burning of gunpowder, but on an explosives timescale, it's really quite slow. The mortar takes about 0.01 seconds, 10 milliseconds, to leave the tube, which we will soon see is an eternity by the standards of some explosions.

△ Terrible accidents have happened when people mistook flash powder for gunpowder and, for example, loaded a ceremonial cannon designed to signal a goal at a sports game with the wrong kind. Flash powder will turn a cannon like this, meant to fire gunpowder, into a thousand shards of hot metal flying at supersonic speed through anyone nearby.

◁ Inside the shell of a common mortar is a much more powerful explosive: flash powder. If for some reason the shell never leaves the tube (for example, if some idiot like me were to intentionally load the shell upside down), the flash powder "break charge" in it will completely destroy the cardboard launch tube. We never found the other half of this one.

This isn't because the total energy released by the break charge is greater than what the lift charge produced, but rather because the flash powder in the break charge burns *much faster* than the gunpowder in the lift charge. The fast reaction means that the gases created don't have time to leave the tube before a huge pressure has built up.

◁ Flash powder is a mixture of powdered aluminum and potassium perchlorate, which reacts very, very fast when confined. But even flash powder is only a low explosive in the same broad category with gunpowder. It is bush-league compared to *really* fast explosives.

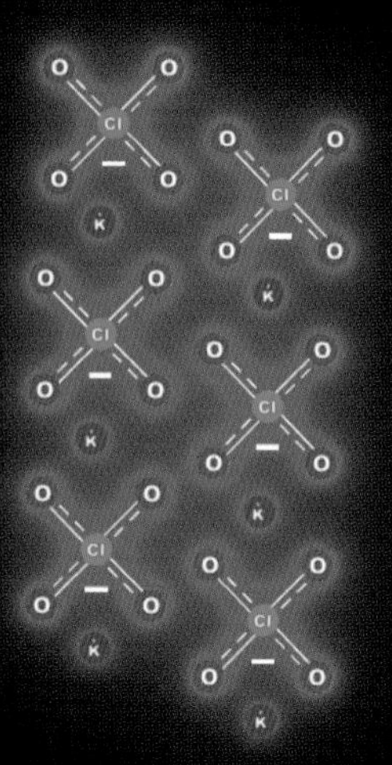

◁ $6\,KClO_4$

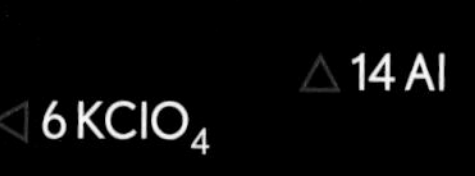

△ $14\,Al$

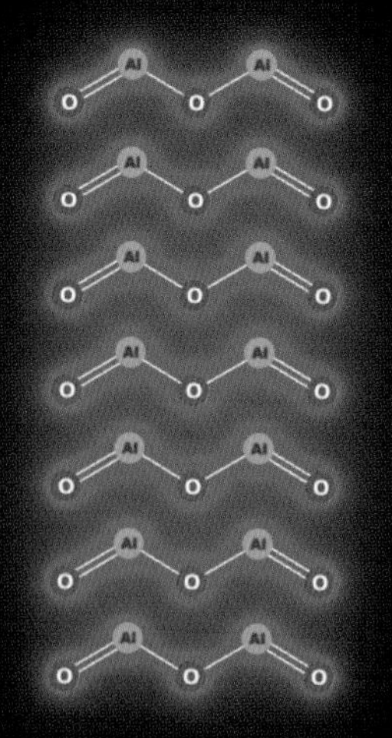

◁ $7\,Al_2O_3$

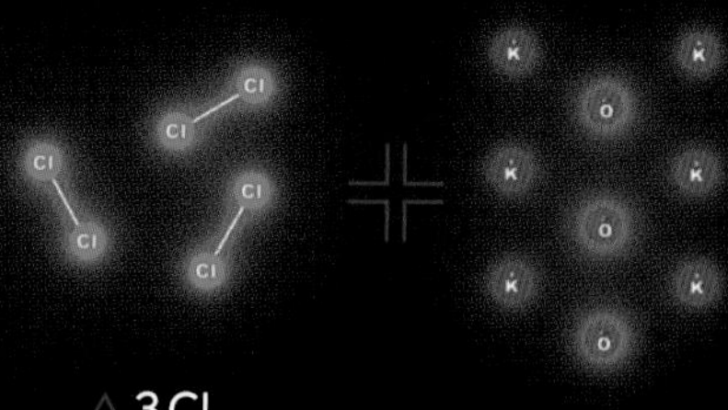

△ $3\,Cl_2$

△ $3\,K_2O$

Really Fast Fire

NITROGLYCERIN IS THE classic example of a *high explosive*. Take a minute to appreciate its brilliant and frightening molecular structure. The oxygen needed for combustion has been moved *right onto the same molecule with the fuel* (the carbon and hydrogen). Nitroglycerin isn't a mixture of two chemicals, one a fuel and the other a source of oxygen. Instead it's a *single molecule* that is both fuel and oxygen source in one.

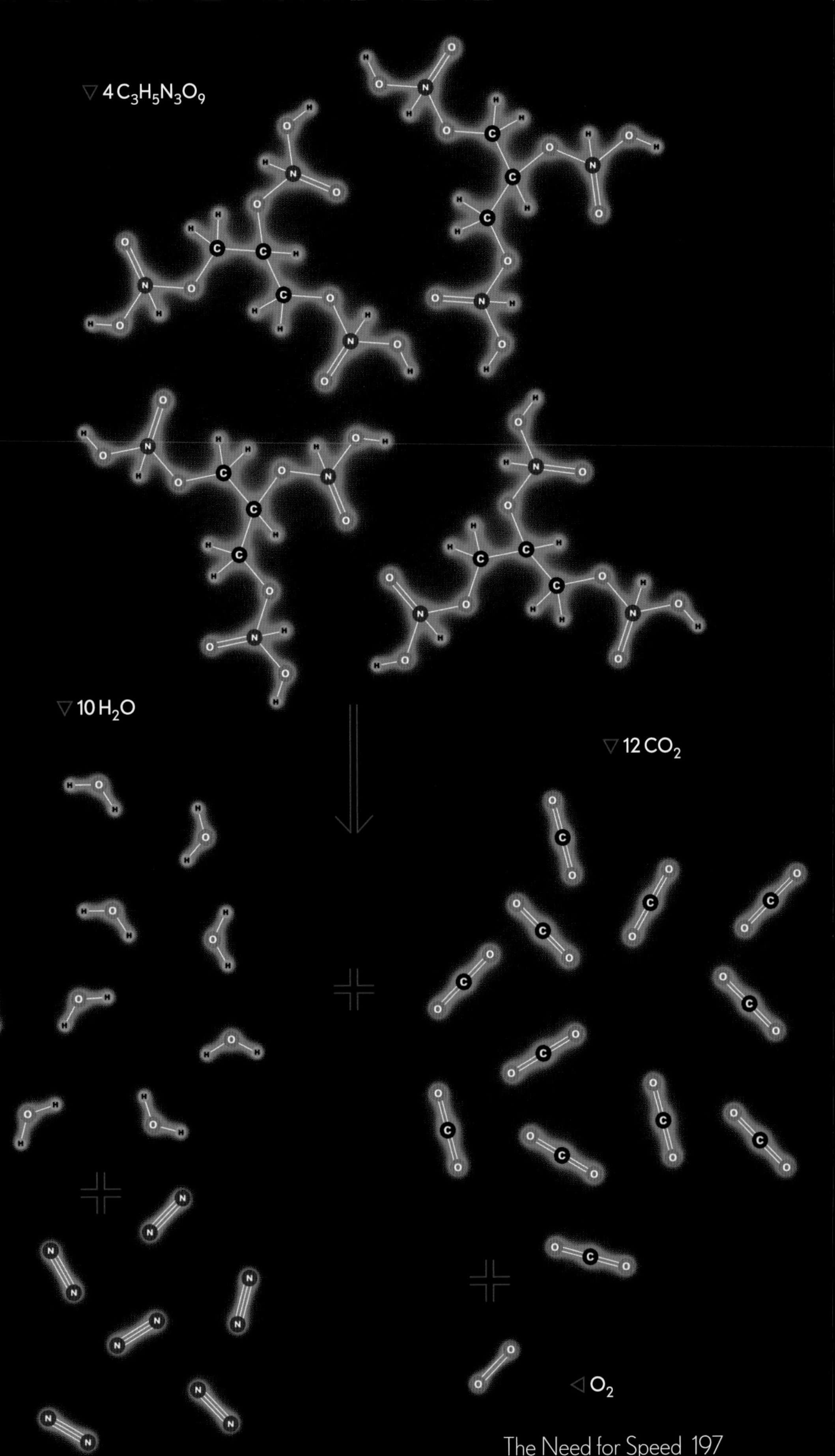

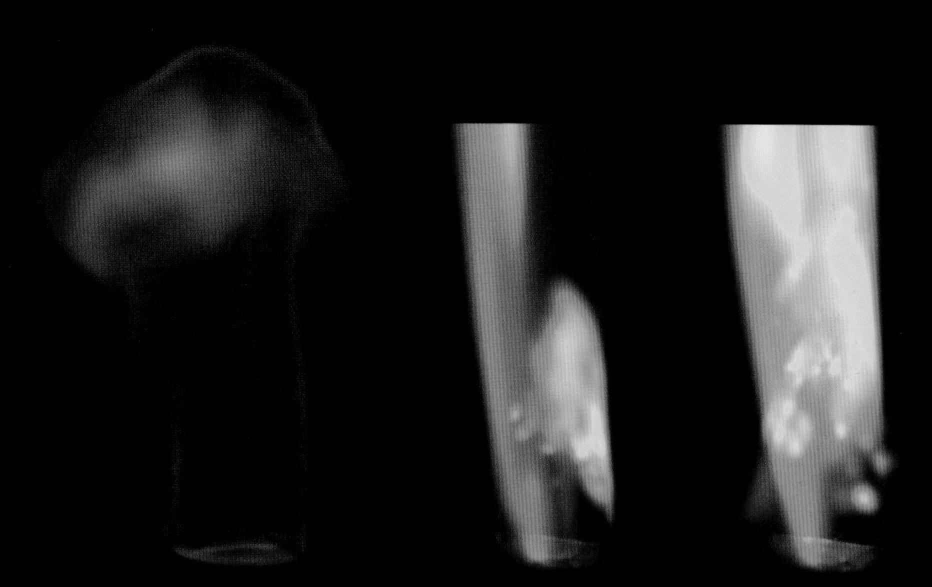

△ If you light a bit of nitroglycerin, it does indeed burn very fast. But, much like gunpowder ignited out in the open (see page 195), it really doesn't *explode*. We've moved the fuel and oxygen as close to each other as they possibly can be, but there is another factor holding back the speed of the reaction.

▷ It takes a certain amount of energy to initiate the disintegration of nitroglycerin molecules. At room temperature, the molecules are stable. (Otherwise how could you have a bottle of them?) But heated up, they start to decompose, which releases heat, which triggers more neighboring molecules to decompose, which releases more heat. This is simply a traditional fire, happening faster than normal, but still fundamentally limited by the speed at which heat can travel through the material. This is fast, but not *FAST*.

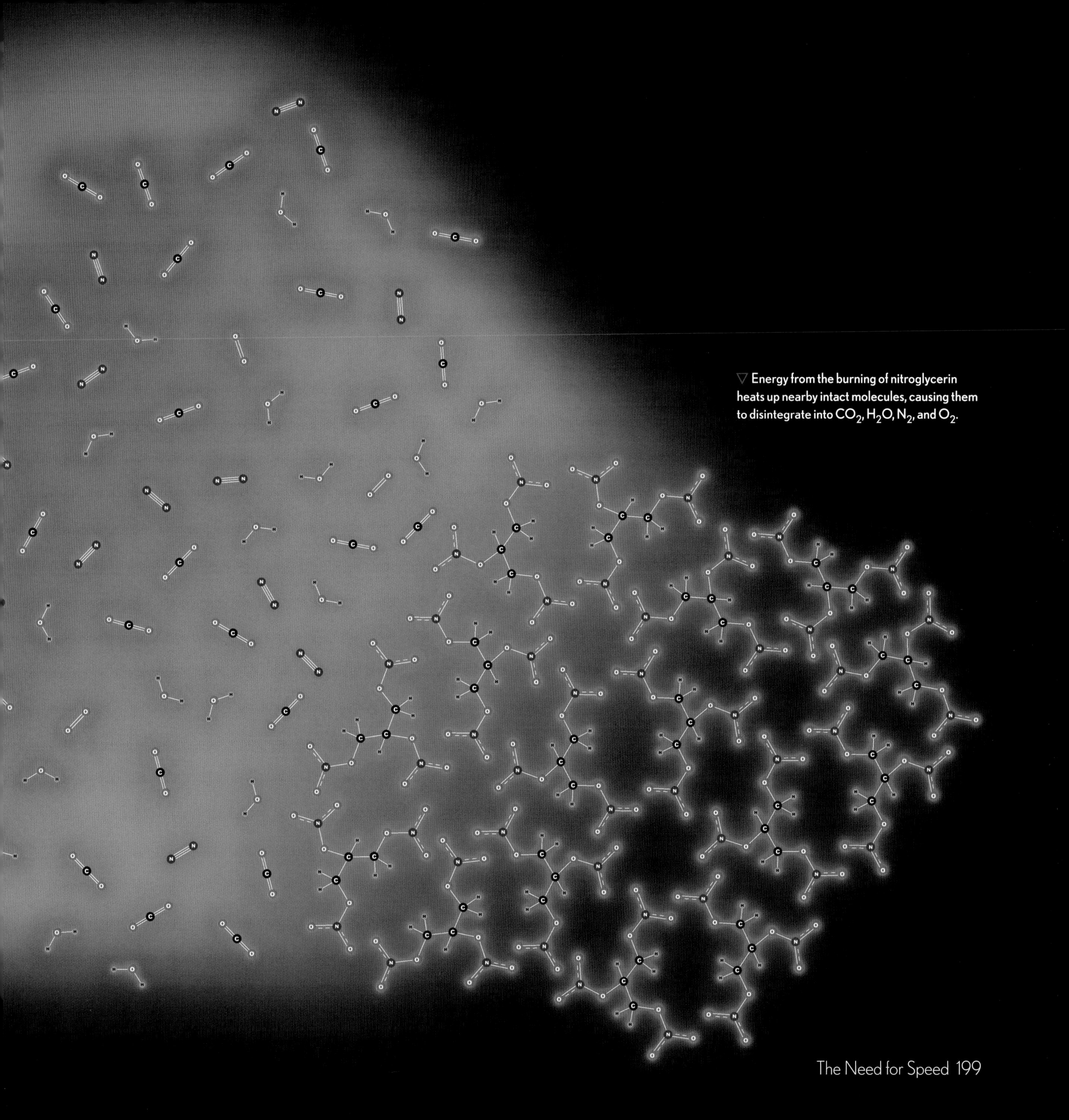

▽ Energy from the burning of nitroglycerin heats up nearby intact molecules, causing them to disintegrate into CO_2, H_2O, N_2, and O_2.

Nitroglycerin burns pretty fast, but watch out because it has a trick up its sleeve. Under the right conditions (you look at it funny), nitroglycerin can *detonate*.

Detonation is a whole other animal from ordinary burning. Instead of relying on the slow transfer of heat, in a detonation a supersonic pressure wave travels through the material at an incredible speed—about 4.7 miles (7.5 kilometers) per second in the case of nitroglycerin—triggering the disintegration of the molecules almost instantaneously as it progresses.

How fast is that? Suppose you had a bottle of nitroglycerin about 4 inches (10 centimeters) tall and you accidentally dropped it. When it hit the floor, a detonation wave would start from the bottom of the liquid. Thirteen microseconds (millionths of a second) later the wave would have reached the top of the liquid, and all the nitroglycerin in the bottle would already have been converted into gas form. This gas (which after only 13 microseconds is probably still inside the bottle) is around 9,000°F (5,000°C) and would like to occupy about 20,000 times its current volume.

A scant few microseconds later the bottle is dust, and perhaps a millisecond (thousandth of a second) later so are you. Now *that's* fast! Despite being one of the oldest, nitroglycerin is still one of the fastest-detonating high explosives known.

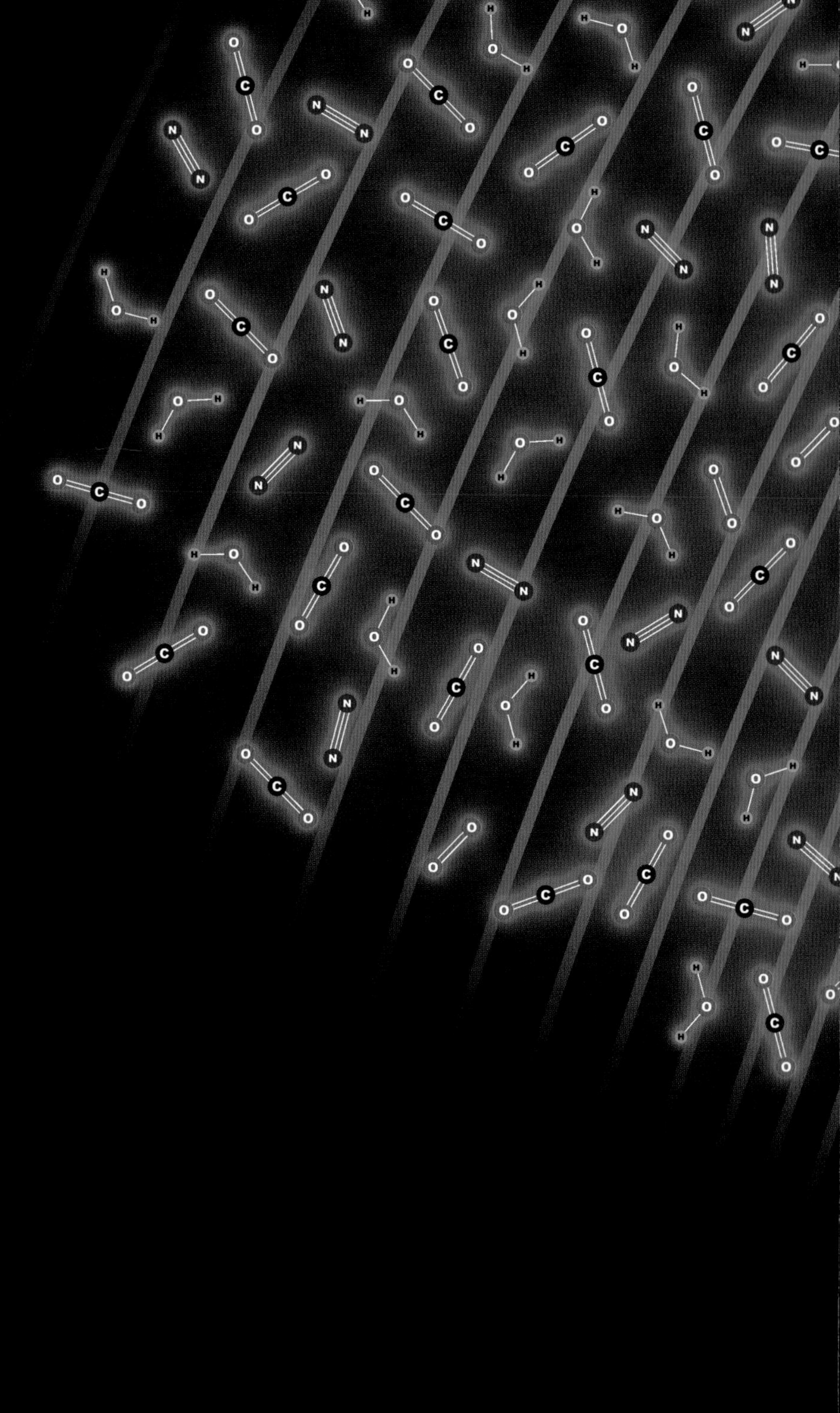

▷ **The disintegration happens so fast that the new gas molecules don't have time to spread out. They create the incredibly high pressure required to sustain the detonation wave.**

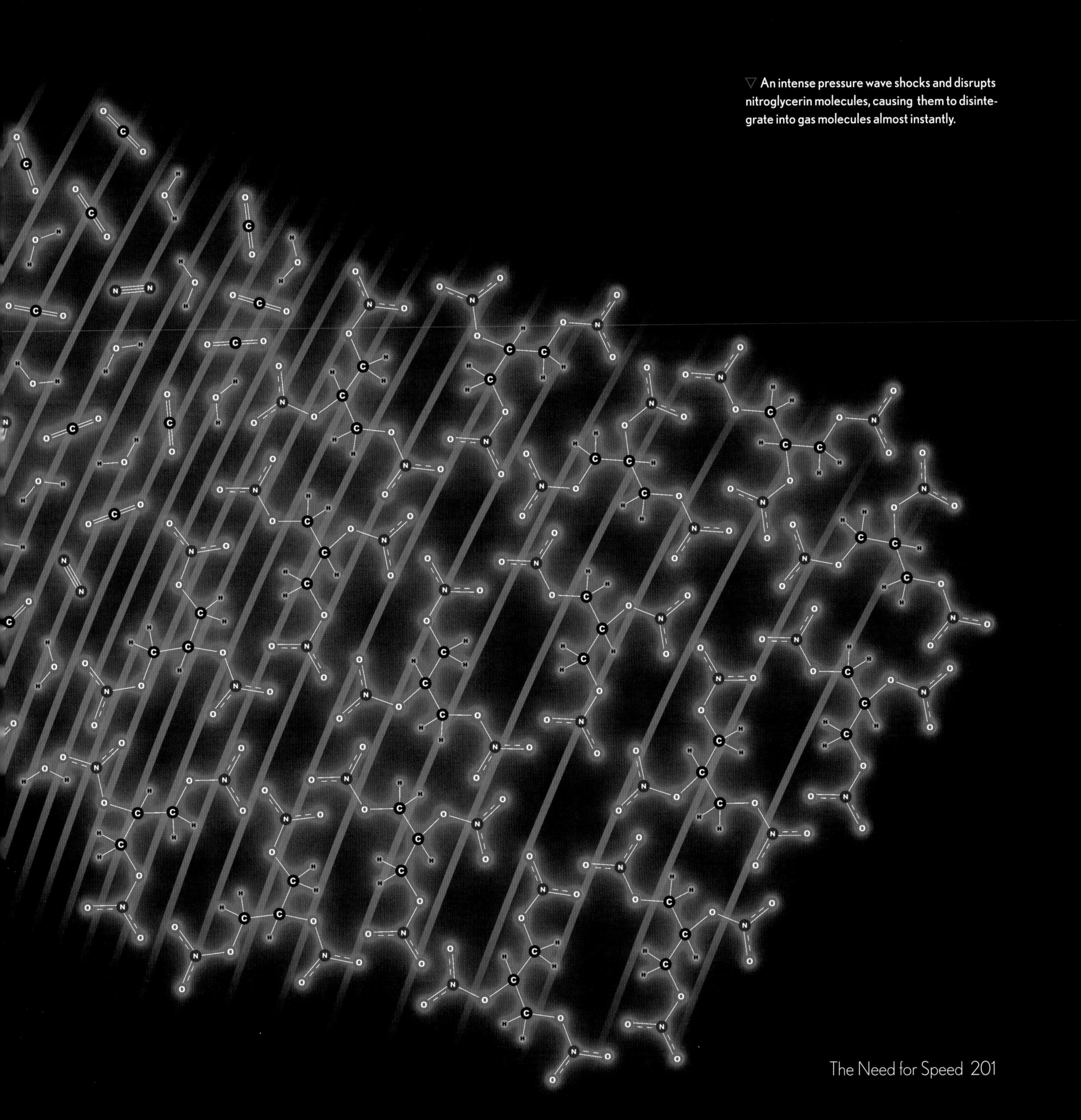

▽ An intense pressure wave shocks and disrupts nitroglycerin molecules, causing them to disintegrate into gas molecules almost instantly.

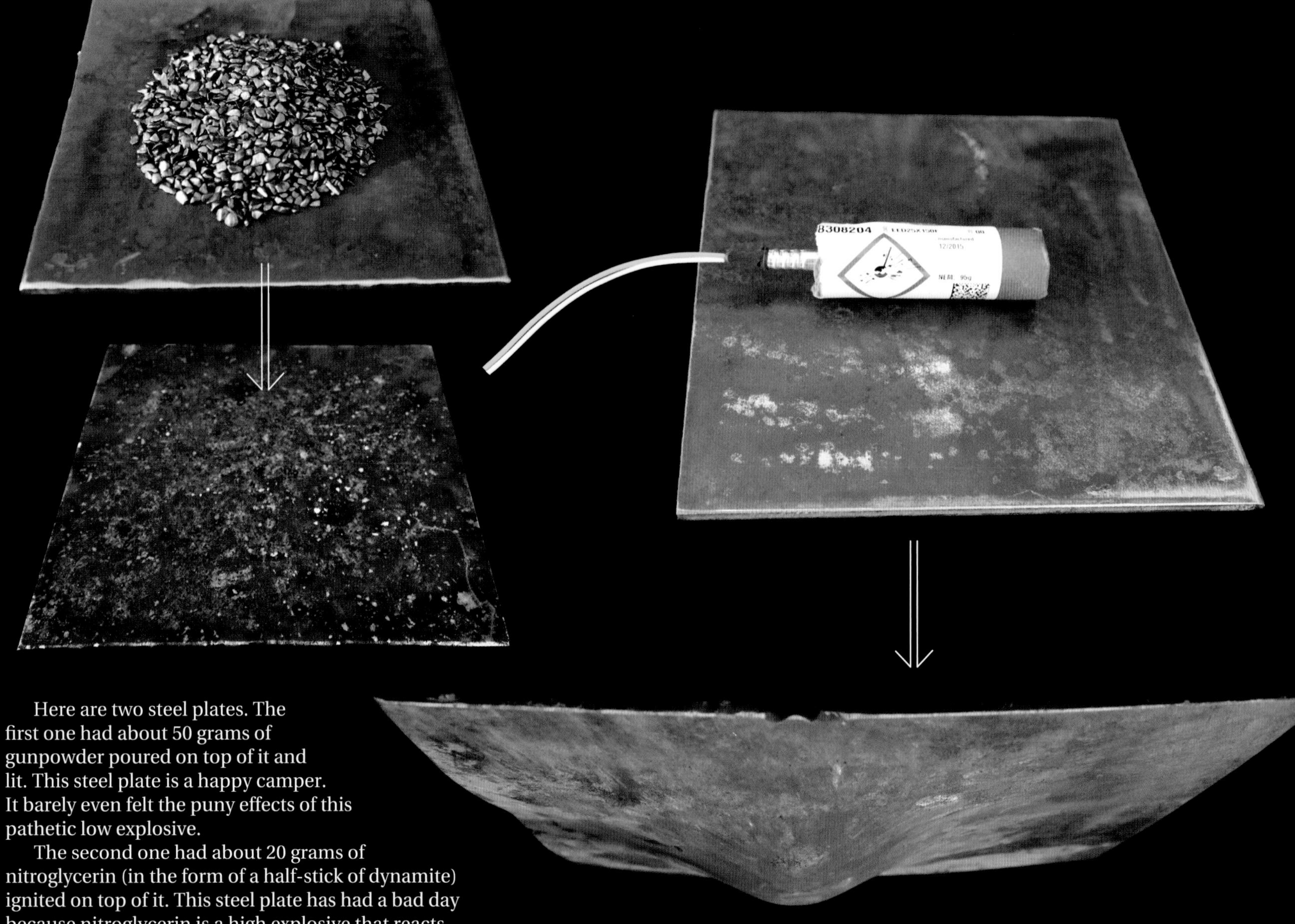

Here are two steel plates. The first one had about 50 grams of gunpowder poured on top of it and lit. This steel plate is a happy camper. It barely even felt the puny effects of this pathetic low explosive.

The second one had about 20 grams of nitroglycerin (in the form of a half-stick of dynamite) ignited on top of it. This steel plate has had a bad day because nitroglycerin is a high explosive that reacts thousands of times faster than gunpowder.

When a low explosive like gunpowder burns in the open air, it simply pushes the air away. The pressure in the area goes up, but not by a whole lot, because at no point does the surrounding air have to move away at faster than the speed of sound (which is the speed at which air is naturally displaced by a pressure wave).

High explosives cross this critical speed-of-sound threshold. They create gas so rapidly that the surrounding air is completely unable to get out of the way fast enough. The inertia of the air molecules around the high explosive creates what is in effect a wall, stronger than the strongest steel.

How can I say this air wall is stronger than steel? Just look at the steel plate: it is dented way out because, when push came to shove, it was literally easier for the explosive gases to push the steel away than to get the air above them to move out of the way any faster.

That's the difference between burning and detonation, between low and high explosives.

The Fastest Reaction of All

THE REACTIONS IN this chapter have been getting faster and faster. Now it's time for the grand finale. The fastest reaction in the whole world. Some kind of superexplosive, perhaps? A black hole imploding? No, it's just water.

As with any such "fastest," "biggest," "tallest" sort of thing, people will quibble. Is that spire on top of your building substantial enough to count as part of its height, or is it just a decoration and my building is actually taller? Does a photo-excited state count as a reaction? I'm not going to get into those arguments because I think the reaction *I* plan to name as the fastest of all is close enough, and it is vastly more important in the scheme of things than any other contender.

My candidate for fastest chemical reaction is the trading of a hydrogen ion between two water molecules.

We encountered the concept of hydrogen bonds when discussing the dissolving of sugar (page 102). As we saw there, water can form hydrogen bonds with itself, creating networks of cross-linked water molecules. The importance of these networks cannot be overstated. They have many subtle effects on the chemistry of water, and one really obvious one: ice floats.

When you cool water down from room temperature, it becomes more and more dense as it gets colder, just like nearly all other liquids. But when it reaches about 40°F (4°C), the internal structures start to become significant and the trend reverses: water becomes *less* dense as it is cooled below that temperature. When water freezes into ice, the structures take over the whole bulk of the water, and the density goes down by a big step. This is very unlike nearly all other substances.

The practical effect is that ice floats. Not only that, very cold, almost-frozen water rises to form a layer above slightly less cold water. This is the only reason that lakes in colder climates don't freeze solid all the way through in winter. It's the only reason freshwater life can survive the winter in these parts of the world.

Why do water molecules come together in these tight structures? Paradoxically, it's because water molecules are constantly tearing each other apart. (Or sharing nicely, depending on your point of view.)

△ The power and majesty of water is impossible to ignore. From a young child taking their first steps into the vast ocean, to an old man hearing echoes of the waves of his youth, we all instinctively know that water is to be cherished.

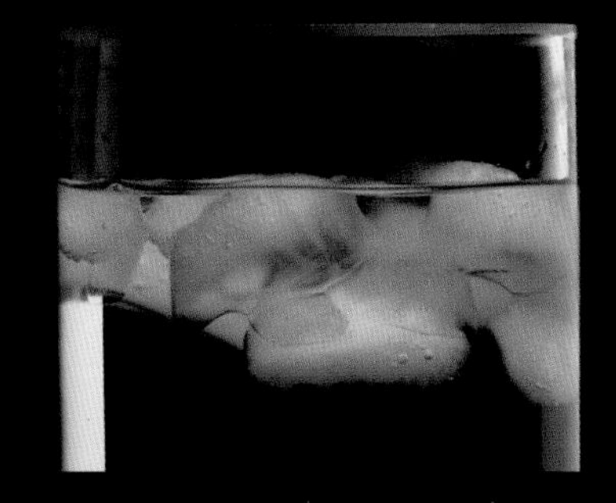

△ Water molecules always have a tendency to align with each other in a way that places a hydrogen atom between two oxygen atoms. At high temperatures, these alignments are very short-lived. But as water gets colder, the molecules begin to spend more and more time locked in semicrystalline rings and small-scale clumps. These structures are slightly less dense than liquid water. (See page 100 for a diagram of what this looks like.)

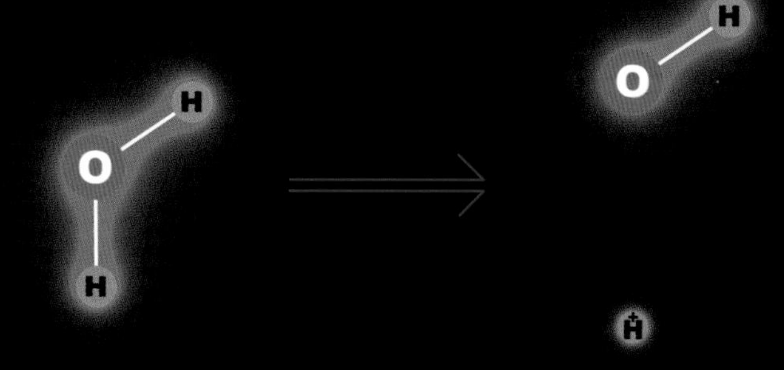

△ A common description of the situation is that water molecules can split apart into two separate ions: a positively charged hydrogen ion (H^+) and a negatively charged hydroxide ion (OH^-). At any given time in a sample of pure water, about one in ten million water molecules will be split apart in this way.

But this is a highly simplified, stylized way of describing the situation. Water molecules don't just throw themselves apart like this diagram implies. They are coaxed.

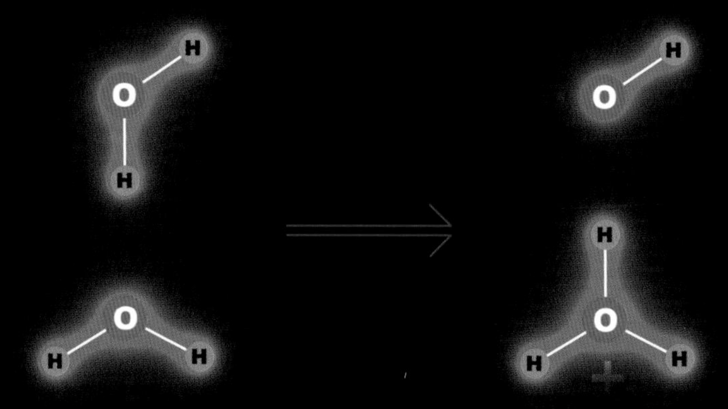

△ This picture is one step closer to reality. It shows a hydrogen ion being *traded* from one water molecule to another, forming a hydroxide (OH^-) ion and what's called a hydronium (H_3O^+) ion. This is a favorite way of representing the situation in textbooks, and a lot of teachers will insist that you never write H^+ to represent a hydrogen ion in water, but rather always write H_3O^+ to emphasize the fact that the hydrogen ion is never just floating freely.

This representation does not capture the full truth, but it does demonstrate a reason for the existence of hydrogen bonds. Imagine the hydrogen ion not just trading once from one water molecule to another, but trading *back and forth* very rapidly. By pulling alternately on one oxygen atom and then the other, the hydrogen ion draws them both toward each other.

This trading happens *incredibly fast*. Hydrogen atoms are the lightest of all atoms, so they can move faster than any others, and hydrogen-oxygen bonds are strong, meaning they pull hard on the lightweight hydrogen atom. The trade from one oxygen atom to the other happens in less than 0.000,000,000,000,05 seconds (50 femtoseconds). That is the fastest of any meaningful chemical reaction. The great speed of this reaction enables much of the unique chemistry of water.

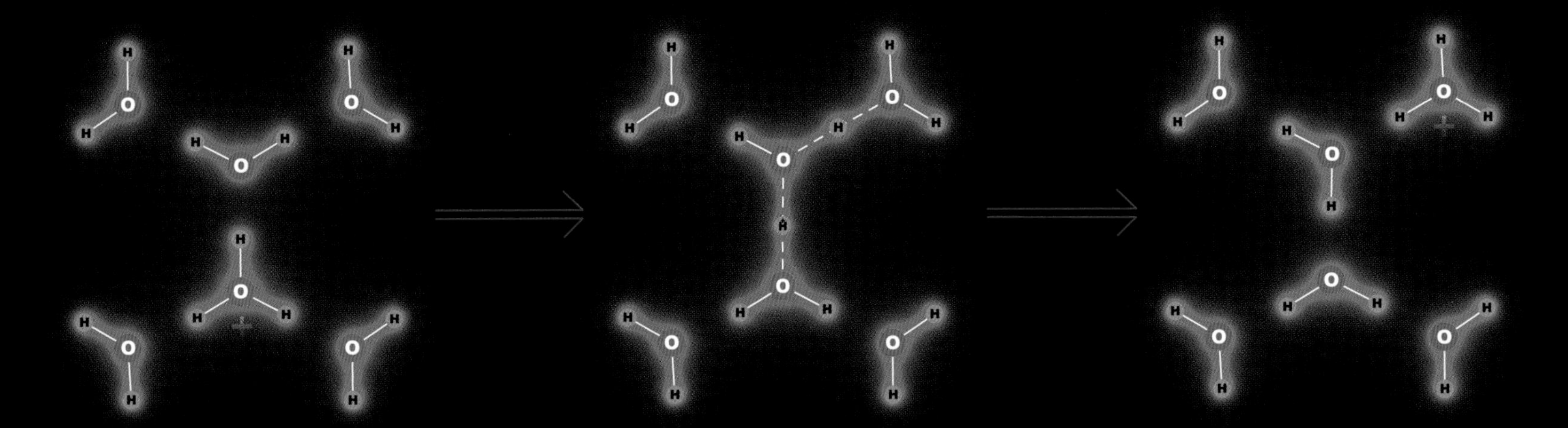

△ The previous diagram makes it look like H_3O^+ is a real thing. That's also the impression you get from a lot of textbooks. But the fact is that no H_3O^+ ion ever exists for long enough to really count as existing. The trading of H^+ ions is almost as fast as the vibrational period of an O–H bond. In other words, hydrogen ions are bouncing back and forth between water molecules pretty much as fast as they are able to.

And not just between one pair of water molecules. When a hydrogen bounces to a second molecule, it's pretty likely that, instead of bouncing back to where it started from, it will stick and knock off one of the other hydrogens on the second molecule. That displaced hydrogen will then bounce onto a third water molecule, and so on.

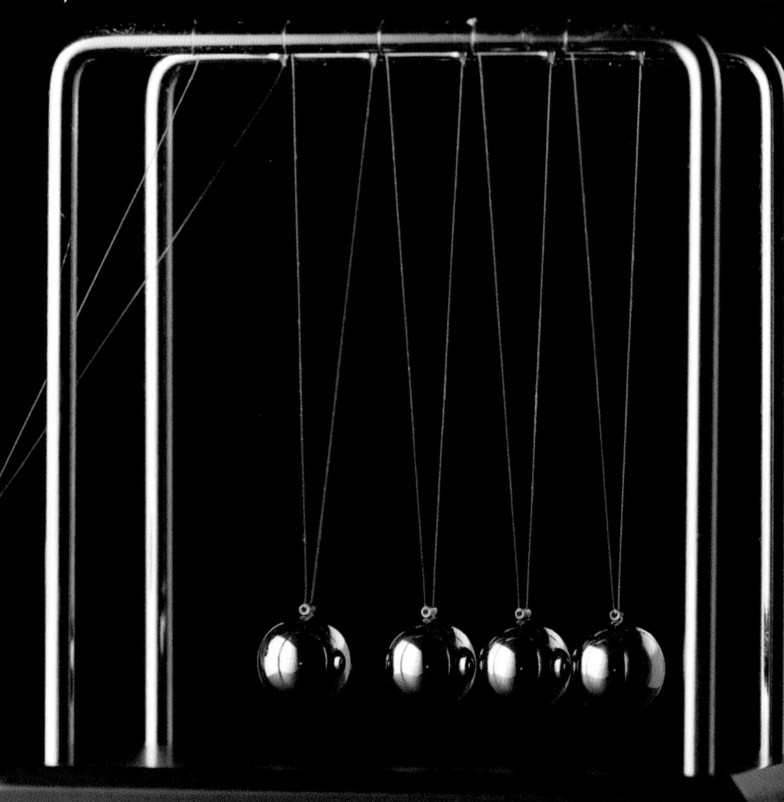

▷ This "Newton's Cradle" toy demonstrates how "bounce energy" can be transferred almost instantly from one end of a chain of balls to another. When the ball hits on the right, a single ball on the left bounces off. (If you haven't seen one of these in action, find a video online: it's fun and surprising if you haven't seen it before.)

Bounce-chains like this happen with hydrogen ion trading as well.

△ When a number of water molecule all happen to line up just right, they form what is called a "proton highway," along which a concerted trade of hydrogen ions can happen almost instantly. The net effect is that you start with an H^+ ion at one end of the highway, and end up with a different H^+ ion at the other end of the highway, without any H^+ ions having actually moved further than the distance from one water molecule to another.

This is just like the Newton's Cradle, where you start with a loose ball on one side and end up with a loose ball on the other side, but no balls moved from one side to the other.

The existence, length, and duration of these proton highways has been studied in computer simulations, and they are a very real thing. They are why certain important reactions happen in water far faster than they would normally be expected to.

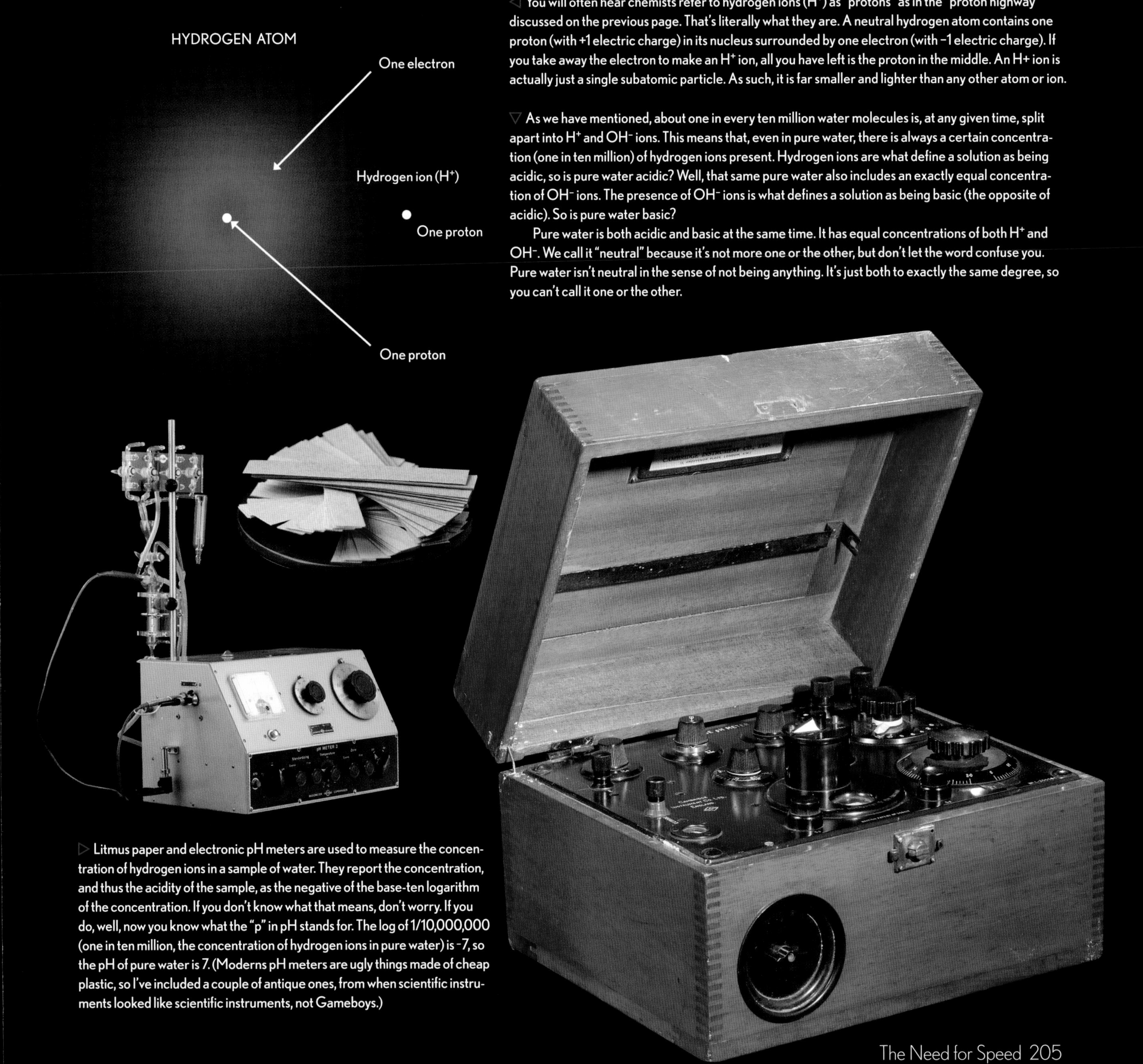

◁ You will often hear chemists refer to hydrogen ions (H^+) as "protons" as in the "proton highway" discussed on the previous page. That's literally what they are. A neutral hydrogen atom contains one proton (with +1 electric charge) in its nucleus surrounded by one electron (with −1 electric charge). If you take away the electron to make an H^+ ion, all you have left is the proton in the middle. An H+ ion is actually just a single subatomic particle. As such, it is far smaller and lighter than any other atom or ion.

▽ As we have mentioned, about one in every ten million water molecules is, at any given time, split apart into H^+ and OH^- ions. This means that, even in pure water, there is always a certain concentration (one in ten million) of hydrogen ions present. Hydrogen ions are what define a solution as being acidic, so is pure water acidic? Well, that same pure water also includes an exactly equal concentration of OH^- ions. The presence of OH^- ions is what defines a solution as being basic (the opposite of acidic). So is pure water basic?

Pure water is both acidic and basic at the same time. It has equal concentrations of both H^+ and OH^-. We call it "neutral" because it's not more one or the other, but don't let the word confuse you. Pure water isn't neutral in the sense of not being anything. It's just both to exactly the same degree, so you can't call it one or the other.

▷ Litmus paper and electronic pH meters are used to measure the concentration of hydrogen ions in a sample of water. They report the concentration, and thus the acidity of the sample, as the negative of the base-ten logarithm of the concentration. If you don't know what that means, don't worry. If you do, well, now you know what the "p" in pH stands for. The log of 1/10,000,000 (one in ten million, the concentration of hydrogen ions in pure water) is -7, so the pH of pure water is 7. (Moderns pH meters are ugly things made of cheap plastic, so I've included a couple of antique ones, from when scientific instruments looked like scientific instruments, not Gameboys.)

▽ Acids and bases are chemicals that increase or decrease the concentration of hydrogen ions (H^+) in water. Acids increase the concentration by releasing H^+ ions when they are dissolved. Bases decrease the concentration either directly by reacting with and absorbing H^+ ions, or by releasing OH^- ions, which react with H^+ ions to form water (thereby decreasing the H^+ concentration).

If you release the same amount of both an acid and a base into the same solution, they will "neutralize" each other. For example, if you dissolve an equal number of molecules of HCl and NaOH in some water, the H^+ ions from the acid will react with the OH^- ions from the base to create a bit more water. All you will have left is some Cl^- ions and same Na^+ ions. This is exactly the same thing you would have if you dissolved salt (NaCl) instead of acid and base. In other words, you've used two dangerous, corrosive chemicals to create harmless salt water.

I said "the H^+ ions from the acid will react with the OH^- ions from the base" but this is not exactly the way it works.

◁ NaOH is a solid at room temperature, so this diagram is closer to reality. But of course, as with all the diagrams in this book, it ignores the fact that all solids are three-dimensional, not flat.

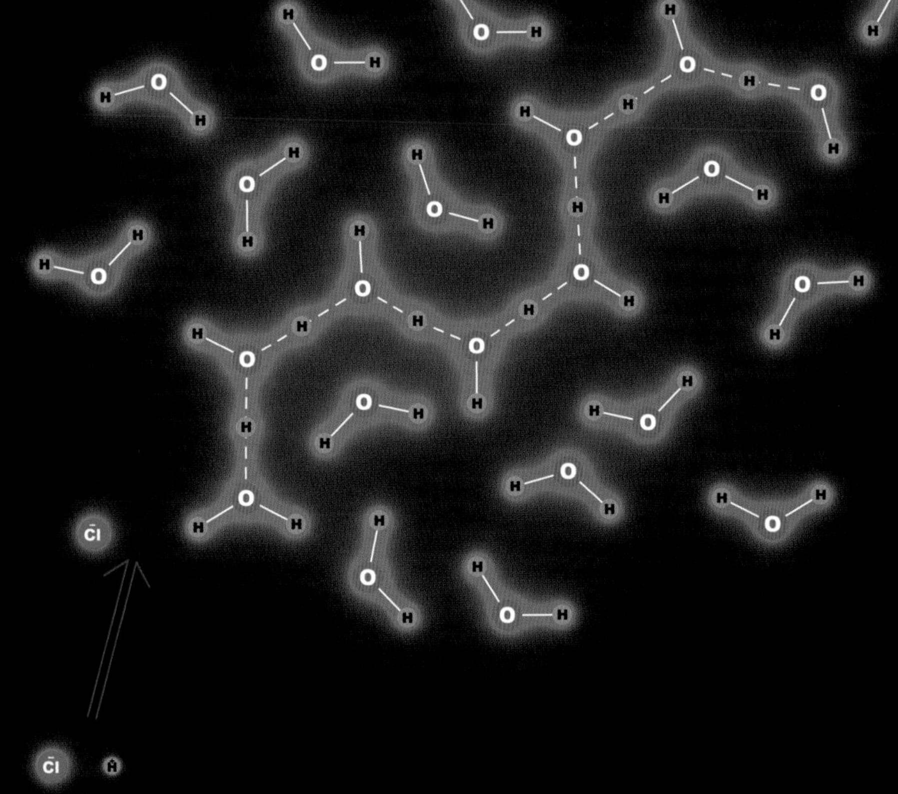

▷ HCl in pure form is actually a gas. It can exist as a liquid at room temperature only when dissolved in water (forming hydrochloric acid). So this diagram should not be taken literally.

△ Proton highways mean that acid-base neutralizing can happen at a distance, without the two substances ever having to meet in person. An excess hydrogen ion from HCl enters the highway at one end, and suddenly there is an excess hydrogen ion at the other end, ready to neutralize an OH^- ion—without any hydrogen ion having traveled the length of the highway.

In this way acids and bases can neutralize each other hundreds of times faster than would otherwise be possible.

The importance to life of the network of hydrogen bonds in water, and between water and other molecules, is so vast that it would be easier to list biologic reactions where it doesn't matter (if there even are any) than trying to count those where it is an important factor.

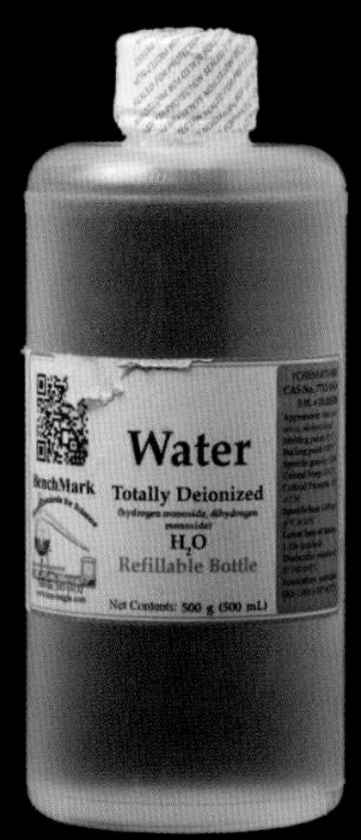

IT IS FITTING TO END our exploration of chemical reactions with water. No other substance is as deeply embedded in our bodies, our minds, our imaginations, and our art as is water. How fitting it is that this substance, known since before there was knowing, revered and honored by all cultures as the giver and sustainer of life, is also so very *chemical* in nature.

Water is both an acid and a base, in the technical meanings of those words. It is a powerful solvent. It participates in all the reactions of life as either a reactant, product, or solvent. Water is the very essence of what it means to be a chemical. How marvelous it is that we can—indeed we must—drink it every day! Water is the ultimate proof that we are made of, depend on, and enjoy chemicals and their marvelous reactions in every aspect of our lives.

and particularly for introducing me to Jerry Bell, whom I worked with as coauthors for a year on the aforementioned book-delaying project. Jerry has tried diligently to keep me honest in my descriptions of chemistry, and finally explained entropy to me in a way that makes sense. He has deepened and sharpened my understanding of chemistry.

My editor, Becky Koh, has been unfailingly supportive, even when I have failed her—first by putting the book off for a year, and then by flaking out until the very last second and no doubt making her wonder if I was ever going to deliver anything. Her support has been greatly appreciated. (She also talked me into changing the title of the second book from *Compounds* to *Molecules*, which is a much better name.)

The crème brûlée that illustrates cooking with danger was made by my fabulous friend Maribel, who is also responsible for restoring my mojo and making the completion of this book possible. Without her, Becky would probably still be waiting for the manuscript.

My daughter Emma, who is now old enough to be really useful, painted me a picture of despair to illustrate the futility of life and made me a drawing of bamboo torture. My son Connor, also useful, lent us his hair to rub a balloon on to illustrate static electricity. And my oldest, Addie, graciously agreed to receive a sewing machine at her dorm room so I could pick it up on a side trip to Scotland. This saved a lot on the shipping cost, though I think if you include the cost of college tuition, it probably would have been cheaper to just have the machine shipped here.

Fiona Barclay was helpful in so many ways, most visibly by finding images for many of the things that we were not able to photograph ourselves (including, for example, historical events), for her chemical expertise, and for introducing us all to Drew Gardner.

Mike Sansom of BrightFire Pyrotechnics went way beyond the call of duty to film burning nitroglycerin and steel plates exploded with gunpowder and dynamite. He also sent hands-down the most beautiful photographs of fireworks I've ever seen.

Drew Berry created the beautiful molecular visualizations of cellulose synthesis. Jonathan Mathews sent molecule structure files (thank you for saving me soooo much work!) for typical coal molecules, which isn't something I thought would actually exist. Braxton Collier invented and implemented a very interesting new technique for rendering atomic and molecular orbitals, used in chapter 2 to illustrate bonding and antibonding electron configurations.

Besides Nick, I've had the privilege of working with several other outstanding photographers, including Mike Walker, who worked with me on my *Popular Science* column for many years, and some of whose images are in this book. Drew Gardener shot the lovely images of thermite welding, and Graham Berry was responsible for many others (see individual photo credits). Chuck Shotwell, who is represented with only a single image, nevertheless looms large in my appreciation of the art of photography.

Matthew Cokeley, who designed and laid out this book and all my others, has been great to work with for many years, and is responsible for much of the beauty of my books.

Thanks to my long-suffering accountant/assistant, Gretchen, for keeping things together, to Bobby Crowe for concrete advice and other assistance making concrete, and to Koatie for generally being there when needed.

For chlorophyll, chloroplast, and chromatophore protein images, thanks to Melih Sener. Thanks to Melih also for a fascinating discussion of the economics of plant growth. Bryan Hanson at DePauw University was essential in providing not only information about, but actual MOL files for, Percy Julian's molecules. Jeff Bryant helped find suitable stars and planets for chapter 4.

Toby, Alexus, Quinton, Brianna, David, and Biscuit did a great job making a monumental mess to illustrate washable kids' paint.

Thanks to Toby and Robert for making the model volcano that illustrates my crushing disappointment in these things. Several other people also went out of their way to make or share things I needed to photograph, some of which I even got to eat. Dannie Otto made an amazing apple pie to illustrate the use of tallow. Gregory Erbach brought in a very nice guitar he had made and finished with nitrocellulose lacquer. Theophilus Jackson spent several hours on a tremendously cold night forging Panamanian beach sand steel into a couple of useful objects.

Finally, I want to thank everyone who worked on the project I mentioned at the start of this section. It was a big project and great work was done, some of which you will find in the pages of this book. These people all distinguished themselves in pursuit of a noble goal: Andy Bull, Anna Evans-Freke, Ashley Cabico, Braxton Collier, Carl Morland, Carole Eychenne, Drew Berry, Drew Gardner, Edward Briffa, Farrell Mackenzie, Fiona Barclay, Graham Berry, Hannah Parry, Ivan Timokhin, Jasper James, Jerry Bell, John Cromie, John Howarth, Matthew Cokeley, Matthew Shribman, Mike Kuiper, Mohsen Ramezanpoor, Nathan March, Nick Mann, Matt Aitken, Tom Weightman, Rob Andrews, Sam Bain, Sam Woolf, Selina Pang, Simon Rice, Stephen Menzie, Mike Sansom, Nick Tall, Milo Shaffer, Bassam Shakhashiri, Deborah Corrigan, Julie Willcott, and Matt Stoltzfus.

NICK MANN ACKNOWLEDGMENTS

I WOULD LIKE to thank Selina, Jackson, Madhura, Andrea, Armando, and my parents for their supportive roles in my life during work on this project. I would also like to thank Theo for the opportunity to work on this book as well as our many other projects over the last decade.

Photo Credits

Unless otherwise noted, all photographs are by Nick Mann, Copyright © 2017 Theodore Gray, and all diagrams and molecule renderings are by Theodore Gray, Copyright © 2017 Theodore Gray.

The following images were produced under the auspices of The Chemistry Project, and are Copyright © 2016 RGB Ltd, credited to the individual photographers named:
vi, vii, 7, 8, 14, 36, 37, 42, 56, 58, 74, 75, 76, 77, 78, 79, 106, 125, 127, 128, 129, 130, 138, 148, 149, 158, 165, 185, 187, 195, 205: Nick Mann.
38, 96, 97: Drew Gardner.
57, 71, 75, 84, 101, 126, 182: Drew Gardner and Graham Berry.
104, 105, 108, 108: Jasper James.
iii, 159, 160, 162, 163: Andrew Berry.

Additional original photographs Copyright © 2016 by Theodore Gray, credited to the photographers named:
36, 37: Nick Mann, courtesy of Touch Press, Gems & Jewels App.
115: Nick Mann, courtesy of The Orchestra app, Amphio, and the Philharmonia Orchestra.
205: Nick Mann, courtesy of the Chemical Heritage Foundation.
22, 83, 89, 92, 114, 176, 90, 184: Theodore Gray.
92: Maribel Odalis Sanchez de Tyler.

Additional original photographs by Nick Mann, Copyright © 2017 Nick Mann:
95, 115, 157, 176, 178.

Additional original photographs copyright by their respective photographers:
8, 19: Max Whitby.
22: Bryan Hanson.
22: Bridget Gourley.
iii, 112, 133, 138: Mike Sansom, courtesy of pyroproductions.co.uk
198, 202: Mike Sansom.

Original renderings by Melih Senar:
50, 51, 52: from A. Hitchcock, C. N. Hunter, and M. Sener. Determination of cell doubling times from the return- on-investment time of photosynthetic vesicles based on atomic detail structural models. J. Phys. Chem. B, 2017.
M. Sener, J. Strumpfer, A. Singharoy, C. N. Hunter, and K. Schulten.
Overall energy conversion efficiency of a photosynthetic vesicle. eLife, page 10.7554/ eLife.09541 (30 pages), 2016.

129: Original renderings of light bulb spectra by Theodore Gray, from data supplied by http://www.designingwithleds.com/light-spectrum-charts-data/

Public domain images courtesy of NASA:
iii, 32, 33, 45, 48, 115, 131, 172, 177, 189, 131 (NASA, ESA, A. Fujii, and Z. Levay, STScI).

Images originally from my Popular Science column, copyright by their respective photographers:
9, 72, 73, 80, 107, 188, 189, 192: Mike Walker.
183: Charles Shotwell.

Additional images:
5: The Three Witches by Daniel Gardner, 1775 (public domain, via Wikimedia Commons).
7: Painting by Pietro Longhi, 1830 (public domain via Wikimedia Commons).
8: Photographer unknown, via Wikimedia Commons.
20: Photographer unknown (public domain via Wikimedia Commons).
22: Photographer unknown (public domain via DePauw University).
31: Copyright © 2003 Adam Block/Mount Lemmon SkyCenter/University of Arizona.
34: ESO (European Southern Observatory)/H. Boffin, Creative Commons Attribution.
73: Gus Pasquarella, public domain via Wikimedia Commons.
86: Copyright Jana Vengels | http://www.dreamstime.com/bluna_info
86: Copyright Fibobjects | http://www.dreamstime.com/fibobjects_info
90: Copyright age fotostock / Alamy Stock Photo.
90: Copyright Dirk Ercken | http://www.dreamstime.com/kikkerdirk_info
91: Copyright Leonid Chernyshev | http://www.dreamstime.com/chernysh_info
93: Copyright Scanrail | http://www.dreamstime.com/scanrail_info
93: Copyright Kuzmichevdmitry | http://www.dreamstime.com/kuzmichevdmitry_info
94: Copyright Antikainen | http://www.dreamstime.com/antikainen_info
95: AP Photo/Bruce Smith.
95: Gretar Ívarsson, public domain via Wikimedia Commons.
105: Bert Chan, CC Share Alike via Wikimedia Commons.
109: Photographer unknown, CC Share Alike via Wikimedia Commons.
110: D. Gordon E. Robertson, CC Share Alike via Wikimedia Commons.
115: Copyright Mitrandir | http://www.dreamstime.com/mitrandir_info
121: Copyright Sharifphoto | http://www.dreamstime.com/sharifphoto_info
123: Copyright Trudywsimmons | http://www.dreamstime.com/trudywsimmons_info
135: Copyright Alison Cornford-matheson | http://www.dreamstime.com/acmphoto_info
148: Artist unknown.
154: Galen Rowell/Mountain Light / Alamy Stock Photo.
163: Copyright Erik1977 | http://www.dreamstime.com/erik1977_info
172: Photographer unknown, Alamy Stock Photo.
172: Keene Public Library and the Historical Society of Cheshire County, public domain.
173: Iain Cameron, CC Attribution.R0349: John Erlandsen, CC Share Alike.
190: Courtesy Photofest
192: AP Photo/The Winchester Star, Scott Mason.
194: Mira / Alamy Stock Photo.
207: Katsushika Hokusai (葛飾北斎), public domain.

Index

Numbers in *italics* indicate photographs or diagrams.

A

B

C

K

L

M

N

O

P

T

U

V

W

X